DYNAMICAL NETWORKS IN PHYSICS AND BIOLOGY

At the Frontier of Physics and Biology

Les Houches Workshop, March 17-21, 1997

Editors

D. BEYSENS
G. FORGACS

Springer

Berlin
Heidelberg
New York
Barcelona
Hong Kong
London
Milan
Paris
Singapore
Tokyo

EDP Sciences

7, avenue du Hoggar
Parc d'Activités de Courtabœuf
BP. 112
91944 Les Ulis Cedex A, France

EDP Sciences
875-81 Massachusetts Avenue
Cambridge, MA 02139, USA

Centre de Physique des Houches

Books already published in this series

1 Porous Silicon Science and Technology
J.-C. VIAL and J. DERRIEN, Eds. 1995

2 Nonlinear Excitations in Biomolecules
M. PEYRARD, Ed. 1995

3 Beyond Quasicrystals
F. AXEL and D. GRATIAS, Eds. 1995

4 Quantum Mechanical Simulation Methods for Studying Biological Systems
D. BICOUT and M. FIELD, Eds. 1996

5 New Tools in Turbulence Modelling
O. MÉTAIS and J. FERZIGER, Eds. 1997

6 Catalysis by Metals
A. J. RENOUPREZ and H. JOBIC, Eds. 1997

7 Scale Invariance and Beyond
B. DUBRULLE, F. GRANER and D. SORNETTE, Eds. 1997

8 New Non-Perturbative Methods and Quantization on the Light Cone
P. GRANGÉ, A. NEVEU, H.C. PAULI, S. PINSKY and E. WERNER, Eds 1998

9 Starbursts Triggers, Nature, and Evolution
B. GUIDERDONI and A. KEMBHAVI, Eds 1998

Book series coordinated by Michèle LEDUC

Editors of "Dynamical Networks in Physics and Biology" (N° 10)
D. Beysens (CEA, Grenoble, France)
G. Forgacs (Clarkson University, Potsdam, USA)

ISBN 3-540-65349-X Springer-Verlag Berlin Heidelberg New York
ISBN 2-86883-382-9 EDP Sciences Les Ulis

Printed in France

PREFACE

The 1997 Les Houches workshop on "Dynamical Network in Physics and Biology" was the third in a series of meetings "At the Frontier between Physics and Biology". Our objective with these workshops is to create a truly interdisciplinary forum for researchers working on outstanding problems in biology, but using different approaches (physical, chemical or biological). Generally speaking, the biologists are trained in the particular and motivated by the specifics, while, in contrast, the physicists deal with generic and "universal" models. All agree about the necessity of developing "robust" models.

The specific aim of the workshop was to bridge the gap between physics and biology in the particular field of interconnected dynamical networks. The proper functioning of a living organism of any complexity requires the coordinated activity of a great number of "units". Units, or, in physical terms, degrees of freedom that couple to one another, typically form networks. The physical or biological properties of interconnected networks may drastically differ from those of the individual units: the whole is not simply an assembly of its parts, as can be demonstrated by the following examples. Above a certain (critical) concentration the metallic islands, randomly distributed in an insulating matrix, form an interconnected network. At this point the macroscopic conductivity of the system becomes finite and the amorphous metal is capable of carrying current. The value of the macroscopic conductivity typically is very different from the conductivity of the individual metallic islands. At the critical concentration, branching polymers in a solution form a macroscopic coupled mesh. The viscosity of the liquid then grows indefinitely, and the system becomes an elastic gel. The macroscopic elastic constants again have very little to do with the elastic properties of the individual polymers. The above networks, extensively studied in the physical and mathematical sciences, are examples of percolation assemblies, lacking an obvious, ordered repetition of the units. An example of an ordered network is a solid crystalline body, being a collection of individual atoms or molecules. As the crystal grows, its physical properties change. Macroscopic attributes not even defined for individual atoms (*i.e.* resistivity, thermal expansion, elasticity) appear.

Interconnected dynamical networks show up at various scales and developmental stages in an organism. The nuclear matrix is a scaffolding made mainly of 9- and 13-nm core filaments. On a slightly larger scale, the cytoskeleton may be viewed as an interconnected network of actin and intermediate filaments, as well as microtubules. On a still larger scale, the extracellular matrix fills the space between various tissues and organs and contains an intricate mesh of interacting macromolecules (*i.e.* collagen, elastin, fibronectin, laminin). At the extreme scale a living system could be viewed as a network of organs coupled through the circulatory or muscular system. Specific components of an organism are often studied in terms of interconnected assemblies. Intensive studies have been carried out in the fields of genetic networks, intracellular protein or enzyme networks, neuronal nets, immunological networks and metabolic networks. The major reason for such an approach is again the realization that the proper understanding of the biological functions performed by these complexes is possible only through the study of the intricate dynamical coupling between a vast number of individual units. It is precisely the dynamical nature of microfilament networks that is employed during cell locomotion: actin subunits are organized into linear polymers, which are crosslinked at the lamella's leading edge. The crosslinked filaments are then disassembled away from the edge, and the resulting subunits are again used for assembly. A fundamental question in developmental biology is the formation of such intra- and extracellular patterns.

Many outstanding questions in both the physical and biological sciences are currently being investigated using networks. Physicists and mathematicians have worked out a number of general methods to study the complex properties of both ordered and random interacting assemblies. Biologists dealing with the networks mentioned above typically build then into in their studies on specificity.

The meeting was attended by 63 scientists. Physicists were brought together with molecular and cell biologists, immunologists and geneticists to discuss the importance of cooperactivity embodied in dynamical networks. Special emphasis was given to the following topics: genetic networks, metabolic networks, cytoskeletal networks, neuronal networks, immune networks, extracellular matrix networks, structure formation in tissues and other complex fluids, percolation and its application in physics, biology and chemistry.

The workshop began, as in the past, with overviews on the general theme from the perspective of a physicist (G. Forgacs, Clarkson University) and a biologist (P. Janmey, Harvard Medical School). 38 talks were given, and each of the above mentioned topics was introduced by an internationally known expert.

The resolution of outstanding problems facing biology today requires more and more accurate measurements and quantitative analysis of the data obtained. At the same time, the changing needs of our societies force physicists to broaden their horizon and venture into territories traditionally not in the realm of physics. The success of the present workshop, which can be measured by the number of collaborations initiated during the meeting, underscores the conviction of the organizers that the Les Houches workshops "At the Frontier of Physics and Biology" fulfill an important role in helping physicists and biologists to better understand each other's scientific approaches and methods and develop fruitful interactions.

Acknowledgments

We wish to kindly acknowledge the generous support of the following agencies, without which this workshop could have not been organized:

- the Commissariat à l'Énergie Atomique,
- the Centre National de la Recherche Scientifique,
- the Direction des Recherches, Études et Techniques du Ministère de la Défense,
- the Ministère des Affaires Étrangères.

We also would like to thank J. Glénat for her help in the administrative organization of the meeting, and the whole team at the Centre de Physique des Houches for making everyday life during the workshop as comfortable as possible.

D. Beysens and G. Forgacs

CONTENTS

INTRODUCTION

LECTURE 1

Cooperative Phenomena in Physical Networks

by G. Forgacs

NETWORKS OF BIOPOLYMERS

LECTURE 2

Complex Networks in Cell Biology

by P.A. Janmey, J.V. Shah and J.X. Tang

LECTURE 3

The Keratocyte Cytoskeleton: A Dynamic Structural Network

by K.I. Anderson

LECTURE 4

Adhesion Mediated Signaling: The Role of Junctional Plaque Proteins in the Regulation of Tumorigenesis

by A. Ben-Ze'ev

LECTURE 5

Of Proteins, Redox States and Living Things

by L. Moldovan and P.J. Goldschmidt-Clermont

LECTURE 6

Tensegrity and the Emergence of a Cellular Biophysics

by D.E. Ingber

LECTURE 7

The Leukocyte Actin Cytoskeleton

by F. Richelme, A.-M. Benoliel and P. Bongrand

LECTURE 8

Branched Polymers and Gels

by M. Daoud

LECTURE 9

Statistical Mechanisms of Semiflexible Polymers: Theory and Experiment

by E. Frey, K. Kroy, J. Wilhelm and E. Sackmann

LECTURE 10

Scale Effects, Anisotropy and Non-Linearity of Tensegrity Structures: Applications to Cell Mechanical Behavior

by S. Wendling, E. Planus, D. Isabey and C. Oddou

CELLULAR AND EXTRACELLULAR NETWORKS

LECTURE 11

Networks of Extracellular Matrix and Adhesion Proteins

by J. Engel

LECTURE 12

Networks of Extracellular Fibers and the Generation of Morphogenetic Forces

by S.A. Newman

LECTURE 13

First Steps Towards a Comprehensive Model of Tissues, or: A Physicist Looks at Development

by J.A. Glazier and A. Upadhyaya

LECTURE 14

Networks of Droplets Induced by Coalescence: Application to Cell Sorting

by D.A. Beysens, G. Forgacs and J.A. Glazier

LECTURE 15

A Monte-Carlo Approach to Growing Solid Non-Vascular Tumors

by D. Drasdo

GENETIC, IMMUNE, MOLECULAR AND METABOLIC NETWORKS

LECTURE 16

Intracellular Communication *via* Protein Kinase Networks

by D.L. Charest and S.L. Pelech

LECTURE 17

Molecular Networks that Regulate Development

by W.F. Loomis, G. Shaulsky, N. Wang and A. Kuspa

LECTURE 18

H-Bond Networks in Stability and Function of Biological Macromolecules

by M.-C. Bellissent-Funel

LECTURE 19

A Network of Cell Interactions Mediated by Frizzled is Essential for Gastrulation and Anteroposterior Axis Determination in *Xenopus* Embryos

by S.Y. Sokol and K. Itoh

LECTURE 20

Fluid Lipid-Bilayer Membranes: Some Basic Physical Mechanisms for Lateral Self-Organization

by L. Miao, P.L. Hansen and J.H. Ipsen

LECTURE 21

Logical Analysis of Timing-Dependent Signaling Properties in the Immune System

by M. Kaufman, F. Andris and O. Leo

LECTURE 22

A Water Channel Network in Cell Membranes of the Filter Chamber of Homopteran Insects

by P. Bron, V. Lagrée, A. Froger, I. Pellerin, S. Deschamps, J.-F. Hubert, C. Delamarche, A. Cavalier, J.-P. Rolland, J. Gouranton and D. Thomas

LECTURE 23

Creative Genomic Webs

by E. Ben-Jacob

NEURONAL NETWORKS

LECTURE 24

Hebbian Learning of Temporal Correlations: Sound Localization in the Barn Owl Auditory System

by J.L. van Hemmen and R. Kempter

APPLICATION OF PHYSICAL MODELS AND PHENOMENA TO BIOLOGICAL SYSTEMS

LECTURE 25

Structures of Supercoiled DNA and their Biological Implications

by T.R. Strick, J.-F. Allemand, A. Bensimon, D. Bensimon and V. Croquette

Introduction

LECTURE 1

Cooperative Phenomena in Physical Networks

G. Forgacs

Department of Physics and Biology, Clarkson University,
Potsdam, NY 13699-5820, USA

1. INTRODUCTION

In the spirit of former winter workshops at the "Frontier Between Physics and Biology" organized at Les Houches, the first talk is an overview of the physics of the chosen topic and is intended to be an introduction into the methods used by physicists.

The present article describes some frequently encountered physical networks and provides an elementary discussion of percolation phenomena (Sahimi, 1994), one of the most widely used approaches to interpret the properties of disordered networks. Special emphasis is put on disordered networks since those found in living organisms mostly belong to this category.

Networks present a ubiquitous form of organization of both inanimate and living systems. They consist of subunits that are interconnected in a way characteristic for the given network. Networks can be highly ordered, as in the case of a perfect crystal, or highly disordered without any apparent spacial regularity of the network forming subunits, as in the case of gel-forming branched polymers. The properties of physical networks may very little to do with the physical characteristics of their subunits: totality is not the simple collection of its parts. The electrical properties of an amorphous conductor are markedly different from those of its granules. Properties defining the network may not even exist for its subunits: the elastic constants of an elastic material reflect the structure and properties of its constituent atoms and molecules, yet such constants have no meaning for these subunits.

Interconnected networks show up at various scales and developmental stages in an organism. As revealed by scanning electron microscopy, the nucleus contains the nuclear matrix, a scaffolding made mainly of 9- and 13-nm core filaments, which contains cell specific proteins (Fey and Penman, 1988; Dworezky *et al.*, 1990). Some of these filaments are of actin-type (Jockusch *et al.*, 1974; Fukui and Katsumara, 1979; Osborn and Weber, 1980), others of intermediate filament-type (Bader *et al.*, 1991; Blessing, *et al.*, 1993). On a slightly larger scale, the cytoskeleton may be viewed as an interconnected network of actin and intermediate filaments, as well as microtubules (Heuser and Kirschner, 1980; Bridgeman and Reese, 1984; Mitchison, 1992; Stossel, 1993). On a still larger scale, the ECM (Birk *et al.*, 1991; McDonald, 1988) fills the space between various tissues and organs and contains an intricate mesh of interacting macromolecules (collagen, elastin, fibronectin, laminin). The physical properties of many of the above macromolecules and methods to study them will be overviewed by Paul Janmey in this volume.

2. ORDERED AND DISORDERED NETWORKS

Figure 1 shows part of the ideal and non-ideal cubic structure of the ordinary kitchen salt (NaCl). The ideal salt crystal, Figure 1A is an atomic network of sodium and chloride atoms occupying the vertices of a cubic unit cell which repeats itself in three dimensions. The known properties of salt stem from the chemical properties of sodium and chloride as well as the physical properties of the network they form. The structure shown in Figure 1A is perfectly regular: each chloride atom is surrounded by six sodium atoms and vice versa. This ideal crystal, however, does not exist in nature. It can be shown (Ashcroft and Mermin, 1976) that for thermodynamic stability imperfections, like substitutional impurities, vacancies (missing atoms), *etc.* must be present. Figure 1B shows some substitutional impurities: K replacing Na and I replacing Cl.

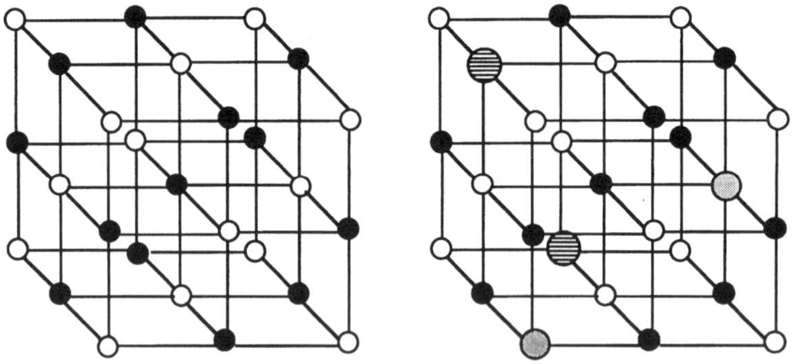

A B

Fig. 1. A. The perfect, ideal cubic lattice of salt, NaCl on the left. Open circles denote Na, full circles denote Cl atoms. B. The non-ideal NaCl structure: dotted circles denote potassium atoms, striped circles denote indium atoms.

With the increase of the impurity concentration the system becomes more and more disordered, although its overall properties may still resemble those of NaCl.

In Figure 2 the schematic view of a conductor is presented. In Figure 2A the conducting spheres form a perfect, triangular lattice. If the system is enclosed in a box with opposing metallic plates then upon applying a potential difference to the plates, current flows through the system as indicated in Figure 2A. In Figure 2B the spheres are made of non-conducting material; the system is a perfect insulator. The completely random mixture, shown in Figure 2C models an amorphous metal or insulator depending on whether the metallic or the insulating subunits form an uninterrupted path between the bounding metallic plates, respectively. In the limit of a very large system (in the macroscopic regime) a sharp transition in the concentration of metallic (or insulating) subunits separates the two regimes.

Let us assume that the metallic (dotted) spheres are made of gold and the insulating (full) spheres are made of wood. Let us further assume that the requirement imposed on the amorphous conductor is that the attached ampmeter show a nonvanishing current and that this objective be achieved at the smallest cost. This requirement can be met using a random mixture, like the one shown in Figure 2C. One needs to determine the minimal concentration of subunits made of gold so that they form an interconnected path or network that spans the system between the electrodes.

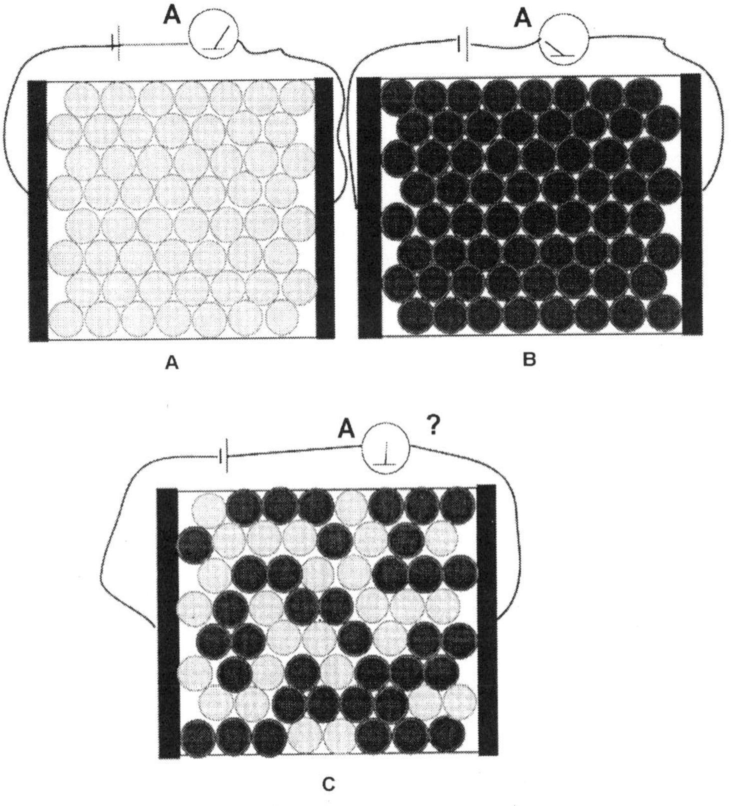

Fig. 2. A. Perfect conductor on a triangular lattice. Dotted spheres denote metallic subunits. B. Perfect insulator. Full spheres denote insulating subunits. C. Mixture of conducting and insulating subunits.

The above examples show that perfectly regular networks are either idealizations that do not exist in nature or are impractical. Most of the interconnected assemblies encountered in physics are imperfect, often disordered networks.

3. PERCOLATION PHENOMENA

Percolation is used to interpret the macroscopic properties of a great variety of systems containing elements distributed randomly or, equivalently, in a way that lacks an obvious repetition of the elements. It gives information on when a system is "macroscopically open" to a given phenomenon (Sahimi, 1994). In such systems connectivity plays an essential role (Sahimi, 1994). When the concentration of "elements" reaches a critical threshold value, customarily denoted by p*, a percolation transition takes place and a global percolation network forms. At this point fundamental changes may occur in the behavior of the system: individual monomers (the "elements") interconnect to form a gel (De Gennes, 1976a,b; Bouchaud *et al.*, 1986); the conducting component (whose atoms are the "elements") of an amorphous conductor-insulator composite spans the specimen and electrical current can flow through the system (Clerk *et al.*, 1990); elastic subunits in an otherwise inelastic amorphous medium interconnect, the system develops macroscopic elastic properties and responds to mechanical signals (Nakayama *et al.*, 1994). Percolation models diverse phenomena including the spread of infectious diseases (Bailey, 1975), or the propagation of forest fire (MacKay and Naeem, 1984). It is used to interpret the peculiar properties of water and other liquids (De Gennes, 1979; Stanley *et al.*, 1981; Xu and Stell, 1988; Campbell and Forgacs, 1990; Utracki, 1991; Sciortino *et al.*, 1991), including microemulsions (Peyrelasse and Boned, 1990) and is extensively employed in the material sciences (Cahn, 1997).

Percolation is an example of a second order or continuous phase transition, similar to a paramagnetic-ferromagnetic transition which takes place at a "critical" (Curie) temperature (Stanley, 1971). In such transitions the significant changes in macroscopic properties occur in a continuous manner. At the threshold, no drastic structural changes can be seen visually in the system. The electrical conductivity and elastic moduli are vanishingly small below and at the threshold, and increase gradually above the threshold. The variations are proportional to $(\Delta p)^{\beta}$, where Δp is the deviation from p* and β is a "critical exponent" corresponding to a given physical quantity (Domb, 1983). (The continuous variation above p* is quite different from what happens in a first order phase transition, like the freezing of water. Here drastic changes can be observed even visually (water turns into ice), and physical parameters may vary in a drastic way, by jumping from one value to another (the specific heat of water is almost twice that of ice at zero Celsius)).

The following examples are meant to illustrate the meaning of the percolation transition. Consider the entire telephone network which connects, for example, Los Angeles and New York, (prior to the age of cellular phones) shown schematically in Figure 3. Such a network consists of a multitude of cables or optical fibers of (varying) finite length, is interconnected and has the appearance of being random (see Fig. 3A, where "cables" drawn with heavy lines belong to the interconnected percolation cluster). It also has redundancy; the signal from Los Angeles can arrive in New York in a number of ways. If this network is subject to a series of disasters, with more and more of the links randomly destroyed, connection between the two cities will still be maintained up to a critical number of finite elements (although the time for the signal to arrive will typically increase). This is illustrated in Figures 3B,C. Telephone service will be definitely interrupted below the percolation threshold, when an interconnected cluster of connections, extending from Los Angeles to New York no longer exists, although many finite clusters of wires may still be present (see Fig. 3D).

The described telephone network is an example of continuous percolation, where no restriction is imposed on the position of the elements in the cluster (*i.e.* cables in Fig. 3). (In discrete percolation the elements are typically confined to sites of a regular discrete lattice.) In this case the value of the percolation threshold strongly depends on the shape of percolating objects. For example, for straight fibers or wires with aspect ratio x = L/d (d: diameter, L: length), in the limit of large system size and large aspect ratio, the percolation threshold is expressed in terms of the volume fraction, v* of fibers or wires as v* = 1/x (Bug *et al.*, 1985).

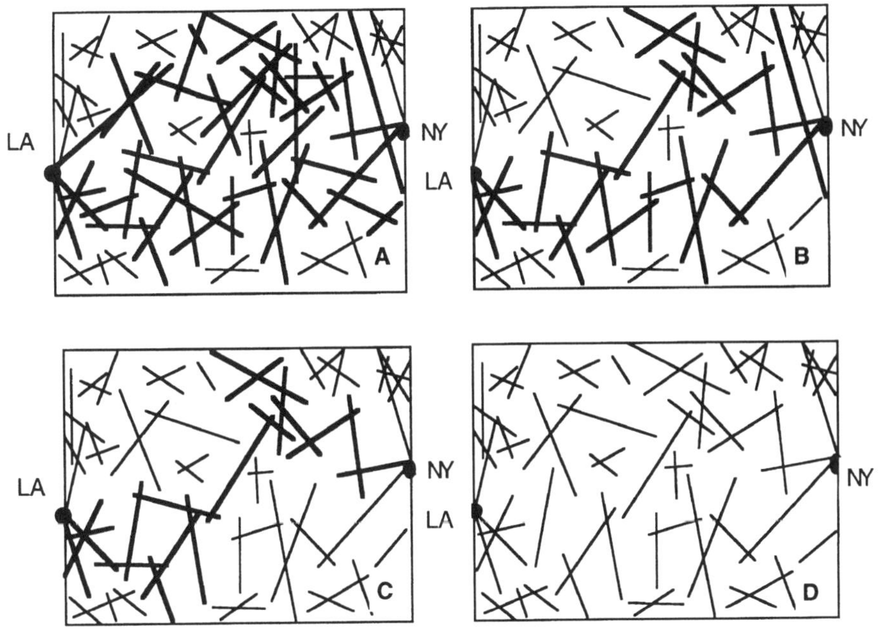

Fig. 3. Schematic representation of the percolation transition. See text for explanation.

As mentioned, at the percolation transition significant changes may occur in the physical properties of the system. In the case of the amorphous conductor shown in Figure 2C, above the transition, when the metallic subunits form a percolating cluster, the electrical conductivity Σ is finite, whereas below the transition it is zero. This is shown schematically in Figure 4. In the vicinity of the transition the conductivity indeed varies according to a power law (Sahimi, 1994).

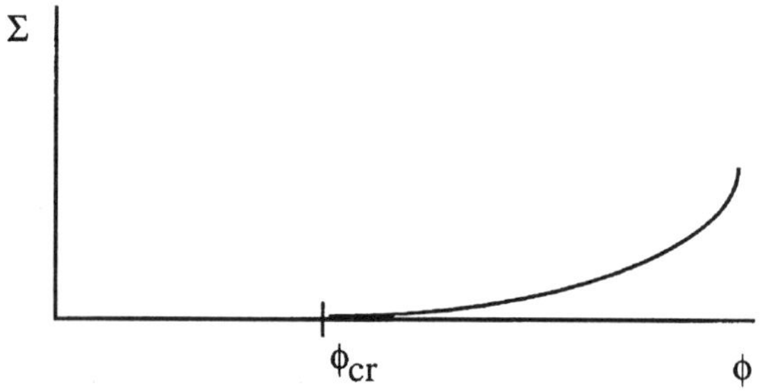

Fig. 4. Electrical conductivity of an amorphous metal as a function of the volume fraction ϕ of the metallic subunits which form an interconnected percolating cluster at ϕ_{cr}.

Another example of the percolation transition is shown in Figure 5. Below a critical concentration of monomers, branching polymers form a sol with finite viscosity as indicated on the left of Figure 5. As polymerization progresses and more and more subunits interconnect, the viscosity rises. Finally, the interconnected cluster spans the entire system. At this moment, which is identified with the onset of percolation, a gel forms with measurable elastic moduli. In the vicinity of this sol-gel transition both the viscosity and the elastic moduli vary according to power laws (De Gennes, 1976a,b; Adam *et al.*, 1985).

Figure 6 is a schematic view of (part of) a lymphocyte membrane. The small discs represent antibody-like receptor molecules, whereas the Y-shape objects symbolize antigens. The process of proliferation and secreting large amounts of antibodies is preceded by cross-linking of antigens through the receptors until a macroscopically large patch is formed on the cell surface. The relationship between this network, which has all the characteristics of a percolation assembly and biological activity is not clear. It is however assumed that network formation provides some sort of simulatory signal to the immune system (Delisi and Perelson, 1976). The complexity of the immune system has been studied using ideas of percolation by several authors (Perelson and Oster, 1979; Stauffer and Weisbuch, 1992; Stauffer and Sahimi, 1993).

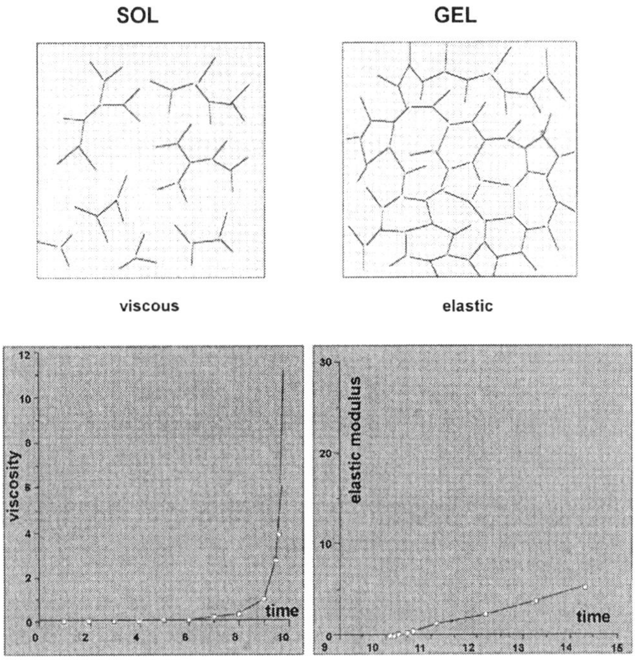

Fig. 5. Left. Polycondensation in the sol phase. The condensation occurs between the Y-shaped monomers. The viscosity (measured in Poise along the y-axis) increases with the degree of polymerization and at the gel point or percolation transition for any practical purpose reaches infinity. Right. The elastic modulus (measured in dyne/cm^2 along the y-axis) is zero below the gel point and rises steadily above the transition to eventually reach a plato. Time is measured in minutes.

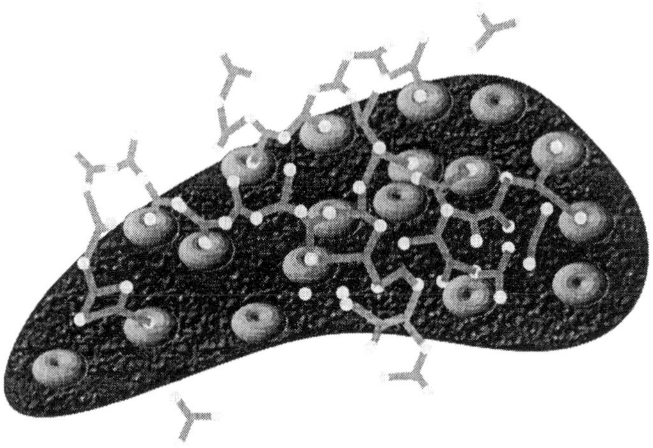

Fig. 6. Lymphocite membrane with antibody molecules (discs) and antigens (Y-shapes).

4. AN ANALYTICALLY TREATABLE EXAMPLE: THE BETHE LATTICE

In practical applications of percolation most of the time one has to resort to a computer in order to evaluate physical properties, like viscosity, elastic moduli, conductivity, *etc.* An amusing and analytically solvable model of percolation is the Bethe lattice shown in Figure 7. Because of its simplicity and flexibility (see below) the Bethe lattice is often used to model complicated phenomena in an exploratory way. For the model in Figure 7 one starts from a central point that has functionality $z = 3$. Each bond ends in another site from which again z bonds emanate one of which connects to the central site, whereas the other $z - 1$ bonds lead to new sites. The branching process resembles the polycondensation shown in Figure 5. If one reaches a site in the interior of the Bethe lattice, then one can go on in $(z - 1)$ other directions in addition to the original direction (no "backward" motion is allowed). The "dangling sites", which do not connect to anything constitute the surface of the system. The important feature of the lattice (which allows for the analytic solution) is that there are no loops.

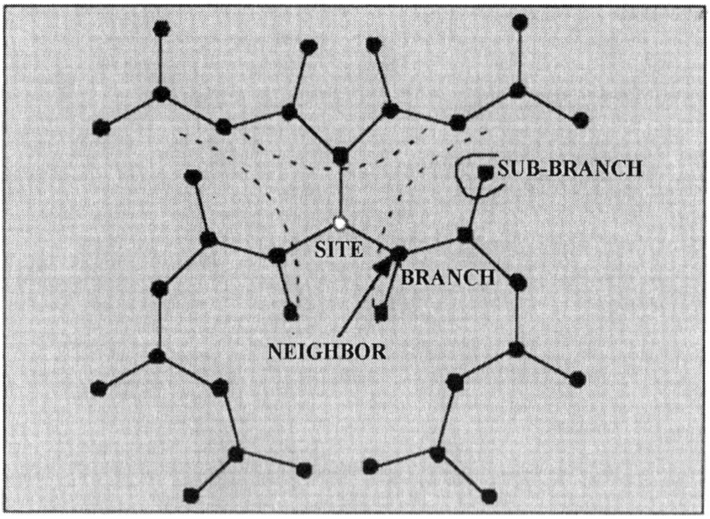

Fig. 7. The Bethe lattice.

The number of sites increases exponentially in this structure. The percolation problem on the Bethe lattice can be formulated in the following way. Assume the central site is chemically active and the other sites are active with probability p (<1). A bond exists between two sites only if both are active. What is the critical value of p at which there will be a path connecting the central site to a site infinitely far away (on the surface).

The average number of occupied neighbors to which we can continue the path from a given site is $(z - 1)p$. If this number is less than one the number of path decreases by this factor at every site and the central site will definitely not be connected to infinity. Therefore, for the percolation threshold one gets $p_{cr} = 1/(z - 1)$. However, even if $p > p_{cr}$, the central site is not necessarily connected to infinity. For example, it may be active but its z neighbors are not (this happens with probability $(1 - p)^{z-1}$). We define the percolation probability P, as the probability that the central site or any other arbitrarily chosen site belongs to the infinite (macroscopic) interconnected cluster. In the case of $z = 3$, P is related to Q, the probability that an arbitrary site is not connected to infinity through one fixed branch originating from that site, by $P = p(1 - Q)^3$. Q is easy to calculate. Consider the central site in Figure 7 (although any other could be chosen) and the site explicitly indicated as neighbor which is on one of the three branches emanating from the central site.

On one hand the central site cannot belong to the infinite cluster if the neighbor is not active (this happens with probability $1 - p$).

On the other hand, even if the neighbor is active (with probability p) but it itself is not connected to infinity (with probability Q since a sub-branch is connected or not connected to infinity with the same probability as the branch itself) the central site will not belong to the infinite cluster with probability pQ^2. (Note that there are two sub-branches emanating from the neighbor, that is why Q^2 appears.) Finally, one obtains $Q = 1 - p + pQ^2$. The non-trivial solution (besides the trivial, $Q = 1$) is $Q = (1 - p)/p$, which determines P. Evaluating the expression for P at p slightly larger than $p_{cr} = 0.5$, one gets $P \sim (p - 0.5)$, indicating that the critical exponent for this quantity is 1.

5. SUMMARY

The purpose of this work is to aquaint biologists with some of the networks encountered in the physical sciences, as well as to provide an elementary introduction into percolation phenomena. We concentrate on percolation since it is one of the most widely used techniques to study disordered networks of which many appear in biological systems.

The examples discussed above illustrate the great redundancy and "flexibility" of percolating assemblies, important characteristics of networks encountered in living organisms. The random structures in Figures 2–4 may assume any shape, with the possibility of one going continuously over to another and still providing global connectivity. Contrary, if the system is allowed to exist only in the form of a highly structured network, its ability to adjust to changes in its environment is strongly limited. Moreover, in percolating systems a phase transition may be induced, in the vicinity of which system properties vary according to well-established quantitative dependencies.

Therefore, it is possible to distinguish this type of networks from others (*e.g.* tensegrity structures) and to analyze and predict the behavior of experimentally measurable quantities.

REFERENCES

Adam M., Delsanti M. and Durand D., *Macromolecules* **18** (1985) 2285.

Ashcroft N.W. and Mermin N.D., Solid State Physics (Holt, Rinehart and Winston, New York, 1976).

Bader B.L., Magin T.M. Freudenmann M., Stumpp S. and Franke W.W., *J. Cell. Biol.* **115** (1991) 1293.

Bailey N., The Mathematical Theory of Infectious Diseases (Hafner, New York, 1975).

Birk D.E., Trelstad R.L. and Silver F.H., In Cell Biology of the Extracellular Matrix, edited by E.D. Hey (Plenum, New York, 1991) p. 221.

Blessing M., Ruther U. and Franke W.W., *J. Cell. Biol.* **120** (1993) 743.

Bouchaud E., Delsanti M., Adam M., Daoud M. and Durand D., *J. Phys. France* **47** (1986) 1273.

Bridgeman P.C. and Reese T.S., *J. Cell. Biol.* **99** (1984) 195.

Bug A.L.R., Safran S.A. and Webman I., *Phys. Rev. Lett.* **54** (1985) 1412.

Cahn R., *Nature* **389** (1997) 121.

Campbell G. and Forgacs G., *Phys. Rev. A* **41** (1990) 4570.

Clerk J.P., Giraud G., Laugier J.M. and Luck J.M., *Adv. Phys.* **39** (1990) 191.

De Gennes P.G., *J. Phys. France* **30** (1976a) 1049.

De Gennes P.G., *J. Phys. France* **37** (1976b) L1-2.

De Gennes P.G., *J. Phys. France* **40** (1979) 783.

Delisi C. and Perelson A.S., *J. Theor. Biol.* **62** (1976) 159.

Domb C., In Percolation structures and processes, edited by G. Deuthcer, R. Zallen and J. Adler, *Ann. Israel Phys. Soc.* **5** (1983) 17.

Dworerzky S.I., Fey E.G., Penman S.J., Lian B., Stein J.L. and Stein G.S., *Proc. Natl. Acad. Sci. USA* **87** (1990) 4605.

Fey E.G. and Penman S., *Proc. Natl. Acad. Sci. USA* **85** (1988) 121.

Fukui Y. and Katsumara H., *Exp. Cell. Res.* **120** (1979) 451.

Heuser J.I. and Kirschner M.W., *J. Cell. Biol.* **86** (1980) 212.

Jockusch B.M. and Becker M., *Exp. Cell. Biol.* **89** (1974) 141.

MacKay G. and Naeem J., *J. Phys. A* **17** (1984) L757.

McDonald J.A., *Ann. Rev. Cell. Biol.* **4** (1988) 183.

Mitchison T.J., *Biol. Cell.* **3** (1992) 1309.

Nakayama T., Yakubo K. and Orback R., *Rev. Mod. Phys.* **66** (1994) 381.

Osborn M. and Weber K., *Exp. Cell. Res.* **129** (1980) 103.

Perelson A.S. and Oster G.F., *J. Theor. Biol.* **81** (1979) 645.

Peyrelasse J. and Boned C., *Phys. Rev. A* **41** (1990) 938.

Sahimi M., Applications of percolation theory (Taylor & Francis, London, 1994).

Sciortino F., Geiger A. and Stanley H.E., *Nature* **354** (1991) 218.

Stauffer D. and Sahimi M., *Int. Mod. Phys.* **C4** (1993) 401.
Stauffer D. and Weisbusch G., *Physica A* **108** (1992) 42.
Stanley H.E., Introduction to phase transitions and critical phenomena (Clarendon Press, Oxford, 1971).
Stanley H.E., Texeira J., Geiger A. and Blumberg R.L., *Physica A* **106** (1981) 260.
Stossel T.P., *Science* **260** (1993) 1086-1094.
Utracki L.A., *J. Rheol.* **35** (1991) 1615.
Xu J. and Stell G., *J. Chem. Phys.* **89** (1988) 1101.

Networks of Biopolymers

LECTURE 2

Complex Networks in Cell Biology

P.A. Janmey, J.V. Shah and J.X. Tang

Experimental Medicine Division, Brigham and Women's Hospital, Harvard Medical School, 221 Longwood Ave., Boston, MA 02115, USA

Networks of filamentous polymers are a common feature of cells and biological tissues. From the cross-linked gels providing strength to blood clots and extra-cellular matrices to the dynamic cytoskeleton that undergoes gel-sol transitions as cells move and the more obscure matrix within the cell nucleus, nearly all cellular surfaces are interconnected by a series of fluctuating but continuous meshworks. In part the function of these networks is mechanical, and their viscoelasticity, together with new methods to observe the dynamics of single polymer strands within the network, has made biological polymers, especially those derived from the cytoskeleton, attractive materials by which to test theories of polymer physics. The unusual viscoelasticity of these networks is crucial to their biological function to provide mechanical continuity throughout cells and tissues. In addition, formation and disassembly of networks composed of the polyanionic filaments comprising the cytoskeleton has many consequences for sequestration of proteins and metabolites and in providing directionality and spatial segregation to biochemical reactions and signal transduction pathways. Concepts related to network formation may transform the way intracellular biochemistry and signaling are understood.

1. INTRODUCTION AND EXAMPLES OF BIOLOGICAL PROTEIN NETWORKS

Much of the structural material in soft tissues is composed of filaments formed by association of proteins into linear polymers. Such filaments contain hundreds to millions of subunits, each of which is itself a polymer of molecular weight typically around 10,000 to 100,000. Because these polymers are generally longer and stiffer than synthetic polymers, and some can form gels in aqueous solutions at very low volume fraction, they offer many possibilities for optical and rheological measurements that illuminate the dynamic behavior of polymers in solution.

1.1. Stable network-forming proteins

Amongst the most common network-forming proteins are collagen and elastin, which account for much of the hydrated elastic material surrounding internal organs, blood vessels, and cells. Other important structural proteins include fibrin, which enables the gelation of blood during clotting and wound healing; and keratins, which form external structures such as skin, nails and hair. Structures such as these are generally formed as stable materials, to be remodeled or removed by proteolytic degradation of the covalent bonds from which they are formed. These proteins are among the oldest known polymers, and their bulk material properties have been extensively measured. Some protein networks, in particular those formed by elastin [26], have viscoelastic properties remarkably similar to those of cross-linked rubbers. In contrast, other proteins, notably collagen and fibrin, are useful models of rodlike or semiflexible polymers, and in the latter case, provide good systems for viscoelastic characterization of topologically constrained but uncrosslinked rodlike polymer networks [1, 6].

1.2. Dynamic biopolymer networks and the cytoskeleton

Another setting in which protein filaments occur in biology is within the cytoplasm of most eukaryotic cells which contain an intricate and dynamic network of polymers called the cytoskeleton. The cytoskeleton provides the mechanical stiffness required for elastic resistance to deformation that allows, for example, blood cells to maintain their shape in the presence of fluid flow, or recover after deformation in capillaries The cytoskeleton also accounts for the resistance of epithelial cells in the skin to mechanical stress, and genetic defects in cytoskeletal proteins can compromise cellular mechanical properties and result in mechanically induced blistering [9].

The fundamental structural unit of the eukaryotic cytoskeleton is a filamentous protein polymer composed of one of three chemically distinct subunits, actin, tubulin, or one of several classes of intermediate filament proteins [35]. The molecular structures of these filaments have been extensively characterized by light and electron microscopy, and in the case of actin filaments, a molecular model based on the crystal structure of the actin monomer is available [27]. The viscoelastic properties of macroscopic samples of each filament types have also been characterized by rheologic techniques [17], and the microscopic elasticity of individual filaments, determined by micromanipulation or analysis of their thermal motions, are available in some cases [11, 16, 20, 23, 33]. The three kinds of cytoskeletal polymers represent the three distinct classes of viscoelastic polymers: rodlike (microtubules), semi-flexible (F-actin) and flexible (intermediate filaments) [22], as illustrated in Figure 1.

Microtubules are among the stiffest polymers known, with contour lengths (L) of many microns, a diameter of 25 nm, and a molecular stiffness characterized by a persistence length (Lp, the length of a polymer over which it appears to be straight, or over which the directions of tangent vectors are correlated) of nearly 1 mm [5, 11, 25, 31, 39]. While MTs are deformable by biologically relevant forces, on typical cellular length scales they are well approximated as straight.

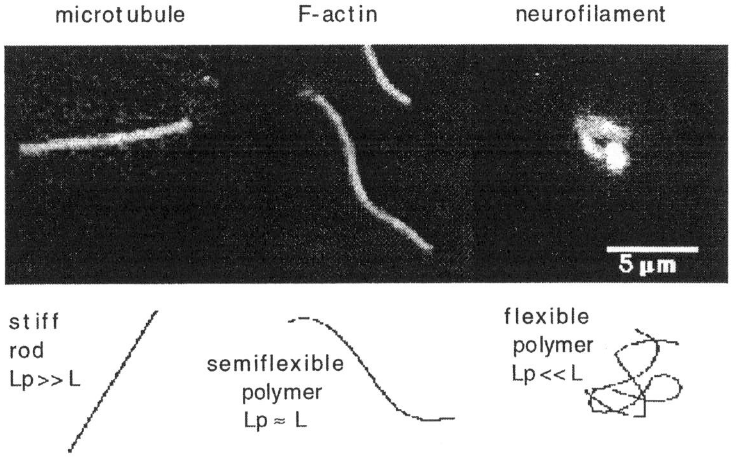

Fig. 1. Structures of cytoskeletal polymers in dilute solution.

Actin filaments (8 nm diameter) are significantly more flexible, with persistence lengths on the order of 1 micron, with different studies reporting values from 0.2 to 17 microns [2, 10, 11, 16, 20, 33, 36]. As a result, on the scale of an entire filament (several microns) or even the mesh size of a cytoplasmic network (several hundred nm) these filaments are highly elongated but exhibit significant flexibility [36].

Intermediate filaments, of all types studied to date, are qualitatively more flexible than F-actin even though they have larger diameter (10 nm). They are better modeled as flexible polymers, although they are far stiffer than the random coil polymers common in many synthetic polymer gels.

2. RHEOLOGY OF CYTOSKELETAL POLYMER NETWORKS

When polymerized in the absence of other proteins, each type of cytoskeletal polymer forms network with distinct viscoelastic properties and some common features. When actin, tubulin, or vimentin (an intermediate filament protein) are polymerized under conditions that maximize filament length, each protein forms aqueous gels at volume fractions below 0.2%. In these materials the storage shear modulus G' is greater than the loss modulus G", implying the formation of highly elastic networks. In each case, the elastic moduli are very strongly dependent on filament length, and modulation of this feature *in vivo* is presumably a mechanism by which cellular stiffness is regulated.

While there are some basic similarities among the networks formed by the three cytoskeletal polymers, there are also large quantitative and qualitative differences. At low strains, F-actin forms the strongest gels, with values of G' approaching 1000 Pa for approximately physiological concentrations of 10 mg/ml (1% volume fraction). Microtubules, the stiffest of the three polymers, form the weakest gels at

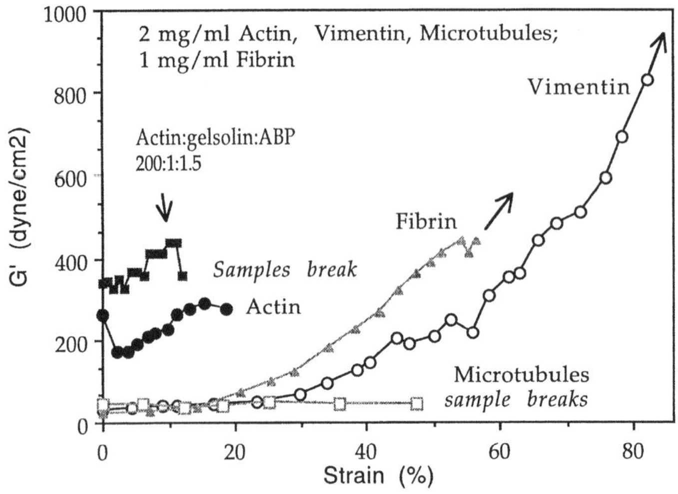

Fig. 2. Strain hardening of fibrin and vimentin and rupture of actin and microtubule networks at high strains.

similar concentrations, presumably because of the absence of entanglements or the tendency of these stiff polymers to form nematic phases and bundles.

One of the most striking differences among the polymers is the dependence of their elastic moduli on the extent of deformation. Figure 2 shows the shear moduli of five biopolymer gels measured over a range of shear strains. F-actin exhibits a high degree of strain hardening (an increase in G' when the strain is increased) over a range of strains from 1 to approximately 20%, followed by an abrupt, irrecoverable decrease at larger strains, attributed to network rupture. In contrast, the shear moduli of vimentin networks, while lower than those of F-actin when measured at very low strains, remain intact even after large deformations (200%), unlike microtubule and actin networks, which are extremely fragile to mechanical deformation. Their shear modulus, G', increases by nearly a factor of 100 over this range. In many respects, the rheology of vimentin networks resembles that of fibrin clots, shown for comparison in Figure 2. The viscoelastic properties of vimentin suggest that this type of intermediate filament may be important for basic cellular integrity, and may provide cytoskeletal networks of some cells with rheologic properties commensurate with those of the extra-cellular environment. In each case, the elastic moduli of purified cytoskeletal polymers, measured at approximately 10 rad/s, is due largely to steric constraints or transient interactions since they all exhibit a steady rate of shear deformation (creep) when subjected to a constant stress for a long time.

In the cell, network structure is modified by a large number of accessory proteins, including some that specifically cross-link individual filaments. Introduction of cross-links between actin filaments strongly increases the elastic moduli of short F-actin solutions *in vitro* [18], and loss of the expression of one such F-actin cross-linker (ABP-280) strongly alters cell shape and inhibits its ability to move [3]. When ABP-280 is added to solutions of short (670 nm) actin filaments regulated by the filament severing protein gelsolin, the initially viscous solution forms a gel with a larger shear modulus than that the same concentration of F-actin polymerized to form filaments with lengths greater than 20 microns (Fig. 2). However, even the presence of cross-links between actin filaments does not prevent the strain-dependent rupture, suggesting that the filaments themselves, and not the contacts between them are altered.

3. MOLECULAR BASIS OF BIOPOLYMER VISCOELASTICITY

Some biopolymers such as elastin and titin [40] are highly flexible and form networks with rubber-like elasticity. In contrast, cytoskeletal and other protein filaments such as fibrin and collagen are rodlike or semi-flexible polymers. Their viscoelasticity is fundamentally different, and the rheological properties of such networks can provide insight into the physics of these rodlike molecular networks. For example, video and electron microscopy reveal that networks of uncross-linked rodlike polymers can exhibit large elastic moduli without any structures resembling blobs or entanglements generally thought to be the basis for the elastic response of uncross-linked flexible polymers [22]. The nature of the elastically effective links between rodlike polymers is not fully defined and a number of alternatives have been suggested [15, 24, 29, 34].

Weak but numerous non-covalent attractive interactions have been proposed to be essential for the viscoelasticity of F-actin, for example, but direct measurement of single filament diffusion in actin networks has failed to reveal such contacts, suggesting that if they are important, their lifetime is very short. Alternatively, the elastic response of the network may depend on the restricted motion of the filaments within the tube formed by their neighbors. Also in this case, alternative models have been proposed that consider both the deformation of the test filament within the virtual tube and the tube itself as contributing to the elastic response [15, 24, 29]. The general importance of steric constraints, as opposed to chemical bonds, is implied by the enormous effect of filament length on biopolymer gel rheology as the filament length L and number concentration c are varied in the range between $1/L^3 << c << 1/dL^2$ where d is the filament diameter [19].

A fundamental question concerning the origin of elasticity in biopolymer solutions and gels in general, and in actin networks in particular, is whether this elasticity results from an entropic mechanism, similar to that which leads to the elasticity of rubbers, or whether the elastic response is due to a purely mechanical effect, such as the bending of the individual filaments. Models have recently been proposed based on both contributions (reviewed in [28]). The ability to manipulate such molecular features as polymer length, stiffness, and cross-linking and ascertain their effects on rheology suggests that biopolymers have much to contribute to development of general theories for polymer physics.

4. FILAMENT DIFFUSION WITHIN NETWORKS

One advantage of some biopolymers is that their enormous lengths and relatively elongated structures cause them to form networks at very low volume fraction, and thus facilitate the interpretation of scattering measurements and their relation to theory [12] (see also the chapter by Frey *et al.* in this volume). In addition, the ability to attach fluorescent labels to cytoskeletal proteins enables single filaments to be visualized in solution by light microscopy [14, 20]. Filament diffusion experiments have been extended to visualize labeled microtubule (MT) and actin filament [21, 22] diffusion in a variety of cytoskeletal polymer networks in the presence and absence of potential cross-linking proteins. If an inter-filament interaction is present, it can be detected by a decrease in long-range diffusive motions.

One example of such an experiment is shown in Figure 3 where the mean square displacement of a microtubule (MT) within a pure actin network is compared to that in a network of actin also containing the neuronal cytoskeletal protein MAP2c, which is thought to mediate interactions between the actin and microtubule systems *in vivo* [4, 32]. The strong inhibition of MT diffusion at long times is clear evidence for a specific molecular link between MTs and F-actin that requires mediation by MAP2c.

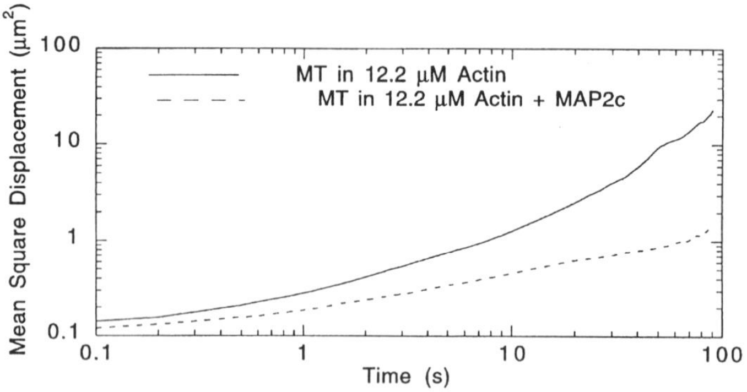

Fig. 3. Mean square displacements of a MT end in an actin network are diminished in the presence of MAP2c.

5. THE POLYELECTROLYTE NATURE OF BIOPOLYMERS

Most filamentous protein biopolymers, including all cytoskeletal elements have large, usually negative, linear charge densities. The polyelectrolyte properties of elastin [41], F-actin [37], and other protein polymers [38] have recently been reported, and they resemble those of the much more thoroughly studied DNA. For instance, the charge spacing on an actin filament is approximately 2.5 Å, significantly less than the Bjerrum length $\lambda_B = 7.1$ Å. As a result distinct populations of multivalent cations surround these filaments but are neither bound to specific sites nor entirely free in solution.

The amount and effect of these condensed ions depend on their valence, and above a critical level of charge neutralization the filaments are predicted to collapse into bundles or toroids whose structure depends on both the inherent stiffness of the filament and the ionic conditions. An example of the generality of this bundling transition is shown in Figure 4 where the formation of bundles is detected by light scattering after addition of cationic lysine oligomers to a variety of anionic filaments. The finding that homo-oligomers of cationic amino acids promote bundle formation by the structurally unrelated anionic filaments formed by F-actin, fd virus, microtubules, vimentin, DNA, and tobacco mosaic virus (TMV) illustrates the strong and general consequences of these polyelectrolyte effects.

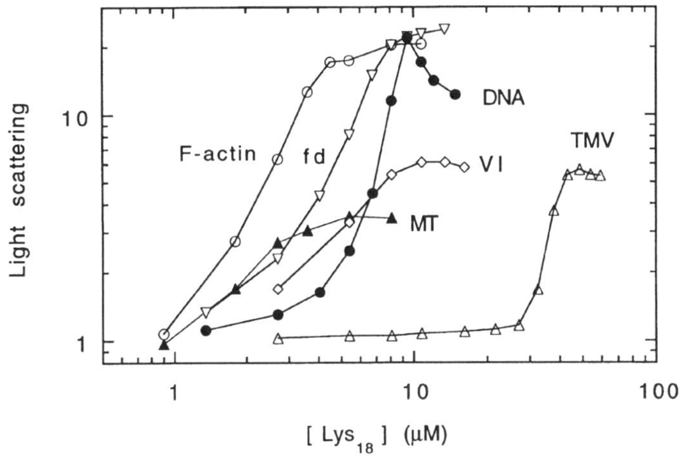

Fig. 4. Structurally distinct anionic filaments can be bundled efficiently by the same cationic ligand.

In addition to promoting bundle formation, condensation of multivalent cations competitively displaces cations of lower valence, and such a result may contribute significantly to the binding of numerous proteins to the cytoskeleton and other filamentous matrices. Experimental evidence for such effects have recently been obtained for a number of protein polymers [37, 38], and a theoretical basis for the attractive interaction facilitated by counterion condensation has been presented [13].

The fact that cytoskeletal networks are not static, but are rapidly assembled and disassembled in response to cellular stimuli, suggests that polyelectrolyte effects may be important not only for mediating protein-protein binding, but also for cellular signaling. The possibility that these anionic filaments can act as conductors of cations such as H^+ or Ca^{2+} has been proposed and awaits experimental verification in a live cell. In a more general sense, the possibility that transient formation of percolating networks spanning the cytoplasm from the cell membrane to the nucleus opens up a vast array of possibilities both for mechanical coupling [30] and transmission of chemical messengers [7, 8].

6. CONCLUSIONS

The unique features of biological polymers and their ability to form networks in very dilute solutions provide many opportunities for studies related both to understanding the physics and chemistry of polymers and to the physiologic functions of these materials. Recent advances in the ability to manipulate the structures of biopolymers and to purify proteins that alter such features as polymer length, stiffness, and interconnections, along with the ability to visualize single filaments in solutions, allow fundamental questions about the way large molecules interact with each other and with their solvent to be addressed. The combination of stable networks that provide mechanical strength to tissues and dynamic networks that form or disappear as cells respond to extra-cellular cues illustrates both the complexity of this biological system and its richness as a source of materials and methods for physical and biological study.

REFERENCES

[1] Bale M.D., Muller M.F. and Ferry J.D., *Biopol.* **24** (1985) 461-482.

[2] Burlacu S., Janmey P.A. and Borejdo J., *Am. J. Physiol. C* (1992) 569-577.

[3] Cunningham C.C., Gorlin J.B., Kwiatkowski D.J., Hartwig J.H., Janmey P.A., Byers H.R. and Stossel T.P., *Science* **255** (1992) 325-327.

[4] Cunningham C.C., Leclerc N., Flanagan L.A., Lu M., Janmey P.A. and Kosik K.S., *J. Cell. Biol.* **136** (1997) 845-857.

[5] Dye R.B., Fink S.P. and Williams R.J., *J. Biol. Chem.* **268** (1993) 6847-6850.

[6] Ferry J.D., *Ann. NY Acad. Sci.* **408** (1983) 1-10.

[7] Forgacs G., *Biochem. Cell. Biol.* **73** (1995) 317-326.

[8] Forgacs G., *J. Cell. Sci.* (1995) 2131-2143.

[9] Fuchs E., *Ann. Rev. Genet.* **30** (1996) 197-231.

[10] Fujime S., Ishiwata S. and Maeda T., *Biophys. Chem.* **20** (1984) 1-21.

[11] Gittes F., Mickey B., Nettleton J. and Howard J., *J. Cell. Biol.* (1993) 923-934.

[12] Gotter R., Kroy K., Frey E., Barmann M. and Sackmann E., *Macromol.* **29** (1996) 30-36.

[13] Gronbech-Jensen N., Mashl R.J., Bruinsma R.F. and Gelbart W.M., *Phys. Rev. Lett.* **78** (1997) 2477-2480.

[14] Harada Y. and Yanagida T., *Cell. Motil. Cytoskeleton* **10** (1988) 71-76.

[15] Isambert H. and Maggs A., *Macromol.* **29** (1996) 1036-1040.

[16] Isambert H., Venier P., Maggs A.C., Fattoum A., Kassab R., Pantaloni D. and Carlier M.F., *J. Biol. Chem.* **270** (1995) 11437-11444.

[17] Janmey P.A., Euteneuer U., Traub P. and Schliwa M., *J. Cell. Biol.* **113** (1991) 155-160.

[18] Janmey P.A., Hvidt S., Lamb J. and Stossel T.P., *Nature* **345** (1990) 89-92.

[19] Janmey P.A., Hvidt S., Peetermans J., Lamb J., Ferry J.D. and Stossel T.P., *Biochemistry* **27** (1988) 8218-8227.

[20] Kas J., Strey H., Barmann M. and Sackmann E., *Europhys. Lett.* **21** (1993) 865-870.

[21] Kas J., Strey H. and Sackmann E., *Nature* **368** (1994) 226-229.

[22] Kas J., Strey H., Tang J.X., Finger D., Ezzell R., Sackmann E. and Janmey P.A., *Biophys. J.* **70** (1996) 609-625.

[23] Kojima H., Ishijima A. and Yanagida T., *Proc. Nat. Acad. Sci. USA* **91** (1994) 12962-12966.

[24] Kroy K. and Frey E., *Phys. Rev. Lett.* **77** (1996) 306-309.

[25] Kurachi M., Hoshi M. and Tashiro H., *Cell. Motil. Cytoskeleton* (1995) 221-228.

[26] Lillie M.A. and Gosline J.M., *Biorheology* **30** (1993) 229-242.

[27] Lorenz M., Poole K.J., Popp D., Rosenbaum G. and Holmes K.C., *J. Mol. Biol.* **246** (1995) 108-119.

[28] MacKintosh F. and Janmey P.A., *Curr. Op. Solid State Mat. Sci.* **2** (1997).

[29] MacKintosh F., Käs J. and Janmey P.A., *Phys. Rev. Lett.* **75** (1995) 4425-4428.

[30] Maniotis A.J., Chen C.S. and Ingber D.E., *Proc. Nat. Acad. Sci. USA* **94** (1997) 849-854.

[31] Mickey B. and Howard J., *J. Cell. Biol.* **130** (1995) 909-917.

[32] Nishida E., Kuwaki T. and Sakai H., *J. Biochem.* **90** (1981) 575-578.

[33] Ott A., Magnasco M., Simon A. and Libchaber A., *Phys. Rev. E* (1993) r1642-r1645.

[34] Satcher R. and Dewey C., *Biophys. J.* **71** (1996) 109-118.

[35] Schliwa M., The Cytoskeleton: An Introductory Survey (Springer-Verlag, Vienna, 1987) p. 326.

[36] Schmidt C.F., Barmann M., Isenberg G. and Sackmann E., *Macromol.* **22** (1989) 3638-3649.

[37] Tang J., Wong S., Tran P. and Janmey P.A., *Ber. Bunsenges Phys. Chem.* (1996) 796-806.

[38] Tang J.X. and Janmey P.A., *J. Biol. Chem.* **271** (1996) 8556-8563.

[39] Venier P., Maggs A.C., Carlier M.F. and Pantaloni D., *J. Biol. Chem.* **269** (1994) 13353-13360.

[40] Wang K., McCarter R., Wright J., Beverly J. and Ramirez-Mitchell R., *Biophys. J.* **64** (1993) 1161-1177.

[41] Winlove C.P., Parker K.H., Ewins A.R. and Birchler N.E., *J. Biomech. Eng.* **114** (1992) 293-300.

LECTURE 3

The Keratocyte Cytoskeleton: A Dynamic Structural Network

K.I. Anderson

Institute of Molecular Biology, Austrian Acadamy of Sciences, Billrothstr. 11, A-5020 Salzburg, Austria

1. INTRODUCTION

Self directed movement is a quality we inherently identify with living things. Among multicellular animals the independent crawling of cells or groups of cells is important during embryonic development, wound healing, the cellular responses of the immune system, and tumor metastasis. As an introduction to my own studies of cell movement I would like to briefly highlight the development of ideas concerning active cell movement due to contractile forces generated inside the cell, and passive cell movement due to surface tension forces generated at the cell surface.

One of the first model systems used to study cell motility was the amoba, in which crawling is accompanied by dramatic cytoplasmic streaming. Early discussion of the mechanism of amoeboid crawling centered on the proposals of Ehrenburg (1830; 1832) and Dujardin (1835; 1839). Ehrenburg proposed that crawling resulted through contraction of the cell body, which forced the cytoplasm forward through weak spots at the front of the cell. Dujardin, on the other hand, proposed that the material comprising the cell was inherently able to both expand and contract; thus motility resulted from the expansion of cell matter at the front and its contraction at the rear. Although contractility was accepted as a feature of living cells, early microscopists failed to identify a structural network within the cell capable of developing and transmitting contractile force (DeBruyn, 1947). This lack of structural data eventually led to the proposal of surface tension based mechanisms of cell motility, which sought to explain ameboid movement in terms of fluid dynamics. These theories were supported by experiments in which aspects of ameboid cytoplasmic streaming and spreading were recreated using carefully chosen fluid systems. For example, Bütchli (1892) found that currents which formed within drops of an oil and soap emulsion suspended in dilute glycerin mimicked cytoplasmic streaming in amoeba.

Ultimately, surface tension based theories of movement failed to explain complex cell behavior (Jennings, 1904) and the large forces generated by moving cells (Mast and Root, 1916). A new generation of ideas followed based on another physical-chemical phenomenon: sol-gel transformation. In these models, the motive force for cell movement was thought to be generated by cyclic conversion of protoplasm between an elastic gel state and a fluid sol state. Hymann (1917), for example, revised Ehrenburgs hernia model and proposed that a cortical layer of elastic protoplasm, the ectoplasm, was a gel which squeezed the inner fluid endoplasm forward and out at the front of the cell.

In 1947 Szent-Gyorgyi demonstrated that the proteins actin and myosin were responsible for muscle contraction. Additional work on protein contraction (Astbury, 1947) suggested that protein folding was a universal biological mechanism for producing force and motion, and led to proposed mechanisms of cell motility along these lines (Goldacre and Lorch, 1950). The hypothesis that muscle contraction and cell motility share common mechanisms (Hill, 1926) was supported by the work of Hoffman-Berling and Weber (1953), who found mechanical and biochemical similarities between contracting muscle and non-muscle cells in culture. The structural basis of cell motility, and the link with muscle contraction, was finally established by Ishikawa (1969), who showed that microfilaments, previously observed by electron microscopy in the cytoplasm of tissue culture and other cells (Beyers and Porter, 1964; Cloney, 1966; Buckley and Porter, 1967), were in fact actin, based on their ability to bind the myosin fragment HMM. The dependence of cell movement upon an intact microfilament network was later established (Schroeder, 1969; Spooner *et al.*, 1970; Wessells *et al.*, 1971) using the drug cytochalasin (Carter, 1967b), which disrupts microfilament networks by blocking the polymerization of actin (Flanagin and Lin, 1980; Brenner and Korn, 1979).

At the same time that contractile elements able to generate the forces required to move cells were being discovered, other investigations led to a reconsideration of surface tension based mechanisms of cell motility. First, Steinberg (1963, 1964) proposed that different strengths of adhesion between different cell types could direct and/or cause cell motility during cell sorting, a process in which cells within a mixed population seggregate. Similarly, Carter (1967a; 1970) proposed that a gradient of adhesion, present either on the substrate or on the cell surface, could drive cell motility from weaker toward stronger adhesion in a process he termed haptotaxis. Experimental investigation failed to substantiate haptotaxis as a mechanism *causing* cell motility (Wolpert *et al.*, 1969; Harris, 1973), although it was confirmed as a mechanism capable of *directing* motility (Harris, 1973). Likewise, experiments performed in the presence of cytochalasin B to inhibit active motility showed that surface tension forces were insufficient to produce normal cell sorting, and instead more likely to function in guiding cell movement (Armstrong and Parenti, 1972; Steinberg and Wiseman, 1972). Since then it has been determined that cell to cell adhesion depends upon a class of protein receptors on the cell surface known as cadherins (Takeichi, 1990), whereas another class of receptors, the integrins, are primarily responsible for cell adhesion to the extracellular matrix (Hynes and Lander, 1992).

Continued investigation of active cell movement has generated a complex picture of a dynamic structural network of actin filaments, the actin cytoskeleton (Lazarides and Weber, 1974; Bershadsky and Vasiliev, 1988), which is responsible for a variety of cell movements, including cell contraction, cell division, cytoplasmic streaming, secretion, and the folding of cell layers during morphogenesis. Actin is a globular protein (Mw = 43 kD) which can polymerize to form polarized filaments capable of directional assembly (Wegner, 1976). Within the cell, actin dynamics are modulated by a multitude of proteins, which control such events as actin polymerization and depolymerization, cross linking of filaments into bundles and meshworks, and the contraction of filament networks (Burridge and Chrzanowska-Wodnicka, 1996). Much current research is focused on how, where, and when these events are regulated within the cell in order to produce directed motility

2. PROTRUSION AND RETRACTION: THE TWO STEPS OF CRAWLING

Recalling the proposals of Ehrenburg and Dujardin, it was early recognized that cell crawling can be broken down into two distinct steps: protrusion, whereby the leading cell boundary advances, and retraction, whereby the trailing cell body advances. Protrusion at the front of the cell is generally associated with a thin layer of active cytoplasm known as the lamellipodium (Loeb, 1922; Abercrombie, 1954; 1970a; 1970b) (Fig. 1). Retraction of the cell body, on the other hand, is a less unified process which varies among different cell types. The level of coordination between these processes determines whether a cell will move smoothly and rapidly or in a slow and erratic manner.

The fish keratocyte is an epithelial cell, popular as a model system for studying cell motility because of its rapid, persistent motility and exaggerated lamellipodium (Goodrich, 1924; Euteneuer and Schliwa, 1984; Kucik *et al.*, 1989; Theriot and Mitchison, 1991, Small *et al.*, 1995). The dependence of keratocyte motility on actin filaments can be demonstrated by the compound effect of cytochalasin B treatment (Fig. 1). The first, rapid effect is to stop protrusion of the lamellipodium. The second effect, inhibition of cell body movement, is delayed until after the cell body has advanced well into the lamellipodium. The first effect demonstrates that protrusion of the leading edge directly depends upon actin polymerization. As I will show, the second effect demonstrates the *indirect* dependence of cell body translocation on actin polymerization, via the filament network actin polymerization generates.

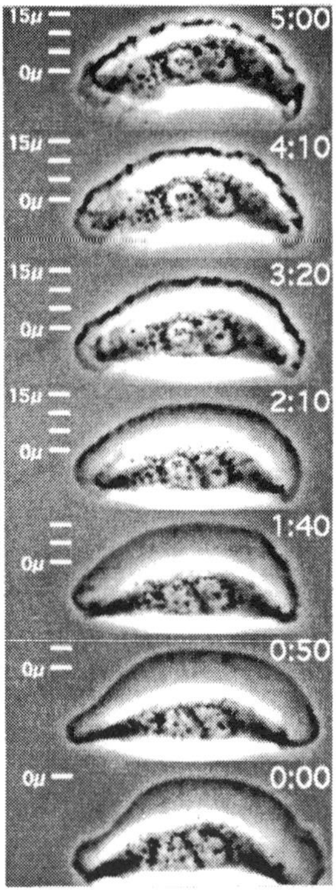

Fig. 1. Time lapse video sequence documenting the arrest of cell movement due to treatment with 200 ng/ml cytochalasin B. Addition of drug at *1:20* rapidly stops protrusion of the lamellipodium by *1:40*, but arrest of the cell body is delayed until it has advanced to the leading edge at *5:00*. Bars mark stationary positions on the substrate separated by 5 μ.

3. ACTIN POLYMERIZATION IN THE LAMELLIPODIUM

How might actin polymerization be involved in cell protrusion? The lamellipodium is a thin layer of cytoplasm comprised of a dense meshwork of actin filaments, oriented with their fast polymerizing ends located at the cell margin (Small *et al.*, 1978; Okabe and Hirokawa, 1989; Small *et al.*, 1995). This organization is consistent with a mechanism of protrusion based on the treadmilling of actin filaments, as first proposed by Wegner for actin filaments in vitro (Wegner, 1976). In this model of actin assembly, new monomers are continuously added to one of the end of a filament and simultaneously removed from its other end. If the actin filaments of the lamellipodium remain stationary relative to the substrate, as is the case in the keratocyte (Theriot and Mitchison, 1991), continuous monomer addition at the cell margin could be coupled to, or even generate protrusion of the leading edge (see below, and Fig. 2). Alternatively, if the leading cell margin remains

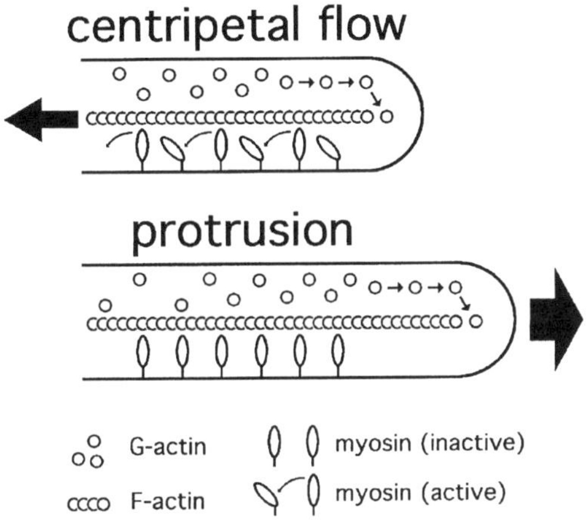

Fig. 2. The treadmilling of actin filaments is compatible with two states of filament dynamics in the lamellipodium. In the first, filament growth at the (+) end is offset by myosin based retraction, which draws actin filaments inward toward the cell body. As a result there is no change in position of the leading edge. In the second case, actin filaments remain stationary relative to the substrate and filament growth results in forward displacement of the leading edge.

stationary, continuous addition of monomers to the filament ends at the cell margin should result in the inward flux of actin through the filaments of the lamellipodium. This flux of actin has been demonstrated by Wang (1985) in stationary fibroblast lamellipodia. Assuming that treadmilling occurs at a constant rate, it has been proposed that different rates of protrusion could arise through the variable coupling of the actin filaments in the lamellipodium to the substrate via a "molecular clutch" mechanism (Mitchison and Kirschner, 1988). Consistent with this hypothesis, Lin and Forscher (1995) have shown that as the rate of protrusion slows in a nerve growth cone, the rate of centripetal flow of its actin filaments rises in inverse proportion.

The juxtaposition of the membrane at the growing filament tips might be expected to block monomer addition and thus inhibit actin polymerization. One proposal for actin based protrusion thus envisions a myosin based motor at, or near the filament tip to push the membrane forward and allow the insertion of new monomers at the filament tip (Small *et al.*, 1993; Cooper, 1991). Alternatively, actin polymerization might provide a mechanism for rectifying the diffusion of the plasma membrane (Peskin, *et al.*, 1993). In this model, oscillations in the position of the plasma membrane due to its thermal energy would generate the spaces necessary for insertion of new monomers at the filament tips. Forward diffusion of the plasma membrane may be aided by the thermal bending of actin filaments pushing against it (Mogiliner and Oster, 1996).

In contrast, other models of protrusion are not strictly based on actin polymerization per se. The nucleation release model of Theriot and Mitchison (1991) envisions the fast growing filament ends to be in the interior of the lamellipodium, distant from actual site of protrusion. The gel-swelling (Janmey *et al.*, 1992) and hydrostatic pressure (Bereiter-Hahn *et al.*, 1981) models of protrusion make protrusion dependent on the influx of water at the leading edge. According to these latter two models, increased pressure within the cell body might be expected to increase the rate of protrusion by driving cytoplasm into the lamellipodium. To test this possibility I used a microneedle to apply constant pressure to the keratocyte cell body for up to 30 seconds (Fig. 3). Despite the increase in extracellular pressure, evidenced by blebbing of the cell membrane, no change in the rate of protrusion of the leading edge was observed.

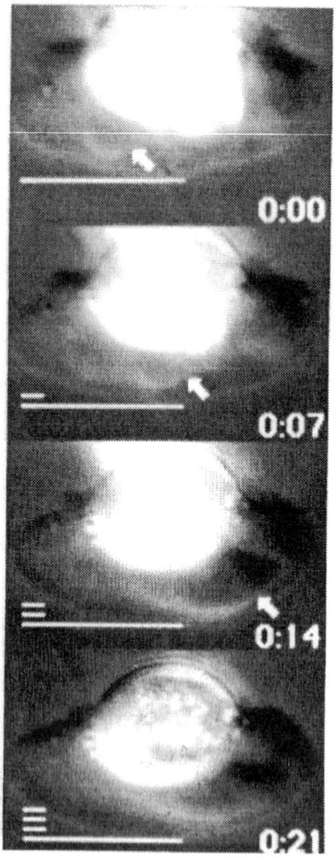

Fig. 3. Time lapse video sequence, in which a microneedle was used to press the cell body against the substrate in the first three frames. (*Arrow*) A bulge spreads across the lamellipodium from left to right, indicating the increase in pressure generated within the cell. This treatment has no effect on the rate of advance of the front edge, marked on the left by cumulative white lines from frame to frame. Normarski optics. Long bar: 15 μ. (Reproduced from *J. Cell. Biol.* **134** (1996) 1208-1218 by copyright permission of The Rockefeller University Press.)

4. CELL BODY TRANSPORT

Forward transportation of the cell body occurs in a variety of ways in moving cells. In the keratocyte, cell body transport must be efficient to keep pace with the rapid advance of the leading edge. Looking for structural clues to the mechanism of transport, we found bundles of actin filaments stretching across the back of the cell body into the lateral lobes of the lamellipodium (Fig. 4a) (see also Heath and Holifield, 1991).

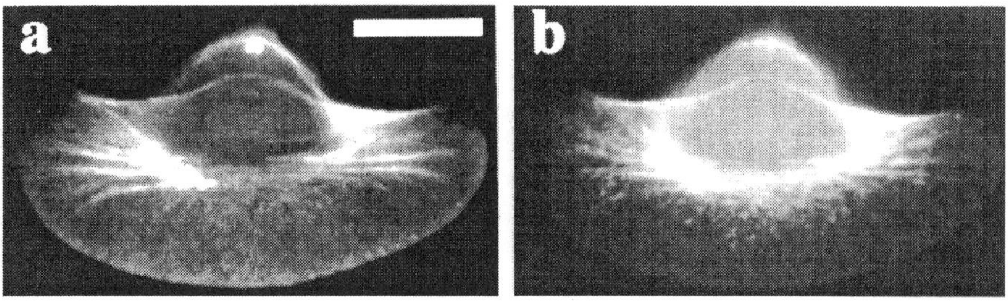

Fig. 4. Immunofluorescence images of a cell labeled with rhodamine-phalloidin (*a*) and antibodies to myosin II (*b*). Bar: 10 μ.

In contrast, at the boundary between the front of the cell body and the rear of the lamellipodium fluorescence intensity scans of rhodamine-phalloidin stained cells indicate a minimum level of actin filaments (Small *et al.*, 1995). Optical sectioning of rhodamine-phalloidin stained keratocytes indicates that the lateral bundles form a cage around the cell body (Fig. 5). Immuno labeling has localized (filamentous) myosin II with the actin bundles connecting the cell body to the lamellipodium (Fig. 4b). The possibility that these bundles may be contractile was tested by microdissection experiments (Anderson *et al.*, 1996). Severing the bundles at one side of the cell body resulted in the rapid recoil of the cell body in the opposite direction, indicating that the cell body itself was under tension (Fig. 6). Our observation of tension across the back of the cell is consistent with the work of Lee *et al.* (1994), who observed keratocytes moving on flexible substrates and obtained the puzzling result that keratocytes primarily generate tension *perpendicular* to their direction of travel. This is contrasted by the direction of force generation during fibroblast motility, which is parallel to the direction of travel (Harris *et al.*, 1981; Oliver *et al.*, 1994) (see below). The organization of the bundles and the lateral tension on the cell body suggest that the cell body is transported forward by actin bundles connecting its sides into the lateral lobes of the lamellipodium. In support of this hypothesis, it was found that the keratocyte cell body rolls during normal motility (Anderson *et al.*, 1996). Cell body rolling means that there can be at best transient connections between the middle of the cell body and the lamellipodium, whereas significant structural connections between the cell body and the lamellipodium must be present the at the sides.

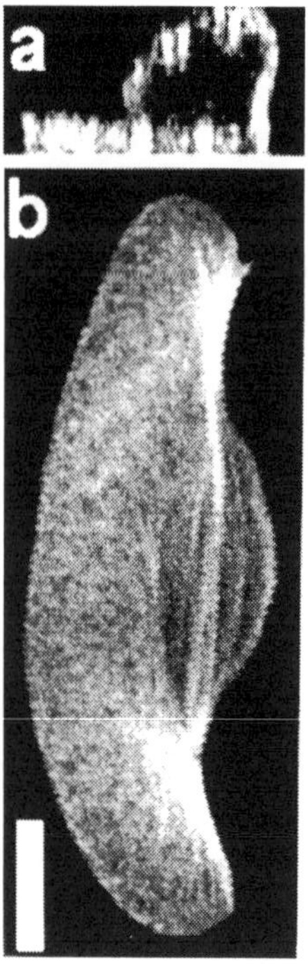

Fig. 5. (*a*) Cross section of a rhodamine-phalloidin stained keratocyte, reconstructed from a series of consecutive confocal images of the cell (*b*) acquired along the Z-axis. Note that the cell body is void of actin filaments. Bar: 10 μ.

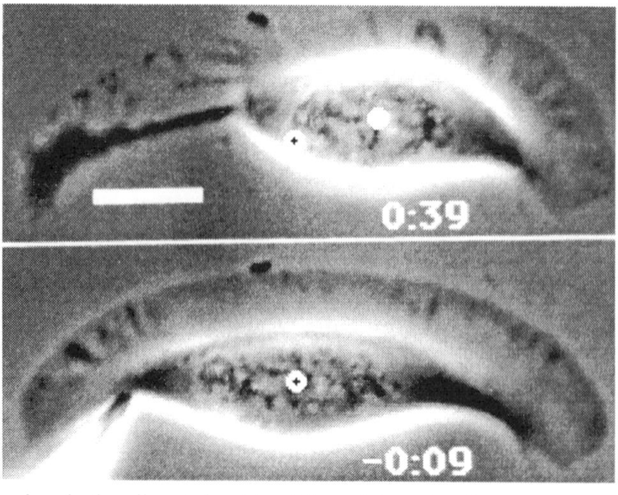

Fig. 6. Severing the bundles at the side of the cell by microdissection causes retraction of the cell body in the opposite direction. (*–0:09*) Cell and microneedle 9 seconds prior to disruption. (*White spot with black point*) initial position of cell body. (*White spot*) position of cell body after retraction. Bar: 10 μ.

5. COORDINATION OF PROTRUSION AND RETRACTION

In fibroblasts, protrusion and retraction of the cell body occur in poorly coordinated cycles (Abercrombie *et al.*, 1970a), in which the net rate of protrusion of the front of the cell is limited by the rate of transport of the cell body. First the front of the cell protrudes as far as it can, while the rear remains firmly attached to the substrate. Then the back of the cell is released and retracted into the cell body, initiating a brief phase of rapid protrusion known as retraction-induced spreading (Chen, 1979; Dunn, 1980). In contrast, keratocytes move at least 10 times faster than fibroblasts while maintaining a constant cell length (average within a sample population = 8.1 ± 0.78 μ, mean ± s.d.), indicating that protrusion and retraction occur at the same rate in these cells.

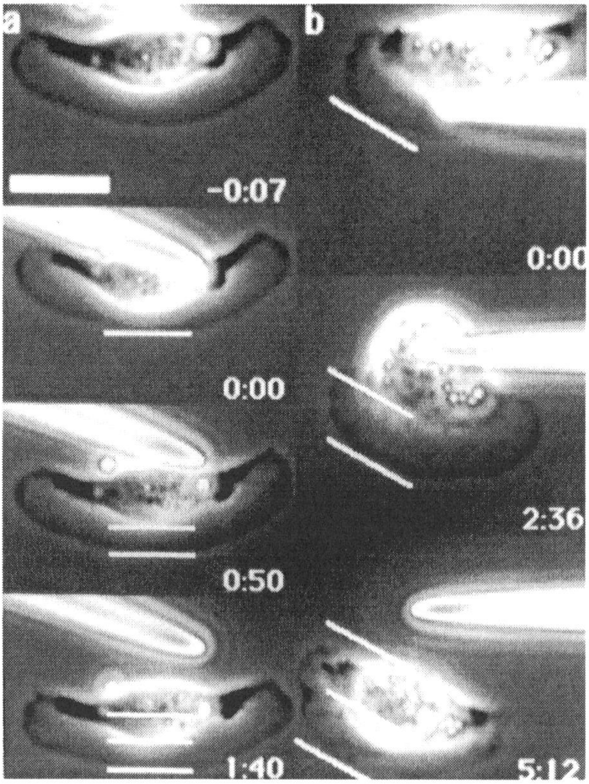

Fig. 7. Mechanical manipulation of the cell body. (*a*) Video sequence showing result of pushing cell body forward from the rear. (*Parallel lines*) Position of front of lamellipodium in subsequent frames. At *–0:07* the needle is positioned behind the cell; at *0:00* it has been thrust 5 μ forward. The time difference between each of the last three frames is the same, as is also the difference in displacement of the lamellipodium. (*b*) Restraint of cell body by glass needle. Otherwise as for *a*. Protrusion rate of lamellipodium is unaffected, as shown by equidistant lines at *5:12*. Phase contrast optics. Bar: 10 μ. (Reproduced from *J. Cell. Biol.* **134** (1996) 1208-1218 by copyright permission of The Rockefeller University Press.)

Significantly, experimental data indicate that in the keratocyte neither protrusion nor retraction directly limits the rate of the other. Manipulation of cell body transport using microneedles has shown that the rate of protrusion of the leading edge is unaffected by the position, and rate of advance of the cell body (Fig. 7), indicating that retraction of the cell body does not limit the rate of protrusion in these cells. On the other hand, inhibition of protrusion by cytochalasin B (Fig. 1) does not immediately inhibit advance of the cell body, indicating that the rate of this process is not limited by the rate of protrusion of the leading edge.

What then is the mechanism of coordination between protrusion of the leading edge and retraction of the keratocyte cell body? We (Anderson *et al.*, 1996) have proposed it is the continuous flow of actin filaments from central regions of the lamellipodium, where they are nucleated and polymerize, out toward the sides of the lamellipodium, where they are bundled, retracted, and pull the cell body forward (Fig. 8). In this model, referred to as the lateral flow model, actin filaments are nucleated and polymerize at the front edge of the lamellipodium.

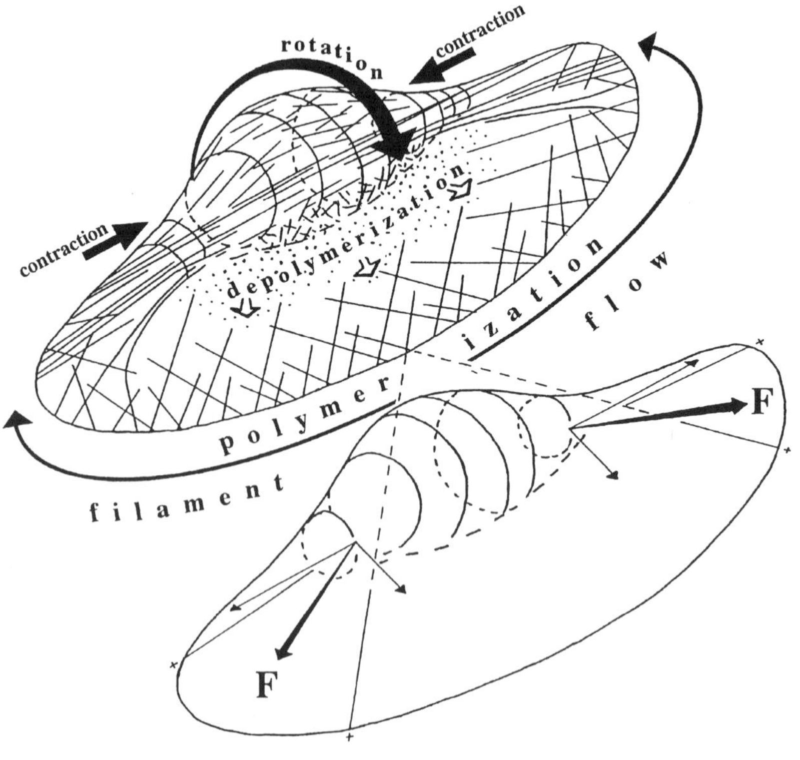

Fig. 8. Proposed model of actin filament dynamics and cell movement in the keratocyte. See text for description. (Reproduced from *J. Cell. Biol.* **134** (1996) 1208-1218 by copyright permission of The Rockefeller University Press.)

Owing to their diagonal orientation, directed filament growth leads to their displacement laterally as the cell moves (*dotted lines*), giving rise to a filament flow towards the lateral flanks of the lamellipodium. Filaments that reach the lateral flanks form bundles at the ends of the cell body and are retracted into the cell body cortex. The interaction of myosin and actin around the cell body develops tension which maintains cell body shape, and leads to a net force (*F*) that pulls the cell body forward at its sides. At the base of the cell body the depolymerization of the trailing ends of the actin filaments of the lamellipodium, and of actin filaments within the cell body, replenishes the actin monomer pool necessary to drive polymerization at the front of the cell. In conclusion, this model suggests a dual function for the lamellipodium: it advances the cell boundary through actin polymerization dependent protrusion, and in the process it generates actin filaments which are subsequently used pull the cell body forward.

ACKNOWLEDGEMENTS

I wish to thank J.V. Small for helpful discussion and critical reading of the manuscript, and Maria Schmittner for excellent photographic assistance.

REFERENCES

[1] Abercrombie M. and Heaysman J.E.M., Observations on the social behavior of cells in tissue culture. II, *Exp. Cell. Res.* **6** (1954) 293-306.

[2] Abercrombie M., Heaysman J.E.M. and Pegrum S.M., The locomotion of fibroblasts in culture. I. Movements of the leading edge, *Exp. Cell. Res.* **59** (1970a) 393-398.

[3] Abercrombie M., Heaysman J.E.M. and Pegrum S.M., The locomotion of fibroblasts in culture. II. "Ruffling", *Exp. Cell. Res.* **60** (1970b) 437-444.

[4] Anderson K.I., Wang Y.-L. and Small J.V., Coordination of protrusion and translocation of the keratocyte involves rolling of the cell body, *J. Cell. Biol.* **134** (1996) 1209-1218.

[5] Armstrong P.B. and Parenti D., Cell sorting in the presence of cytochalasin B, *J. Cell. Biol.* **55** (1972) 542-553.

[6] Astbury W.T., The Croonian Lecture: On the structure of biological fibres and the problem of muscle, *Proc. R. Soc. London* **134** (1947) 303-328.

[7] Bereiter-Hahn J., Strohmeier R., Kunzenbacher I., Beck K. and Vöth M., Locomotion of *Xenopus* epidermis cells in primary culture, *J. Cell. Sci.* **52** (1981) 289-311.

[8] Bershadsky A.D. and Vasiliev J.M., Cytoskeleton. In Cellular Organelles, edited by P. Siekevitz (Plenum Press, New York, 1988).

[9] Beyers B.a.P., K.R., Oriented microtubules in elongating cells of the developing lens rudiement after induction, *Proc. Nat. Acad. Sci. USA* **52** (1964) 1091-1099.

[10] Brenner S.L. and Korn E.D., Substoichiometric concentrations of cytochalasin D inhibit actin polymerization, *J. Biol. Chem.* **254** (1979) 9982-9985.

[11] Buckley I.K. and Porter K.R., Cytoplasmic fibrils in living cultured cells, *Protoplasma* **64** (1967) 349-380.

[12] Burridge K. and Chrzanowska-Wodnicka M., Focal adhesions, contractility, and cell signaling, *Ann. Rev. Cell. Dev. Biol.* **12** (1996) 463-519.

[13] Bütschli O., Untersuchungen über mikroskopische Schäume und das Protoplasma (W. Engelman, Leipzig, 1982).

[14] Carter S.B., Haptotaxis and the mechanism of cell motility, *Nature* **213** (1967a) 256-260.

[15] Carter S.B., Effects of cytochalasins on mammalian cells, *Nature* **213** (1967b) 261-264.

[16] Carter S.B., Cell movement and cell spreading: A passive or an active process? *Nature* **225** (1970) 858-859.

[17] Chen W.-T., Induction of spreading during fibroblast movement, *J. Cell. Biol.* **81** (1979) 684-691.

[18] Cloney R.A., Cytoplasmic filaments and cell movements: Epidermal cells during ascidian metamorphosis, *J. Ultrastruct. Res.* **14** (1966) 300-328.

[19] Cooper J.A., The role of actin polymerization in cell motility, *Ann. Rev. Physiol.* **53** (1991) 585-605.

[20] De Bruyn P.P.H., Theories of amoeboid movement, *Quart. Rev. Biol.* **22** (1947) 1-24.

[21] Dujardin F., Recherches sur les organismes inférieurs, *Ann. Sci. Nat. Zool.* **4** (1835) 343-377.

[22] Dujardin F., Mémoire sur l'organisation des Infusiores, *Ann. Sci. Nat. Zool.* **10** (1839) 230-315.

[23] Dunn G.A., Mechanisms of fibroblast locomotion. In Cell Adhesion and Motility, edited by A.S.G. Curtis and J.D. Pitts (Cambridge University Press, Cambridge, 1980).

[24] Ehrenberg C.G., Organisation, Systematik und geographisches Verhaltniss der Infusionsthierchen (F. Dümmler, Berlin, 1830).

[25] Ehrenberg C.G., Zur Erkenntniss der Organisation in der Richtung des kleinsten Raumes (F. Dümmler, Berlin, 1832).

[26] Euteneuer U. and Schliwa M., Persistent, directional motility of cells and cytoplasmic fragments in the absence of microtubules, *Nature* **310** (1984) 58-61.

[27] Flanagan M.D. and Lin S., Cytochalasins block actin filament elongation by binding to high affinity sites associated with F-actin, *J. Biol. Chem.* **255** (1980) 835-838.

[28] Goldacre R.J. and Lorch I.J., Folding and unfolding of protein molecules in relation to cytoplasmic streaming, amoeboid movement and osmotic work, *Nature* **166** (1950) 497-500.

[29] Goodrich H.B., Cell behavior in tissue cultures, *Biol. Bull.* **46** (1924) 252-262.

[30] Harris A., Behaviour of cultured cells on substrata of variable adhesiveness, *Exp. Cell. Res.* **77** (1973) 285-297.

[31] Harris A.K., Stopak D. and Wild P., Fibroblast traction as a mechanism for collagen morphogenesis, *Nature* **290** (1981) 249-251.

[32] Heath J. and Holifield B., Actin alone in lamellipodia, *Nature* **352** (1991) 107-108.

[33] Hill A.V., The viscous elastic properties of smooth muscle, *Proc. R. Soc. London* **100** (1926) 108-115.
[34] Hoffman-Berling H. and Weber H.H., Vergleich der Motilität von Zellmodellen und Muskelmodellen, *Biochem. Biophys. Acta.* **10** (1953) 629-630.
[35] Hyman L.H., Metabolic gradients in amoeba and their relation to the mechanism of amoeboid movement, *J. Exp. Zool.* **24** (1917) 55-99.
[36] Hynes R.O. and Lander A.D., Contact and adhesive specificities in the associations, migrations, and targeting of cells and axons, *Cell.* **68** (1992) 303-322.
[37] Ishikawa H., Bischoff R. and Holtzer H., Formation of arrowhead complexes with heavy meromyosin in a variety of cell types, *J. Cell. Biol.* **43** (1969) 312-328.
[38] Janmey P.A., Cunningham C.C., Oster G.F. and Stossel T.P., Cytoskeletal networks and osmotic pressure in relation to cell structure and motility. In Mechanics of Swelling. edited by T.K. Karalis (Springer-Verlag, Berlin, 1992) pp. 333-346.
[39] Jennings H.S., Contributions to the study of the behavior of lower organisms, 6. The movements and reactions of amoeba (Carnegie Inst. Wash. Publ., 1904).
[40] Kucik D.F., Elson E. and Sheetz M.P., Forward transport of glycoproteins on leading lamellipodia in locomoting cells, *Nature* **340** (1989) 315-317.
[41] Lazarides E. and Weber K., Actin antibody: the specific visualization of actin filaments in non-muscle cells, *Proc. Natl. Acad. Sci. USA* **71** (1974) 2268-2272.
[42] Lee J., Leonard M., Oliver T., Ishihara A. and Jacobson K., Traction forces generated by locomoting keratocytes, *J. Cell. Biol.* **127** (1994) 1957-1964.
[43] Lin C.-H. and Forscher P., Growth cone advance is inversely proportional to retrograde f-actin flow, *Neuron* **14** (1995) 763-771.
[44] Loeb L., Amoeboid movement, tissue formation and consistency of protoplasm, *Am. J. Psyiol.* **56** (1922) 140-167.
[45] Mast S.O. and Root F.M., Observations on ameba feeding on rotifers, nematodes, and ciliates, and their bearing on the surface tension theory, *J. Exp. Zool.* **21** (1916) 33-49.
[46] Mitchison T.J. and Kirschner M.H., Cytoskeletal dynamics and nerve growth, *Neuron* **1** (1988) 761-772.
[47] Mogilner A. and Oster G., Cell motility driven by actin polymerization, *Biophys. J.* **71** (1996) 3030-3045.
[48] Okabe S. and Hirokawa N., Incorporation and turnover of biotin-labeled actin microinjected into fibroblastic cells: An immonoelectron microscopic study, *J. Cell. Biol.* **109** (1989) 1581-1595.
[49] Oliver T., Lee J. and Jacobson K., Forces exerted by locomoting cells, *Semin. Cell. Biol.* **5** (1994) 139-147.
[50] Peskin C., Odell G. and Oster G., Cellular motions and thermal fluctions: The Brownian ratchet, *Biophys. J.* **65** (1993) 316-324.
[51] Schroeder T.E., The role of "contractile" ring filaments in dividing *Arbacia* egg, *Biol. Bull.* **137** (1969) 413-414.
[52] Small J.V., Isenberg G. and Celis J.E., Polarity of actin at the leading edge of cultured cells, *Nature* **272** (1978) 638-639.

[53] Small J.V., Rohlfs A. and Herzog M., Actin and cell movement. In Cell Behaviour: Adhesion and Motility, edited by G. Jones, C. Wigley and R. Warn (The Company of Biologists Limited, Cambridge, 1993) pp. 57-71.

[54] Small J.V., Herzog M. and Anderson K., Actin filament organization in the fish keratocyte lamellipodium, *J. Cell. Biol.* **129** (1995) 1275-1286.

[55] Spooner B.S. and Wessels N.K., Effects of cytochalasin B upon microfilaments involved in morhpogenesis of salivary epithelium, *Proc. Nat. Acad. Sci.* **66** (1970) 360-364.

[56] Steinberg M.S., Reconstruction of tissues by dissociated cells, *Science* **141** (1963) 401.

[57] Steinberg M.S., The problem of adhesive selectivity in cellular interactions, *Symp. Soc. Study Dev. Growth* **22** (1964) 321.

[58] Steinberg M.S. and Wiseman L.L., Do morphogenetic tissue rearrangements require active cell movements? *J. Cell. Biol.* **55** (1972) 606-615.

[59] Szent-Gyorgyi A., Chemistry of Muscle Contraction (Academic Press, New York, 1947).

[60] Takeichi M., Cadherins: A molecular family important in selective cell-cell adhesion, *Ann. Rev. Biochem.* **59** (1990) 237-352.

[61] Theriot J.A. and Mitchison T.J., Actin microfilament dynamics in locomoting cells, *Nature* **352** (1991) 126-131.

[62] Wang Y.-L., Exchange of actin subunits at the leading edge of living fibroblasts: Possible role of treadmilling, *J. Cell. Biol.* **101** (1985) 597-602.

[63] Wegner A., Head to tail polymerization of actin, *J. Mol. Biol.* **108** (1976) 139-150.

[64] Wessells N.K., Spooner B.S., Ash J.F., Bradley M.O., Luduena M.A., Taylor E.L., Wrenn J.T. and Yamada K.M., Microfilaments in cellular and developmental processes, *Science* **171** (1971) 135-143.

[65] Wolpert L., Macpherson I. and Todd I., Cell spreading and cell movement: An active or a passive process? *Nature* **223** (1969) 512-513.

LECTURE 4

Adhesion Mediated Signaling: The Role of Junctional Plaque Proteins in the Regulation of Tumorigenesis

A. Ben-Ze'ev

Department of Molecular Cell Biology, The Weizmann Institute of Science, Rehovot 76100, Israel

1. INTRODUCTION

Cell adhesion to neighboring cells and to the extracellular matrix (ECM) plays important roles in cell motility, growth, differentiation and survival (Ben-Ze'ev, 1997; Geiger *et al.*, 1995; Gumbiner, 1996). The molecular interactions at cell adhesion sites include transmembrane integrin-type receptors that link cells to the ECM, and cadherin receptors at cell-cell contact sites that are associated with submembranal plaque proteins that bridge between the cytoskeleton and the adhesion receptors (Geiger *et al.*, 1995). Major junctional plaque proteins are vinculin, α-actinin and the catenins. In addition, recent studies have amply indicated the localization in the submembranal plaque of a large number of regulatory molecules involved in signal transduction (Fig. 1, and Geiger *et al.*, 1995; Gumbiner, 1996). The submembranal plaque area is now viewed not only as a structural link that mediates cell adhesion, but also as an important component in the control of signal transduction regulating cell behavior. Tumor cells are often characterized by altered adhesion, disorganized cytoskeletal assembly and impaired adhesion-mediated signaling (Ben-Ze'ev, 1985, 1992, 1997). Many cancer cells are "anchorage independent" and less susceptible to cell density-dependent inhibition of growth. Our previous studies have demonstrated that both the organization and expression of junctional plaque proteins is modulated during growth activation, differentiation and transformation (Ben-Ze'ev, 1991). Here, I summarize our studies on modulating the expression of the adhesion plaque proteins vinculin and α-actinin, and the cell-cell junctional plaque proteins plakoglobin and β-catenin, to define their role in cell behavior, with special emphasis on their effect on the tumorigenic ability of cells.

2. RESULTS AND DISCUSSION

2.1 Induction of vinculin and α-actinin expression in growth-activated cells *in vitro* and *in vivo*

Vinculin and α-actinin are major junctional plaque proteins that link actin stress fibers to areas of cell adhesion to the ECM (Fig. 1). While these proteins are among the most abundant, constitutive cellular proteins, their expression is extensively modulated in response to growth stimulation of quiescent 3T3 cells by serum factors. When such cells are treated with serum, a dramatic transient increase in the level of vinculin synthesis was observed (Ben-Ze'ev *et al.*, 1990; Bellas *et al.*, 1991). The synthesis of α-actinin is also induced, but follows a different kinetics of induction (Glück *et al.*, 1992). These changes in vinculin and α-actinin synthesis are accompanied by parallel changes in the levels of the corresponding mRNAs. For vinculin, a rapid and transient induction of vinculin gene transcription precedes the increase in vinculin synthesis in serum stimulated cells (Ben-Ze'ev *et al.*, 1990; Bellas *et al.*, 1991). Posttranscriptional regulation of vinculin expression is also implicated in this process: in 3T3 cells stably expressing a transfected vinculin under a constitutive viral promoter, there is a similar, yet more moderate, elevation in the synthesis of the transfected vinculin after serum stimulation (Geiger *et al.*, 1992).

→

Fig. 1. A scheme depicting the interactions between signaling by focal adhesions and growth factor receptors. Actin filaments are linked to the extracellular matrix via transmembrane integrin receptors by talin (Tal) and α-actinin (α-Act). Vinculin (V) has talin- and α-actinin-binding sites and can stabilize the link between these structural components of focal adhesions. Vinculin can also fold into a conformation where its binding sites are masked (V*). Zyxin (Zyx) is an α-actinin-binding protein that can bind to vasodilator phosphoprotein (VASP) which can bind to actin filaments. The assembly of focal adhesions depends on the activation of Rho that is activated by a variety of growth factors. Rho has multiple targets in the cell, some of which are directly related to assembly of focal adhesions and actin bundles. It activates Rho-kinase that phosphorylates (and hence inactivates) a phosphatase of myosin II light chain. This increases contraction (tension) that is central for actin bundles and focal adhesion formation. Rho also activates phosphatidyl inositol-4-phosphate-5 kinase (PIP5-kinase) to generate phosphatidyl inositol-4-5 bisphosphate (PIP_2) from phosphatidyl inositol-4-phosphate (PIP). PIP_2 can interact with actin-binding proteins to promote actin polymerization, and can stimulate the transition of vinculin from a non-active conformation (V*) to one capable of talin- and actin-binding. Protein tyrosine kinases are also localized at focal adhesions. These include: focal adhesion kinase (FAK) which binds the cytoplasmic domain of β integrin, Src and Src-family kinases, Csk (a negative regulator of Src kinases), and Abl (binds actin). Other components of focal adhesions, including paxillin (Pax), tensin (Ten), and $p130^{cas}$ (p130), are substrates for tyrosine phosphorylation. Upon adhesion of cells to the ECM, tyrosine phosphorylation of FAK is induced, and this increases its tyrosine kinase activity. This is followed by tyrosine phosphorylation of paxillin, tensin and $p130^{cas}$. Then, the Ras-MAP kinase signaling pathway is induced by a Ras-dependent mechanism as follows: Tyrosine phosphorylated FAK binds the SH2/SH3 adaptor protein Grb2, and tyrosine phosphorylated paxillin, binds another SH2/SH3 adaptor, Crk. Tyrosine-phosphorylated $p130^{cas}$ can bind both Crk and Grb2. The SH3 domains of both Crk and Grb2 bind to the guanine nucleotide exchange factors SOS and C3G, which can activate Ras and thus stimulate the MAP kinase pathway. Another adaptor protein, Sch, can couple via caveolin a class of integrins with the MAP kinase pathway. A cross-talk between Ras and Rho may also exist. (Reproduced from Ben-Ze'ev, 1997).

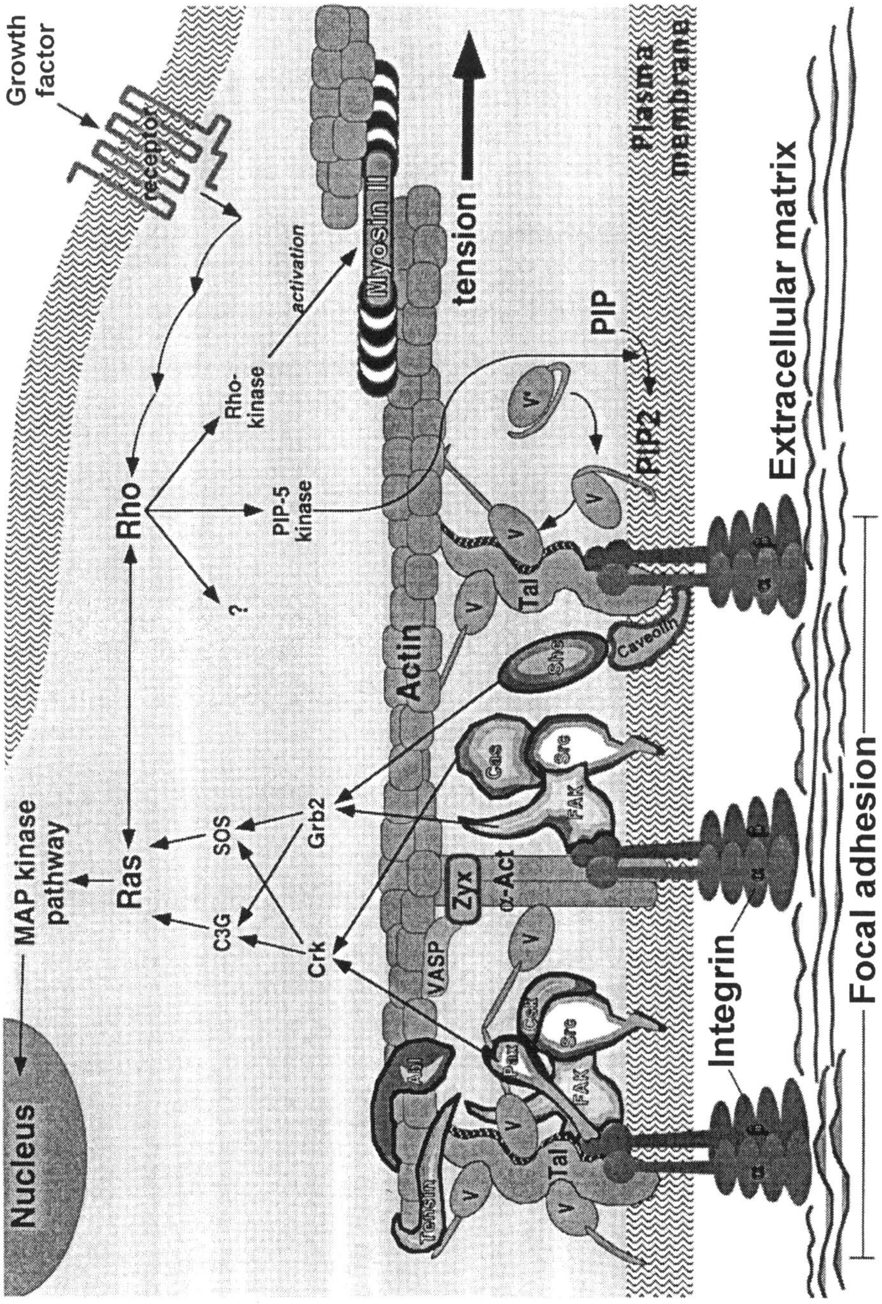

Nucleus
MAP kinase pathway
Growth factor
receptor
Ras
Rho
C3G
SOS
Grb2
Crk
?
PIP-5 kinase
Rho-kinase
activation
Myosin II
tension
Actin
Zyx
VASP
α-Act
Tal
V
FAK
Src
Cas
Csk
Pax
Abl
Tensin
Shc
Caveolin
PIP
PIP2
Plasma membrane
Integrin
α
β
Extracellular matrix
Focal adhesion

The possible physiological relevance of these changes in vinculin and α-actinin expression in response to growth stimulation was examined in regenerating liver, a system widely used as an *in vivo* model for studying gene expression at early stages of induced cell proliferation. Northern blots of RNA isolated at different times after 2/3 hepatectomy showed a transient elevation in vinculin RNA levels, and latter in the abundance of α-actinin RNA that persisted into the S-phase (Glück *et al.*, 1992).

2.2 The effects of vinculin and α-actinin overexpression

To investigate the role of these modulations in vinculin and α-actinin expression on cell behavior, 3T3 clones stably expressing these transgenes were isolated. Since cultured 3T3 cells express high levels of these proteins, in stably transfected cells the transgenes were expressed at levels usually less than 50% of the endogenous protein level. Nevertheless, overexpression of vinculin by only about 20% was sufficient to confer phenotypic and functional changes in 3T3 cells. Such cells assembled more abundant stress fibers, terminating in larger vinculin-containing plaques (Geiger *et al.*, 1992).

Cells overexpressing either vinculin or α-actinin also displayed a decrease in cell motility, measured as shorter phagokinetic tracks on colloidal gold-coated substrates, or as reduced ability to close an artificial wound created in a confluent monolayer of cells (Rodríguez Fernández *et al.*, 1992a; Glück and Ben-Ze'ev, 1994). These results suggest that relatively moderate changes in vinculin or α-actinin expression may have a major impact on cell morphology, cytoskeletal assembly, and the motile abilities of cells.

2.3 The effects of suppressing vinculin and α-actinin levels and targeted inactivation of the vinculin gene

To study the effect of forced reduction in vinculin and α-actinin levels, 3T3 cells were transfected with antisense cDNA constructs of these genes and clones stably expressing reduced levels of either vinculin or α-actinin were isolated. Antisense vinculin cDNA transfection generated clones displaying vinculin levels equal to only 10% – 30% of control levels (Rodríguez Fernández *et al.*, 1993). Such cells showed a marked change in morphology, with reduced capacity to spread on the substrate, and small vinculin-containing plaques at the cell periphery. Moreover, the suppression in vinculin expression resulted in enhanced cell motility (Rodríguez Fernández *et al.*, 1993). These changes in the morphological and motile properties of cells correlated with the degree of suppression in vinculin expression. Only cells displaying over 70% reduction in vinculin levels showed increased motility and reduced spreading on the substrate. Cells displaying a more moderate decrease in vinculin, and revertants of clones originally showing effective vinculin suppression, also reverted to normal motile characteristics.

Vinculin expression was completely eliminated in the F9 embryonal carcinoma and embryonal stem (ES) cells by targeted disruption of the vinculin gene after homologous recombination (Coll *et al.*, 1995). Vinculin deficient F9 cells had unchanged levels of α-actinin, paxillin and talin (Volberg *et al.*, 1995).

Vinculin-null cells showed a significantly reduced ability to spread on the substrate, and a slower rate of initial adhesion (Coll *et al.*, 1995), reminiscent of the phenotype of antisense vinculin transfected 3T3 cells (Rodríguez Fernández *et al.*, 1993). Vinculin-deficient F9 cells displayed increased locomotory capacity, in agreement with the results obtained with antisense vinculin transfection. This increase in the motility of vinculin-null F9 cells was also maintained when these cells were induced to differentiate into the more motile parietal endodermal (PE) cells. The vinculin null F9 cells could be induced to express the differentiation markers in the absence of vinculin (Coll *et al.*, 1995), and their ability to form stress fibers and assemble focal adhesions containing α-actinin, paxillin, talin and phosphotyrosinated components was not impaired (Volberg *et al.*, 1995). These findings suggest that there are multiple molecular mechanisms for focal adhesion formation in the absence of vinculin, including one that is based on α-actinin bridging between integrin receptors and actin (Fig. 1).

2.4 Vinculin and α-actinin levels and the transformed phenotype

Suppression of vinculin expression decreased the cell adhesion rate and cell spreading, and conferred anchorage independent growth on 3T3 cells (Rodríguez Fernández *et al.*, 1993). In antisense α-actinin transfected 3T3 clones, when α-actinin levels were reduced by 40% – 75%, these cells became tumorigenic after injection into nude mice (Glück and Ben-Ze'ev, 1994). The appearance of tumors correlated with the level of α-actinin suppression, and cells expressing lower levels of the protein caused faster developing tumors. Furthermore, revertants of such clones that regained control levels of α-actinin, became nontumorigenic. These results support the notion that vinculin and α-actinin may have a tumor suppressive activity.

In agreement with this view, we found that SV40-transformed 3T3 cells (SVT2) display diminished levels of vinculin and α-actinin (Glück *et al.*, 1993; Rodríguez Fernández *et al.*, 1992b), and vinculin was absent in a highly metastatic adenocarcinoma cell line (ASML) that expresses high levels of α-actinin and talin (Raz *et al.*, 1986).

2.5 Suppression of tumorigenicity by transfection with vinculin and α-actinin

To examine the effect of restoring vinculin and α-actinin levels in transformed cells on their tumorigenic phenotype, 3T3 fibroblasts transformed by SV40 (SVT2) and the highly metastatic ASML epithelial cells were transfected with either full length vinculin or α-actinin. Clones stably expressing different levels of these transgenes were isolated. High levels of exogenous vinculin expression resulted in cells with more abundant stress fibers and larger vinculin-positive plaques (Rodríguez Fernández *et al.*, 1992b). The tumorigenicity of cells overexpressing vinculin was dramatically affected, and cells expressing the transfected vinculin at levels similar to nontransformed 3T3 cells, have completely lost their tumorigenic ability in mice (Rodríguez Fernández *et al.*, 1992b). Similarly, the overexpression of α-actinin in SVT2 cells, resulted in suppression of their tumorigenic capacity, and correlated with the level of the transfected α-actinin in the different clones (Glück *et al.*, 1993).

The expression of the viral (SV40) T-antigen in SVT2 cells however, and that of mutant p53 molecules accumulating in these cells was not affected in the vinculin- and α-actinin-transfected clones. This implies that these junctional molecules have used alternative route(s) to influence the tumorigenic ability of SVT2 cells.

The effect of vinculin overexpression on the metastatic spread of tumor cells was examined in a highly metastatic adenocarcinoma (ASML) cell line that does not express vinculin. Expression of high levels of the transgene suppressed the malignant metastatic ability of these cells, while low levels of vinculin only partially altered the number of lung metastases that formed with these clones (Rodríguez Fernández *et al.*, 1992b).

2.6 Plakoglobin and β-catenin: cell-cell junctional molecules involved in signaling and regulation of tumorigenesis

The most direct effect of cell-cell adhesion is on morphogenesis, i.e. the assembly of individual cells into highly ordered tissues through cell-cell adhesion junctions (Gumbiner, 1996). These interactions among cells involve transmembrane cell adhesion receptors of the cadherin family, which link cells to each other. Effective adhesion requires, in addition, an association of the transmembrane receptors with the cytoskeleton that is mediated by junctional plaque proteins in the cytoplasm including plakoglobin, β-catenin and α-catenin (Fig. 2, and Geiger *et al.*, 1995; Gumbiner, 1996; Ben-Ze'ev, 1997). β-Catenin can associate in addition to cadherins at cell-cell junctions, with a variety of other molecules including the EGF receptor, and a major tumor suppressor molecule APC whose mutation and subsequent loss are involved in human colon cancer (Fig. 2, and Gumbiner, 1996; Ben-Ze'ev, 1997; Peifer, 1997). The interaction between β-catenin and APC in the cytoplasm regulates β-catenin degradation by the ubiquitin-proteasome system (Aberle *et al.*, 1997).

——————————▶

Fig. 2. A model for signaling by cell-cell adhesion molecules of adherens junctions that can affect gene expression and morphogenesis. Signals generated by cell-cell adhesion via cadherins and their assembly with catenins can modulate the level of β-catenin (β), or plakoglobin (Pg) that is available for interaction with the APC (Adenomatosis Polyposis Coli) tumor suppressor molecule, or with transcription factors such as Tcf-Lef (T cell factor-Lymphoid enhancer factor). The complex between the transcription factor and β-catenin, or plakoglobin, can translocate into the nucleus and directly bind to the 5' end of the E-cadherin (E-CAD) gene, and other genes, to regulate their expression. Signaling that involves β-catenin can also be elicited by the Wnt signaling pathway, that includes the Wnt receptor (Dfz2), then *Dsh* (a submembrane phosphoprotein), and glycogen synthase kinase (GSK), that can transmit a signal to the nucleus to regulate gene expression. The association between the tumor suppressor APC and β-catenin, or plakoglobin, and GSK can regulate the stability of β-catenin and plakoglobin by phosphorylation (P), and thus controls their translocation to the nucleus. The association of APC with microtubules (Mt) may, in addition, transduce a negative effect on cell growth. An association between receptor tyrosine kinases such as the EGF (epidermal growth factor) receptor (EGFR) and ErbB2 with β-catenin and plakoglobin may also affect signaling. Note the presence of protein kinases such as Src and protein kinase C (PKC) in the junctional plaque, together with structural plaque proteins vinculin (V), α-actinin (α-Act), radixin (Rad), α-catenin (α) and zyxin (Zyx). (Reproduced from Ben-Ze'ev, 1997).

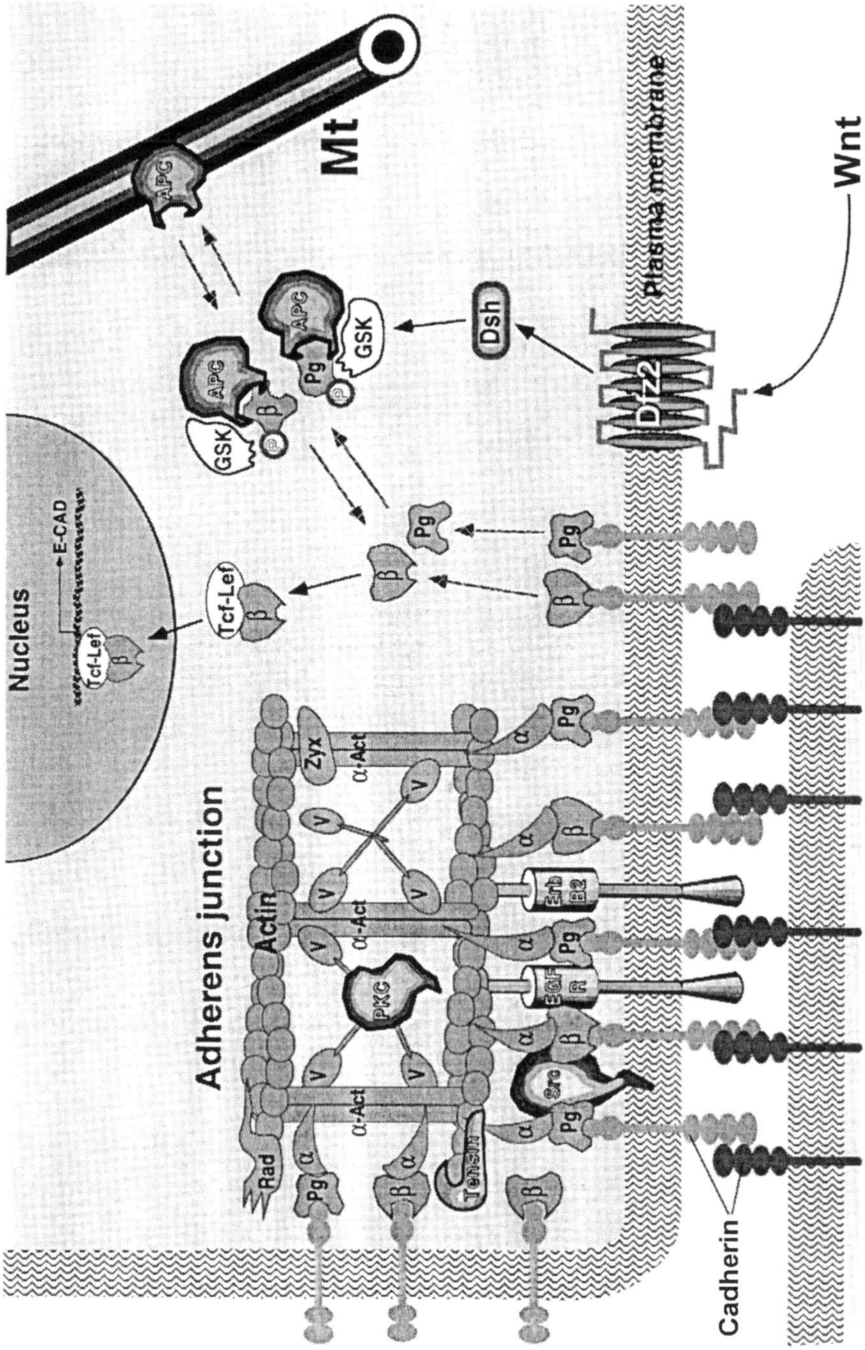
Nucleus
E-CAD
Tcf-Lef
Adherens junction
Actin
α-Act
Zyx
PKC
Rad
Src
Pg
EGF R
Erb B2
APC
GSK
Mt
Dsh
Plasma membrane
Wnt
Cadherin

Recent studies have shown that the homologues of β-catenin in *Drosophila* and *Xenopus* can function in signal transduction: the *Drosophila* homologue of β-catenin, armadillo, is a segment polarity gene whose level is increased by the wingless/wnt secreted glycoprotein and it regulates the transcription of target genes in the nucleus (Fig. 2, and Peifer, 1997; Nusse, 1997). In *Xenopus*, β-catenin can determine cell fate and dorso ventral axis formation when overexpressed (Gumbiner, 1996). We have studied the signals conveyed by changes in the level of plakoglobin, that is closely related to β-catenin, and found that a variety of tumor cells display reduced levels or complete loss of plakoglobin while expressing high levels of β-catenin (Simcha *et al.*, 1996). Overexpression of plakoglobin in such cells from mouse and human tumors resulted in suppression of their tumorigenic ability (Simcha *et al.*, 1996). Furthermore, in cells lacking a cadherin and catenin system, overexpression of plakoglobin resulted in the nuclear accumulation of plakoglobin, and a dose dependent suppression of their tumorigenic capacity (Simcha *et al.*, 1996). Moreover, in tumor cells lacking plakoglobin, but expressing β-catenin, when high levels of plakoglobin were induced, β-catenin was displaced from cell-cell junctions and rapidly degraded (Salomon *et al.*, 1997).

This could be of major importance for tumorigenesis regulation since high levels of β-catenin can be oncogenic, owing to constitutive transactivation of target genes by β-catenin (Peifer, 1997). The competition between plakoglobin and β-catenin for cadherin binding (Fig. 2) thus could be an important regulatory pathway for limiting the level of β-catenin in the cell.

3. CONCLUSIONS

Our earlier studies have shown modulations in the expression of both the organization and expression of cytoskeletal and junctional molecules during growth activation, differentiation and cell transformation. The studies summarized here strongly imply that such changes in expression may have important long term effects on cell behavior. This was demonstrated by the moderate overexpression of vinculin or α-actinin that dramatically influences cell motility, and the restoration of normal levels of expression of these molecules in transformed cells that effectively suppresses their tumorigenicity. Recent studies on the involvement of plakoglobin and β-catenin in both signaling and adhesion, and their additional role in the transactivation of genes, provide another exciting direction for future studies on deciphering the molecular mechanisms by which adhesion mediated signaling is regulating tumorigenesis.

ACKNOWLEDGMENTS

I would like to express my gratitude to all my colleagues and collaborators who have participated in the studies summarized in this manuscript. To J.L. Rodríguez Fernández, U. Glück, I. Simcha, D. Salomon, T. Volberg, B. Geiger, M. Zöller, R. Pankov, J.L. Coll and E. Adamson. These studies were supported by grants from the Pasteur-Weizmann Joint Research Program, the USA-Israel Binational Fund, and by a grant from the Leo and Julia Forchheimer Center for Molecular Genetics. A. B-Z holds the Lunenfeld-Kunin Professorial chair in Genetics and Cell Biology.

REFERENCES

[1] Aberle H., Bauer A., Stappert J., Kispert A. and Kemler R., *EMBO J.*, in press.
[2] Ben-Ze'ev A., *Biochim. Biophys. Acta.* **780** (1985) 197-21.
[3] Ben-Ze'ev A., *Crit. Rev. Eukaryotic Gene Expr.* **2** (1992) 265-281.
[4] Ben-Ze'ev A., *Curr. Opin. Cell. Biol.* **9** (1997) 99-108.
[5] Ben-Ze'ev A. *BioEssays* **13** (1991) 207-212.
[6] Ben-Ze'ev A., Reiss R., Bendori R. and Gorodecki B., *Cell. Regul.* **1** (1990) 621-636.
[7] Bellas R.E., Bendori R. and Farmer S.R., *J. Biol. Chem.* **266** (1991) 12008-12014.
[8] Coll J-L., Ben-Ze'ev A., Ezzell R.M., Rodríguez Fernández J.L., Baribault H., Oshima R.G. and Adamson E.D., *Proc. Nat. Acad. Sci. USA* **92** (1995) 9161-9165.
[9] Geiger B., Ginsberg D., Ayalon O., Volberg T., Rodríguez Fernández J.L., Yarden Y. and Ben-Ze'ev A., *Cold Spring Harbor Symp. Quant. Biol.* **57** (1992) 631-642.
[10] Geiger B., Yehuda-Levenberg S. and Bershadsky A.D., *Acta Anat.* **154** (1995) 46-62.
[11] Glück U. and Ben-Ze'ev A., *J. Cell. Sci.* **107** (1994) 1773-1782.
[12] Glück U., Kwiatkowski D.J. and Ben-Ze'ev A., *Proc. Nat. Acad. Sci. USA* **90** (1993) 383-387.
[13] Glück U, Rodríguez Fernández J.L., Pankov R. and Ben-Ze'ev A., *Exp. Cell. Res.* **202** (1992) 477-486.
[14] Gumbiner B.M., *Cell.* **84** (1996) 345-357.
[15] Nusse R., *Cell.* **89** (1997) 321-323.
[16] Peifer M., *Science* **275** (1997) 1752-1753.
[17] Raz A., Zöller M. and Ben-Ze'ev A., *Exp. Cell. Res.* **162** (1986) 127-141.
[18] Rodríguez Fernández J.L., Geiger B., Salomon D. and Ben-Ze'ev A., *Cell. Motil. Cytoskel.* **22** (1992a) 127-134.
[19] Rodríguez Fernández J.L., Geiger B., Salomon D. and Ben-Ze'ev A., *J. Cell. Biol.* **122** (1993) 1285-1294.
[20] Rodríguez Fernández J.L., Geiger B., Salomon D., Sabanay I., Zöller M. and Ben-Ze'ev A., *J. Cell. Biol.* **119** (1992b) 427-438.
[21] Simcha I., Geiger B., Yehuda-Levenberg S., Salomon D. and Ben-Ze'ev A., *J. Cell. Biol.* **133** (1996) 199-206.
[22] Volberg T., Geiger B., Kam Z., Pankov R., Simcha I., Sabanay H., Coll J.-L., Adamson E. and Ben-Ze'ev A., *J. Cell. Sci.* **108** (1995) 2253-2260.
[23] Salomon D., Sacco P.A., Guha Roy S., Simcha I., Johnson K.R., Wheelock M.J. and Ben-Ze'ev A., *J. Cell. Biol.* **139** (1997) 1325-1335.

LECTURE 5

Of Proteins, Redox States and Living Things

L. Moldovan and P.J. Goldschmidt-Clermont

The Heart and Lung Institute at the Ohio State University 420 W, 12th Avenue Columbus, Ohio 43210, USA

"It is a gamble to bet on science for moving ahead, but it is, in my view, the only game in town."

Lewis Thomas

1. INTRODUCTION

Proteins are amazing structures. For a long while, we have trusted them to be these inoffensive, carefully crafted, highly organized particles, coded for by individual genes (by all means the culprits), and contributing largely to the harmonious activity of cells and tissues. With the recent discovery that proteins can transmit such obscure diseases as Creutzfeld-Jakob neurodegeneration, capable of twisting the programming of invaded cells, the prion proteins, and by analogy, all proteins, have forever lost our trust.

Made of more than 400 amino acids in average, human proteins are coded for by nearly 100,000 genes, the deoxyribonucleic acid blueprint to all proteins. While some rare diseases are caused by a major alteration in the structure of a single protein, most diseases are due to the conspiracy of multiple minor genetic factors (such as the slight variations in the structure of proteins that we call polymorphisms), with environmental factors. While we know some about environmental factors that cause human diseases, we know nearly nothing about the protein (or genetic) contribution. Actually we know one thing: the maximal number of genes that could contribute to most polygenic diseases, such as cancers and heart diseases, has to be comprised between one and 100,000. If you conclude that we, the biologists, know little, then be kind to the doctors who deal with their newly developed scientific branch called medicine. It is a miracle that humans have come up with such efficient therapies as vaccinations, antibiotics and thrombolytics. It is still possible that these "progresses", when considered at the cosmic level, might not reflect true advances. Microorganisms rapidly adapt to environmental changes through hypermutation (as evidenced by the development of multi-drug resistant staphylococcus strains), and in the long run, these human interventions might not impact that much on the fate of the species. Nevertheless, it is great that lives can be saved for the time being. However, there is no drug available to treat protein-caused diseases. Thus, it is likely that we will need the input of physicists, as we sought in the past for other crisis, to develop the relevant antagonists for prions, a ample justification for having a meeting between physicists and biologists, in our view.

2. REDOX STATE AND PROTEIN STRUCTURE

The structure of proteins can oscillate between two extreme conformations from unfolded to optimally folded, with a spectrum of statistically relevant states, usually stabilized by a given set of environmental factors. For example, when exposed to certain detergents, proteins can lose their compact structure and are unable to associate with other proteins, lipids, or nucleic acids and, therefore, provide their function. When exposed to ammonium sulfate within a "hanging drop," they can crystallize in such a way that they self-associate within a unit of space called "cell", and become suitable for X-ray diffraction analysis. Certain proteins can also self-associate in more physiological conditions. Actin is such a protein, which, upon self-association, can form a sophisticated network, further elaborated by an additional 30+ families of actin binding proteins. While many protein-protein interactions are highly transient, some interactions are more sustained. When such interactions result from the oxidation of sulfhydryl groups, linking cysteine residues through disulfide bridges, they require relatively large activation energy to be resolved. Therefore, cells are equipped with an army of anti-oxidant systems. In the absence of these anti-oxidants, cells would be frozen into a rigid network of cross-linked proteins, almost independently of the "probability of space occupation" for individual proteins.

To illustrate the magnitude of the effect of redox interactions among cellular proteins, a simple assay can be performed. Proteins extracted from cells can be sieved in their native oxidized state in a gel of increasing concentrations of acrylamide (gradient gel), upon exposure to an electrical field. The proteins will migrate to a position at which the energy required to move them through the holes of the acrylamide network is larger than the strength of the electric field that mediates their migration. Then, the resulting strip of protein bands can be placed on top of a second acrylamide gradient gel and re-exposed to the same electric field, but this time after reducing all of the disulfide bridges. Three outcomes can be observed:

(1) If there were no disulfide bridges, all proteins would line up along a diagonal (hence, the name of this technique, diagonal two-dimensional gel electrophoresis), as their migration in the first and second gels would be exactly identical.

(2) Disulfide bridges could have been disrupted that were linking proteins together, and in this case, the isolated proteins would migrate further in the second gel than in the first one, thus away from the diagonal distribution.

(3) In contrast, if disulfide bridges were linking distant domains within a single protein, reduction would affect the packing of this protein, thereby increasing the span of the reduced protein, and the energy required to force it through the holes of the acrylamide network, limiting its migration in the second gel. When all proteins of cells are exposed to such a diagonal electrophoresis, few proteins actually line up exactly on the diagonal line, with most proteins dispersed above and below the line, indicating a high degree of oxidation for proteins inside cells. Oxidation of proteins can also involve non-cysteine residues, a process that has been linked to protein aging, and resides beyond the focus of this review. We realize that, to physicists, such experiments might have more to do with alchemy than with science. Remember that, while the discipline of physics has grown to maturity, biology, in its own way, is growing up too and at best, we can now reliably account for empirical evidence.

Recently, we have become interested in processes that control the generation of reactive oxygen species (ROS), highly reactive molecules that contribute largely to the oxidized state of proteins. A recurrent quest in the field of biology is the deciphering of the mechanisms that explain how signals originating from outside cells can trigger specific cellular behaviors. One protein that corresponds to a cornerstone for many signaling processes is called $p21^{ras}$ or ras. We and others have recently discovered that ras can regulate cascades of signaling reactions through the production of oxygen radicals by specialized enzymes. This review will next focus on how ras and ROS contribute to the organization of the actin cytoskeleton.

3. THE ACTIN NETWORK AS AN INTEGRATOR OF CELLULAR FUNCTIONS

Comprising up to 10% of the total cellular protein, there is no wonder that actin is one of the most studied proteins, and that the mechanisms that control the organization of actin within cells are under active investigation. About half of the actin pool in any non-muscle cell is unpolymerized (G-actin, Pollard and Cooper, 1986). Upon polymerization, actin monomers associate head-to-tail and form filaments (F-actin), which associate in various higher-level structures. Actin filaments form a network throughout the cell, however the highest density is in the cortex, which is the region just beneath the plasma membrane (Alberts *et al.*, 1994).

Many quiescent cultured cells, such as endothelial cells (the cells lining the inner surface of blood vessels) and fibroblasts (the cells responsible for wound healing), display bundles of actin filaments associated with myosin and α-actinin, called stress fibers (SF) (Kreis and Birchmeyer, 1980; Wong *et al.*, 1982; Byers and Fujivara, 1982; Gabbiani *et al.*, 1983). At one or both ends SF come in tight contact with specialized domains of the plasma membrane, named focal adhesions (FA), through which cells adhere to the extracellular matrix. These are complex structures, the main component of which are transmembrane proteins of the integrin family (Tamkun *et al.*, 1986). Integrins have an extracellular domain which specifically binds components of the extracellular matrix, such as fibronectin, and a cytoplasmic domain, connected to actin stress fibers through interaction with several ligand proteins: vinculin, talin, α-actinin, tensin, radixin (for review, see Jockusch *et al.*, 1995). Focal adhesions are highly dynamic structures, as they continuously remodel, and their distribution is responsive to environmental signals (Davies *et al.*, 1994).

A dramatic change in the appearence of the actin cytoskeleton occurs when the cells are challenged. Reendothelialization following injury to the endothelium (Gottlieb, 1990), growth factors (*e.g.*, PDGF, Herman and Pledger, 1985), tumor promoters (Schliwa *et al.*, 1984), as well as other physiologic or experimental stimuli, induce the redistribution of actin filaments, responsible for the motile responses to these factors, including the intense membrane activity known as "ruffling", shape changes and directional locomotion (Cooper, 1991). Advancing cells extend in the direction of their movement a flat process, the lamellipodium, or thin, cylindrical projections, named filopodia (Laufenburger and Horwitz, 1996). Actin fibers are continuously assembled at the plasma membrane in these regions, while depolymerization occurs at the opposite end, and vinculin and other FA proteins are redistributed (Schliwa *et al.*, 1984; Theriot and Mitchison, 1992). The organization of actin filaments within these structures is different from that in SF, and, accordingly, the actin-binding proteins which contribute to the stabilization of these structures are different: while in SF filaments are arranged with opposite polarities and the loose packing is maintained by α-actinin cross-bridges, fimbrin contributes to the tight association of actin filaments in filopodia, and filamin helps formation of a three-dimensional network (Alberts *et al.*, 1994; Bretscher, 1991).

Although the term "actin cytoskeleton" has a connotation of rigidity, suggesting that it mainly contributes to the mechanical support of the cell, this is not entirely correct. The actin cytoskeleton is highly dynamic, since only such dynamic interactions between actin and binding proteins can support cellular motile responses on time scales ranging from sub-seconds to hours. Moreover, there is one facet of actin biology that has been often missed: actin not only confers "plasticity" to cells, thus allowing them to move and divide, but it is also contributing to the orchestration of key metabolic pathways. It has long been known that the enzymes of the glycolytic pathway associate with actin, that their kinetic parameters are specifically modulated by this interaction, that such regulation depends on whether actin is in monomeric or filamentous form, and whether it is associated or not with actin-binding proteins (*e.g.*, Clarke and Masters, 1975; Walsh *et al.*, 1977; Kuo *et al.*, 1986; Poglazov and Livanova, 1986). A more recent report (Bereiter-Hahn *et al.*, 1995) suggests that the increase in glycolytic activity during the G_2 phase[1] is correlated with the dissolution of SF, and with the formation of a looser network of actin fibers, which provides an increased surface for association with glycolytic enzymes. In this respect, actin fibers may be considered as surface catalysts (Friend, 1993) for these enzymes (Crawford *et al.*, 1994).

In addition to the enzymes of the glycolytic pathway, other enzymes were found to bind to actin, such as 17β-estradiol dehydrogenase in porcine endometrial cells

---

[1] The cell cycle comprises several phases: G_1, when the cell simply grows, S, when the DNA is duplicated, G_2, when the cell becomes committed to division, and mitosis, when the cell divides.

(Adamsky *et al.*, 1993), possibly contributing to the correct positioning of the enzyme within the cell. Moreover, actin filaments regulate the activity of sodium channels in myeloid leukemic cells, as their disruption with cytochalasin D leads to a marked increase of channel activity (Negulayev *et al.*, 1996). Some investigators have even suggested that the actin and tubulin networks drive metabolic fluxes within cells (Aon *et al.*, 1996).

4. COMPLEX CONTROL OF THE ACTIN CYTOSKELETON ORGANIZATION AND DYNAMICS

The nature of the signalling pathways which control actin reorganization was long sought. However, despite efforts in many laboratories, the biochemical steps connecting the extracellular signals to the reorganization of the actin cytoskeleton are not yet completely characterized. There are many points of control along these pathways that are often redundant, and the factors involved dictate *when, where,* and *how fast* the changes should occur.

The first step in the formation of actin filaments is nucleation. It involves the association of actin monomers to form dimers, and then trimers, which then act as templates for elongation. This is an extremely unfavorable reaction (high activation energy), which minimizes the probability of random formation of new filaments (Pollard, 1988). It is the rate-limiting step in actin polymerization, and it is also an important point of control for actin assembly. Further limitations in the generation of new filaments include the binding of actin monomers to protein ligands. G-actin binds stoichiometrically to thymosin β_4 (Cassimeris *et al.*, 1992), and other G-actin-sequestering proteins. When the cell is stimulated, actin is released, the rate of exchange of ADP/ATP increases (Goldschmidt-Clermont *et al.*, 1992a), and F-actin is formed. The steady-state of actin assembly is controlled by the interaction of F-actin with capping proteins (Carlier and Pantaloni, 1994).

There are numerous other proteins that interact with F-actin, thereby controlling different steps of the process of actin fibers formation, assembly, and disassembly. Although specific, many of these interactions are weak, in the range of 10^{-5} M (Pollard, 1988). An obvious advantage of such weak interactions resides in the rate with which they are formed and broken, which, in turn, is essential for the dynamic properties of the actin cytoskeleton and for the promptitude with which cells respond to stimuli. As a corollary, one would expect that signalling pathways exert control upon the interactions between actin and its regulatory proteins (Pollard, 1988), and indeed, this proved to be the case.

We will elaborate on two prototypical proteins that regulate the organization of actin: gelsolin and profilin. Gelsolin is a paradigm for a family of filament-severing proteins that also includes villin, severin, and adseverin. In the presence of micromolar concentrations of calcium, it severs actin filaments, thus reducing their length (Yin and Stossel, 1979). Gelsolin also has an apparently paradoxical effect: it binds actin monomers and oligomers, forms new nucleation sites and thereby promotes polymerization (Matsudaira and Janmey, 1988).

Profilin binds to actin monomers in a 1:1 complex, thereby inhibiting the spontaneous nucleation of actin filaments (Pollard and Cooper, 1984), and also catalyses the exchange of adenosine nucleotides bound to actin monomers (Mockrin and Korn, 1980; Goldschmidt-Clermont *et al.*, 1992a,b), an effect that might potentiate filament formation in physiological conditions. Calcium regulates the binding of gelsolin, as well as other actin filament-severing proteins, to actin filaments (Yin and Stossel, 1979). Interestingly, the effect of calcium on actin-bundling proteins (α-actinin, fimbrin, 30-kd bundling protein) is to promote dissociation, therefore reversal of bundling (Janmey, 1994). Both profilin and gelsolin are regulated by plasma membrane molecules, namely the inositol phospholipids (PPIs). The best characterized signaling role for PPIs is as a source of inositol trisphosphate (IP_3) and diacylglycerol (DAG) through the hydrolysis of phosphatidylinositol 4,5-bisphosphate (PIP_2) by phospholipase C (PLC). Binding of profilin and gelsolin to PPIs appears to reverse the interaction of these proteins with actin, thus regenerating these ligands for more interaction with new actin molecules (Lassing and Lindberg, 1985; Janmey *et al.*, 1989). Moreover, binding of profilin to PPIs can alter the turnover rate of these phospholipids (Goldschmidt-Clermont, *et al.*, 1990; Goldschmidt-Clermont and Janmey, 1991). Thus, it seems that both profilin and gelsolin are at the nodes of a network of signalling pathways, which means that they are under multiple control systems, and that, in turn, they participate in the control of these pathways. Such reactions are likely to contribute to the synchronization of signalling reactions at the membrane level and reorganization of the submembranous actin cytoskeleton. These reactions have been reviewed previously and therefore will not be the subject of this discussion.

5. REGULATION OF ACTIN BY SMALL GTP-BINDING PROTEINS

An important step in understanding the regulation of the actin cytoskeleton was the discovery of the involvement of small GTP binding proteins of the $p21^{ras}$ (ras) family. Ras is the product of the cellular protooncogene *ras*. The proteins from this family are regulated by the binding, hydrolysis and release of guanosine triphosphate (GTP). Guanosine diphosphate (GDP) is bound in the quiescent state; when activated by exchange factors, GDP is released and replaced by GTP. GTP is then hydrolyzed to GDP and phosphate, the inorganic phosphate is released and the protein becomes again inactive. The K_d for guanine nucleotide binding is several orders of magnitude smaller than the concentration of the nucleotide within cells, such that the small GTP binding proteins are not controlled by the cellular level of GTP (Bourne *et al.*, 1991). Instead, the interaction with the guanine nucleotides is controlled by associated proteins: guanine nucleotide exchange factors (GEFs), GTPase-activating proteins (GAPs), and GDP-dissociation inhibitors (GDIs) (Boguski and McCormick, 1993). When activated, some of these proteins associate with the plasma membrane, which is also a prerequisite for the transforming activity of ras (Lowy *et al.*, 1993). This association requires posttranslational modifications at the C-terminus of the protein, which include methylation and isoprenylation (Hancock *et al.*, 1990).

The analysis of the thermodynamic cycles of actin and ras reveals striking similarities. Both bind in the inactive state to nucleotide diphosphates (NDP), ADP for actin and GDP for ras. Both are activated by the release of NDP and binding of the respective nucleotide triphosphate (NTP), with high affinity. Both have an NTPase activity, but the spontaneous hydrolysis of NTP is slow. Nevertheless, it can be accelerated by interaction with other proteins (ras) or with itself (actin within filaments). This step is irreversible, forcing the cycle in one direction. The rate-limiting step is the dissociation of NDP, which can be accelerated by regulatory factors (exchangers) (Goldschmidt-Clermont *et al.*, 1992b).

GTP binding proteins from a ras-related family, the rho proteins, which include rhoA, rac1, rac2, and cdc42, were found to coordinate the spatial and temporal changes in the actin cytoskeleton that lead to cellular movements (Nobes and Hall, 1995). Microinjection experiments showed that rho acts downstream of PDGF and other growth factors, as well as lysophosphatydic acid (LPA), on the pathway leading to the assembly of stress fibers and focal adhesions in Swiss 3T3 fibroblasts (Ridley and Hall, 1992). In contrast, rac1 induces a strong pinocytotic and membrane ruffling activity, accompanied by redistribution of actin fibers at the periphery of the cells (Ridley *et al.*, 1992). At the same time, the increase in stress fibers triggered by PDGF, but not by LPA, is also dependent on rac, which acts upstream of rho (Ridley *et al.*, 1992). Finally, cdc42 promotes the formation of filopodia and it also induces the activation of rac (Nobes and Hall, 1995).

Since the first reports on the function of GTP binding proteins in fibroblasts, their involvement in actin reorganization has been confirmed in other systems as well: mast cells (Norman *et al.*, 1994), intestinal epithelial cells (Nusrat *et al.*, 1995), cytokinesis in *Dictyostelium* (Larochelle *et al.*, 1996), and border cells in *Drosophila* (Lee and Montell, 1997). Moreover, some of the downstream effectors of rac were uncovered. Peppelenbosch *et al.* (1995) suggest that growth factors-induced stress fibers formation in Swiss 3T3 cells is mediated by arachidonic acid release. They observed that rac-induced stress fiber formation in these cells is inhibited by inhibitors of leukotriene synthesis, and they hypothesize that leukotrienes act downstream of rac in the triggering rho-mediated stress fibers formation.

Furthermore, two groups (Coso *et al.*, 1995; Minden *et al.*, 1995) concomitantly reported that rac1, rac2 and cdc42 have roles not only in the regulation of cellular morphology, but also in activating gene expression. Constitutively activated forms of these proteins, expressed in either HeLa and NIH 3T3 cells (Minden *et al.*, 1995), or COS-7 cells (Coso *et al.*, 1995), activated c-Jun N-terminal kinases (JNK), through the activation of JNK-kinase, which triggered increased c-Jun transcriptional activity.

A probable downstream effector of rac on the pathway of membrane ruffles formation was also described, as a rac1-interacting protein (Van Aelst *et al.*, 1996). This protein, named POR1, has no homology with known proteins, except a 25% homology with a protein of unknown function from *Caenorhabdites elegans*, and contains a putative leucine zipper sequence. This protein binds preferentially to GTP-bound rac.

A novel family of serine/threonine kinases, which are activated by the interaction with GTP binding proteins from the rho family, has been recently described (see, *e.g.*, Manser *et al.*, 1994; Bagrodia *et al.*, 1995). These proteins are mammalian homologs of the *Saccharomyces cerevisiae* Ste20 kinase, involved in the pheromone response pathway. One of the members of this family, mPAK-3, contains potential SH_3 domain-binding motifs[1], and it does bind *in vitro* to the SH_3 domains of phospholipase C-γ (PLC) and Nck. Since PIP_2, which is a substrate of PLC, is, as already discussed, a regulator of profilin, gelsolin and other actin binding proteins, it is suggested that mPAK-3 is one of the missing links between rac and cdc42, on one hand, and actin cytoskeleton reorganization, on the other hand (Bagrodia *et al.*, 1995). However, when rac mutants that were able either to bind PAK3 and activate JNK activity, but not POR1, or to bind POR1 but not PAK3, were assayed functionally, the result was that stimulation of JNK and mPAK3, and induction of membrane ruffling through POR1, were independent pathways (Joneson *et al.*, 1996; Westwick *et al.*, 1997). Thus, the complete sequence of events linking small GTP binding proteins to actin reorganization continues to resist unveiling.

6. RAC AS AN ACTIVATOR OF PHAGOCYTIC CELL NADPH OXIDASE

While rac control of membrane ruffling in fibroblasts was being described, two groups reported the involvement of the same protein in an apparently unrelated process (Abo *et al.*, 1991; Knaus *et al.*, 1991). The killing of microorganisms by phagocytic cells (neutrophils, macrophages) requires the production of superoxide and other reactive oxygen species (ROS) (for reviews, see, *e.g.*, Thrasher *et al.*, 1994; Robinson and Badwey, 1995). Responsible for the production of superoxide is a membrane-bound enzyme, NADPH oxidase, which catalyses the reaction:

$$\mathrm{NADPH} + 2\mathrm{O}_2 \rightarrow 2\mathrm{O}_2^- + \mathrm{NADP}^+ + \mathrm{H}^+ \cdot$$

By dismution of O_2^-, superoxide dismutase promotes hydrogen peroxide formation:

$$\mathrm{O}_2^- + \mathrm{O}_2^- + 2\mathrm{H}^+ \rightarrow \mathrm{H}_2\mathrm{O}_2 + \mathrm{O}_2 \cdot$$

The active enzyme is a complex of a membrane flavocytochrome b, and several cytosolic factors: p47phox, p67phox and p40phox (Dagher *et al.*, 1995). Upon activation, p47phox becomes phosphorylated and all three proteins are recruited at the plasma membrane. However, an absolute requirement for a GTP-binding protein in the activation of the NADPH oxidase was observed (Bokoch and Prossnitz,

---

[1] The protein $pp60^{v\text{-}src}$ (src) is the prototype of a large family of protein-tyrosine kinases, which have in common several regions of homology. One of these is named src homology domain 3 (SH_3), and its function is to mediate the interaction with guanine nucleotide exchange factors and GTP-ase activating proteins.

1992), and the search for this protein identified both rac1 (Abo *et al.*, 1991) and rac2 isoforms (Knaus *et al.*, 1991). In resting neutrophils, rac is found in the cytosolic fraction, while in activated phagocytes it is translocated to the plasma membrane. The process is dependent of rac isoprenylation and requires that the bound nucleotide is GTP (Bokoch, 1994).

In an inherited disease characterized by recurrent infections and inflammatory reactions with granuloma formation, therefore called chronic granulomatous disease (CGD), phagocytes are unable to produce superoxide, due to defects in components of the NADPH oxidase. Interestingly, when fibroblasts from patients with CGD were analyzed for their ability to produce superoxide, it was observed that this was not impaired, and that the enzymatic system producing superoxide was comparable to the phagocytic NADPH oxidase (Meier *et al.*, 1991). However, it was recognized that the fibroblast and phagocytic oxidases were immunologically and genetically distinct (Meier *et al.*, 1993; Emmendorffer *et al.*, 1993). These findings suggested that more than one NADPH oxidase system exists within the body, of which one produces superoxide in non-phagocytic cells (Irani and Goldschmidt-Clermont, 1997).

Several questions were raised by these findings: What is the role of this non-phagocytic NADPH oxidase? What are the protein components of this enzyme? Where is superoxide produced, extracellularly, intracellularly, or within phagolysozomes? In what circumstances is this enzyme activated and by what pathway? Does its activation also implicate rac? And maybe the most intriguing question: how can rac be involved in the regulation of two apparently so different processes, such as superoxide production and actin cytoskeleton reorganization?

7. COULD ROS BE MEDIATORS OF ACTIN REORGANIZATION?

A large body of work developed over the last decade changed our view about ROS. Previously, they were thought of as the damaging oxidants involved in aging (Nohl, 1993), malignant transformation (Toyokuni, 1996; Dreher and Junod, 1996), reperfusion injury (Flitter, 1993), nervous tissue degeneration disorders (Borlongan *et al.*, 1996; Jenner, 1996), etc. In other words, the toll eukaryotes had to pay for their aerobic existence. The cytoplasmic environment of cells is essentially reductive, and molecular oxygen undergoes not only tetravalent reduction by the mitochondrial electron transport chain, but also univalent reduction, in which the superoxide anion, H_2O_2, and hydroxyl radicals are generated (Davies, 1995).

More recently, a new concept gradually gained acceptance: ROS may play an important role in the homeostasis of the cell, as well. They are found to be involved in several signaling pathways, of which some of the best documented, albeit still controversial, are the regulation of the transcriptional activators nuclear factor kappa-B (NF-κB) and AP1 (Sen and Packer, 1996), and the regulation of apoptosis by p53 (Polyak *et al.*, 1997). NF-κB contains a cysteine in its DNA-binding site, which is essential for DNA binding. This cysteine has to be maintained in a reduced form in order for NF-κB to be active, and it sensitizes NF-κB to the redox status of the cell (Sen and Packer, 1996). AP1 is a transcriptional activator formed of homo- or heterodimers of two proteins: c-jun and c-fos. Their activity *in vitro* is also regulated by the oxidation status of a cysteine in the DNA binding domain (Abate *et al.*, 1990), however, the *in vivo* pathway of redox regulation in less well understood.

Several other processes were found to involve, at some point, control by ROS: cellular proliferation (Burdon, 1995), lymphocyte activation (Goldstone *et al.*, 1995, 1996), signal transduction (Sundaresan, 1995; Suzuki *et al.*, 1997), cellular transformation (Irani *et al.*, 1997), and the list is growing. We raised the question: are the two pathways, in which rac plays a role, superoxide generation and regulation of actin, divergent, or are they, in fact, interconnected? We have several lines of evidence that support a connection.

NIH 3T3 fibroblasts, which either transiently expressed a constitutively activated form of ras (EJ-ras) (Sundaresan *et al.*, 1996), or had been stably transformed with this gene (Irani *et al.*, 1997), produced increased amounts of superoxide. The production was suppressed by coexpression of dominant negative forms of either ras or rac1. This indicates that rac1 acts downstream of ras on this pathway. Interestingly, the superoxide production was also inhibited by treatment of cells with diphenylene iodonium, a flavoprotein inhibitor, which suggests the activation of a NADPH oxidase (Irani *et al.*, 1997). In EJ-ras-transformed cells the amount of stress fibers decreased, while the amount of actin fibers at the margins of cells, within membrane ruffles, increased (Heldman *et al.*, 1996).

These findings imply the following sequence of events: ras is activated by extracellular signals, such as growth factor receptors occupancy. Consequently, it activates rac, which in turn activates a non-phagocytic NADPH oxidase. This produces superoxide, and membrane ruffles, containing polymerized actin fibers, form. These steps may follow directly each other, or may be separated by a number of additional, unknown, steps (Fig. 1).

Furthermore, if superoxide is indeed involved in actin cytoskeleton reorganization, then superoxide scavengers should reverse this process, while exposing cells to oxidants should trigger the actin depolymeryzation-repolymerization cycle. Reoxygenation after hypoxia is a condition that occurs *in vivo* during reperfusion of tissues previously exposed to ischemia. It is known to be accompanied by the production of ROS, in particular superoxide (Zweier *et al.*, 1994). In *in vitro* experiments on human aortic endothelial cells (HAEC) maintained in hypoxic conditions for 4 hrs, re-addition of oxygen induces the reorganization of the actin cytoskeleton, which is similar to the changes observed in EJ-ras transformed cells: increase of the total filamentous actin and translocation of the filaments to the periphery of the cells (Crawford *et al.*, 1996). If these cells are infected with replication incompetent adenovirus, containing the superoxide dismutase (SOD) gene[1], they will overexpress SOD, and in these cells the actin

[1] The adenovirus was engineered such that it is not able to replicate within these cells. By further engineering, this replication deficient virus may be used as a vector, for the introduction within cells of different cDNAs (a cDNA corresponds to the coding domain of a gene).

response is suppressed. Similar observations are made if HAEC are overexpressing the constitutively activated form of rac, racV12, after infection with an adenoviral vector carrying racV12 cDNA. Increase of total actin occurs, concomitantly with prominent formation of membrane ruffles, processes which are both inhibited by the overexpression of SOD (unpublished results).

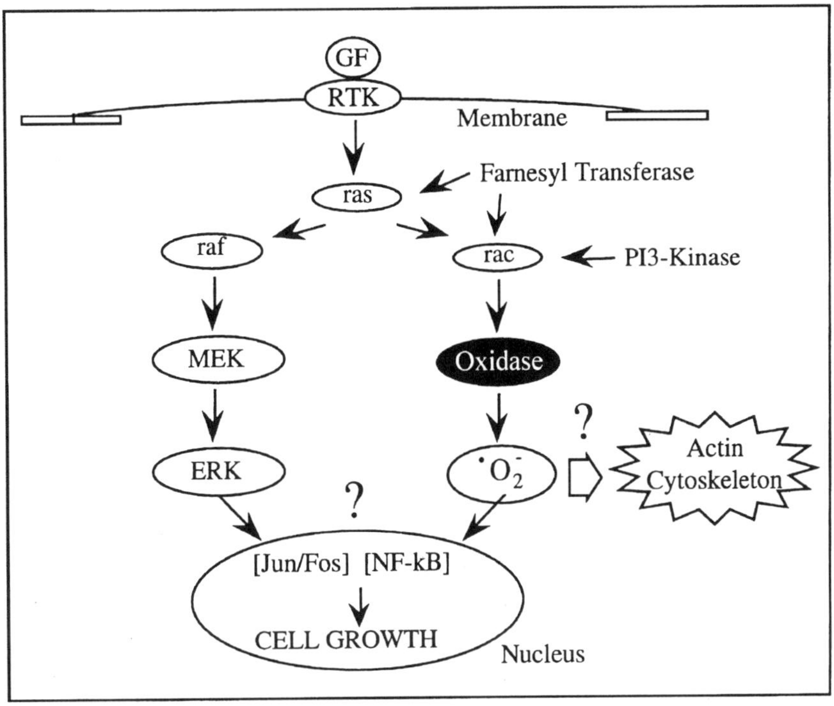

Fig. 1. Ras, Superoxide and Signal Transduction. An alternative effector pathway for Ras is shown. The pathway includes Rac, an NADPH oxidase and superoxide. Specifically how Ras, Rac and superoxide regulate mitogenesis, cell growth, and actin cytoskeleton, remains essentially uncharacterized (question marks).

Recently, the laboratory of Bert Vogelstein has initiated a series of experiments aiming at investigating the downstream targets for the ubiquitous tumor suppressor protein p53 (Polyak *et al.*, 1997). Mutations/deletions of p53 represent a common step in the development of many human cancers. Using the powerful "serial analysis of gene expression" technique (SAGE) to identify target genes for this DNA transcription regulatory protein, they found that most strongly responsive genes code for proteins that regulate the redox state of the cell. While this finding opens new avenues in cancer research to elucidate the role of these gene products in inhibiting the development of cancer cells, it strongly suggests that ROS are directly involved in the unchecked proliferation of cells, through the regulation of either apoptosis or mitosis.

8. MECHANISMS AND CONCLUSIONS

How is it be possible that such short-lived, highly reactive and supposedly harmful molecules to surrounding cytoplasmic structures, can function as second messengers? Cellular components are never exclusively exposed to oxidants. Instead, they are exposed to oxidants in a context where large excess of reducing agents are also present, such as reduced glutathione (GSH) which functions as a true sulfhydryl buffer. Due to the very strong activity of the enzyme glutathione reductase, the oxidized conformation of glutathione (GSSG) is rapidly converted to GSH, such that the ratio of GSH to GSSG in most cells is maintained ≥ 500. Therefore, the oxidized conformation of targeted proteins is likely to be rapidly reduced by the abundant cellular anti-oxidants. Moreover, intracellular oxidants are promptly inactivated by dismutases, catalases and other cellular anti-oxidants. Hence, their oxidizing effects are expected to be limited to a site directly surrounding their producing units. In addition, cells also contain large amounts of chaperone proteins, such as heat shock proteins. These protein-chaperones are known to protect their ligands against the damaging action of various stresses, including heat shock (Buchner, 1996). Thus in the absence of chaperones, superoxide (and other ROS) might oxidize proteins to result in their denaturation, which would be followed rapidly by their degradation (Prinsze *et al.*, 1990). Instead, in the presence of chaperones, the titrated oxidation of targeted proteins like actin or other protein/enzymes might result in conformational changes that could contribute to their activation (de Crouy-Chanel *et al.*, 1995; Pratt, 1997) (Fig. 2). Alternatively, the relative sensitivity of proteins to oxidants in a given system might result in the disruption of steady state, resulting in substantial reorganization of the affected system (Beckmann *et al.*, 1992).

The role ascribed to superoxide and derived oxidants in biology is clearly expanding. By analogy with nitric oxide, whose activity at low concentration is to transduce signals within vessels and neurons, while high concentrations produce damage to cells and microorganisms (Yun *et al.*, 1997), superoxide and probably other oxidants function as messengers at low concentration, while larger amounts are required for bactericidal activity. While the production of oxidants by phagocytes has been traditionally implicated as the main source of superoxide and other ROS in injured tissues, the discovery of the widespread use by cells of oxidants as signalling molecules is likely to improve our understanding of the contribution of such molecules to the biology of normal as well as injured tissues. The precise orchestration of the targeted production of oxygen radicals at specific sites of cells is likely to be timed, as many cellular processes are, by the hydrolysis of GTP, bound to the triphosphatases of the Ras family.

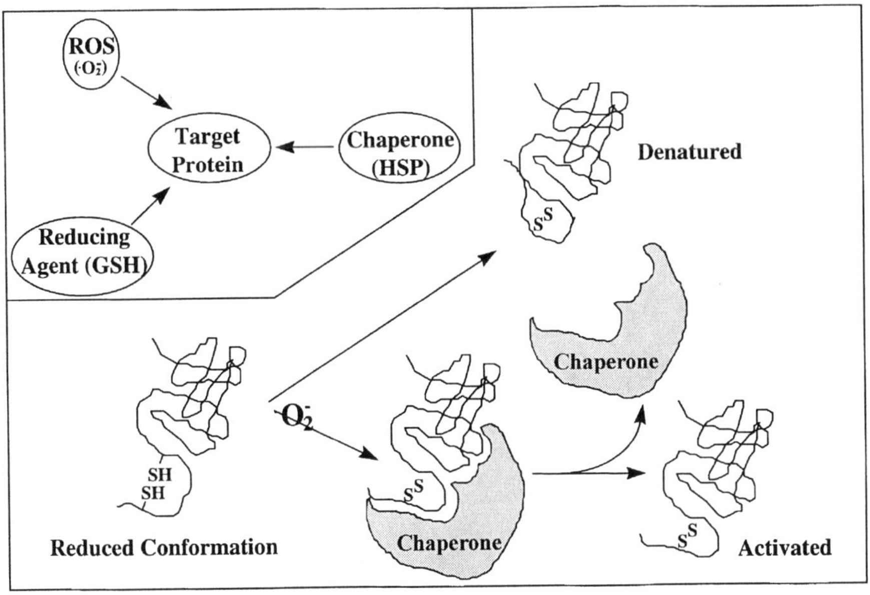

Fig. 2. Superoxide, Chaperones and Protein Conformation. Due to the presence of a large excess of reducing agents (GSH), as well as dismutases, catalases and cellular anti-oxidants, the intracellular oxidants have limited effects. One of the effects of protein oxidation might be their denaturation, and hence inactivation. However, in the presence of chaperones different conformational changes might occur, contributing to protein activation.

ACKNOWLEDGMENT

We dedicate this review to Yves Goldschmidt-Clermont, who taught at the School of Physics of Les Houches.

REFERENCES

Abate C., Patel L., Rauscher F.J. and Curran T., *Science* **249** (1990) 1157-1162.

Abo A., Pick E., Hall A., Totty N., Teahan C.G. and Segal A.W., *Nature* **353** (1991) 668-670.

Adamski J., Husen B., Thole H.H., Groeschel-Stewart U. and Jungblut P.W., *Biochem. J.* **296** (1993) 797-802.

Alberts B., Bray D., Lewis J., Raff M., Roberts K. and Watson J.D., In Molecular Biology of the Cell (Garland Publishing, Inc., 1994).

Aon M.A., Cáceres A. and Cortassa S., *J. Cell. Biochem.* **60** (1996) 271-278.

Bagrodia S., Taylor S.J., Creasy C.L. Chernoff J. and Cerione R.A., *J. Biol. Chem.* **270** (1995) 22731-22737.

Beckmann R.P., Lovett M. and Welch W.J., *J. Cell. Biol.* **117** (1992) 1137-1150.

Bereiter-Hahn J., Stübig C. and Haymann V., *Exp. Cell. Res.* **218** (1995) 551-560.

Bokoch G.M., *Curr. Opin. Cell. Biol.* **6** (1994) 212-218.

Bokoch G.M. and Prossnitz V., *J. Clin. Invest.* **89** (1992) 402-408.
Boguski M.S. and McCormick F., *Nature* **366** (1993) 643-654.
Borlongan C.V., Kanning K., Poulos S.G., Freeman T.B., Cahill D.W. and Sanberg P.R., *J. Fla. Med. Assoc.* **83** (1996) 335-341.
Bretscher A., *Ann. Rev. Cell. Biol.* **7** (1991) 337-374.
Buchner J., *FASEB J.* **10** (1996) 10-19.
Burdon R.H., *Free Rad. Biol. Med.* **18** (1995) 775-794.
Bourne H.R., Sanders D.A. and McCormick F., *Nature* **349** (1991) 117-127.
Byers H.R. and Fujiwara K., *J. Cell. Biol.* **93** (1982) 804-811.
Carlier M.F. and Pantaloni D., *Semin. Cell. Biol.* **5** (1994) 183-191.
Cassimeris L., Safer D., Nachmias Z.T. and Zigmond S.H., *J. Cell. Biol.* **119** (1992) 1261-1270.
Clarke F.M. and Masters C.J., *Biochim. Biophys. Acta* **381** (1975) 37-46.
Cooper J.A., *Ann. Rev. Physiol.* **105** (1991) 585-605.
Coso O.A., Chiarello M., Yu J.C., *et al.*, *Cell.* **81** (1995) 1137-1146.
de Crouy-Chanel A., Kohiyama M. and Richarme G., *J. Biol. Chem.* **270** (1995) 22669-22672.
Crawford L.E, Tucker R.W., Heldman A.W. and Goldschmidt-Clermont P.J., In Actin: Biophysics, Biochemistry, and Cell Biology, edited by J.E. Estes and P.J. Higgins (Plenum Press, 1994) pp. 105-112.
Crawford L.E., Milliken E, Irani K., *et al.*, *J. Biol. Chem.* **271** (1996) 26863-26867.
Dagher M.C., Fuchs A., Bourmeyster N., Jouan A. and Vignais P.V., *Biochimie* **77** (1995) 651-660.
Davies P.F., Robotewskyj A. and Griem M.L. *J. Clin. Invest.* **93** (1994) 2031-2038.
Davies K.J., *Biochem. Soc. Symp.* **61** (1995) 1-31.
Dreher D. and Junod A.F., *Eur. J. Cancer.* **32A** (1996) 30-38.
Emmendorffer A., Roesler J., Elsner J., Raeder E., Lohmann-Matthes M.L. and Meier B., *Eur. J. Haematol.* **51** (1993) 223-227.
Flitter W.D., *Brit. Med. Bull.* **49** (1993) 545-555.
Friend C.M., *Sci. Am.* **268** (1993) 74-79
Gabbiani G., Gabbiani F., Lombardi D. and Schwartz S.M., *Proc. Nat. Acad. Sci. USA* **80** (1983) 2361-2364.
Goldschmidt-Clermont P.J., Machesky L.M., Baldassare J.J. and Pollard T.D., *Science* **247** (1990) 1575-1578.
Goldschmidt-Clermont P.J. and Janmey P.A., *Cell.* **66** (1991) 419-421.
Goldschmidt-Clermont P.J., Furman M.I., Wachsstock D., Safer D., Nachmias V.T. and Pollard T.D., *Molec. Biol. Cell.* **3** (1992a) 1015-1024.
Goldschmidt-Clermont P.J., Mendelsohn M.E. and Gibbs J.B., *Curr. Biol.* **2** (1992b) 669-671.
Goldstone S.D., Fragonas J.C., Jeitner T.M. and Hunt N.H., *Biochim. Biophys. Acta* **1263** (1995) 114-122.
Goldstone S.D., Milligan A. and Humt N.H., *Biochim. Biophys. Acta* **1314** (1996) 175-182.
Gotlieb A.L., *Toxicol. Pathol.* **18** (1990) 603-617.
Heldman A.W., Kandzari D.E., Tuckcer R.W., *et al.*, *Circ. Res.* **78** (1996) 312-321.
Hancock J.F., Paterson H. and Marshall C.J., *Cell.* **63** (1990) 133-139.

Herman B. and Pledger W.J., *J. Cell. Biol.* **100** (1985) 1031-1040.
Irani K. and Goldschmidt-Clermont P.J., *Biochem. Pharmacol.* **55** (1998) 1339-1346.
Irani K., Xia Y., Zwlier J.L., *et al.*, *Science* **275** (1997) 1649-1652.
Janmey P.A., *Ann. Rev. Physiol.* **56** (1994) 169-191.
Janmey P.A. and Stossel T.P., *J. Biol. Chem.* **264** (1989) 4825-4831
Jenner P., *Pathol. Biol.* **44** (1996) 57-64.
Jockusch B.M., Bubeck P., Giehl K., *et al.*, *Ann. Rev. Cell. Dev. Biol.* **11** (1995) 379-416.
Joneson T., McDonough M., Bar-Sagi D. and Van Aelst L., *Science* **274** (1996) 1374-1376.
Knaus U.G., Heyworth P.G., Evans T., Curnutte J.T. and Bokoch G.M., *Science* **254** (1991) 1512-1515.
Kreis T.E. and Birchmeier W., *Cell.* **22** (1980) 555-561.
Kuo H.J., Malencick D.A., Liou R.S. and Anderson S.R., *Biochemistry* **25** (1986) 1378-1286.
Larochelle D.A, Vithalani K.K. and De Lozanne A., *J. Cell. Biol.* **133** (1996) 1321-1329.
Laufenburger D.A. and Horwitz A.F., *Cell.* **84** (1996) 359-369.
Lassing I. and Lindberg K., *Nature* **314** (1985) 472-474.
Lee T. and Montell D.J., *Develop. Biol.* **185** (1997) 25-33.
Lowy D.R., Johnoson M.R., DeClue J.E., *et al.*, *Ciba Found. Symp.* **176** (1993) 67-84.
Manser E., Leung T., Salihuddin H., Zhao Z.S. and Lim L., *Nature* **367** (1994) 40-46.
Matsudaira P., and Janmey P.A., *Cell.* **54** (1988) 139-140.
Meier B., Cross A.R., Hancock J.T., Kaup F.J. and Jones O.T.G., *Biochem. J.,* **275** (1991) 241-245.
Meier B., Jesaitis A.J., Emmendorfer A., Roesler J. and Quinn M.T., *Biochem. J.* **289** (1993) 481-486.
Minden A., Lin A., Claret F.X., Abo A. and Karin M., *Cell.* **81** (1995) 1147-1157.
Mockrin S.C. and Korn E.D., *Biochemistry* **19** (1980) 5359-5362.
Negulayev Y.A., Vedernikova E.A. and Maximov A.V., *Mol. Biol. Cell.* **7** (1996) 1857-1864.
Nobes C.D. and Hall A., *Cell.* **81** (1995) 53-62.
Nohl H., *Br. Med. Bull.* **49** (1993) 653-667
Norman J.C., Price L.S., Ridley A.J., Hall A. and Koffer A., *J. Cell. Biol.* **126** (1994) 1005-1015.
Nusrat A., Giry M., Turner J.R., *et al.*, *Proc. Nat. Acad. Sci. USA* **92** (1995) 10629-10633.
Peppelenbosch M.P., Qiu R.G., de Vries-Smits A.M.M., *et al.*, *Cell.* **81** (1995) 849-856.
Poglazov B.F. and Livanova N.B., *Adv. Enzyme Regul.* **25** (1986) 297-305.
Pollard T.D., In Signal Transduction in Cytoplasmic Organization and Cell Motility (A.R. Liss, Inc., 1988) pp. 263-269.
Pollard T.D. and Cooper J.A., *Biochemistry* **23** (1984) 6631-6641.

Pollard T. and Cooper J.A., *Ann. Rev. Biochem.* **55** (1986) 987-1035.
Polyak K., Xia Y., Zwier J.L., Kinzler K.W. and Vogelstein B., *Nature* **389** (1997) 300-305.
Pratt W.B., *Ann. Rev. Pharmacol. Toxicol.* **37** (1997) 297-326.
Prinsze C., Dubbelman T.M. and Van Steveninck J., *Biochim. Biophys. Acta* **1038** (1990) 152-157.
Ridley A.J. and Hall A., *Cell.* **70** (1992) 389-399.
Ridley A.J., Paterson H.F., Johnston C.L., Diekmann D. and Hall A., *Cell.* **70** (1992) 401-410.
Robinson J.M. and Badwey J.A., *Histochemistry* **103** (1995) 163-180.
Schliwa M., Nakamura T., Porter K.R. and Euteneuer U., *J. Cell. Biol.* **99** (1984) 1045-1059.
Sen C.K. and Packer L., *FASEB J.* **10** (1996) 709-720.
Sundaresan M., Yu Z.X., Ferrans V.J., Irani K. and Finkel T., *Science* **270** (1995) 296-299.
Suzuki Y.J., Forman H.J. and Sevanian A., *Free Rad. Biol. Med.* **22** (1997) 269-285.
Tamkun J.W., DeSimone D.W., Fonda D., Patel R.S., Buck C., Horwitz A.F. and Hynes R.O., *Cell.* **46** (1986) 271-282.
Theriot J.A. and Mitchison T.J., *J. Cell. Biol.* **119** (1992) 367-377.
Thrasher A.J., Keep N.H., Wientjes F. and Segal A.W., *Biochim. Biophys. Acta* **1227** (1994) 1-24.
Toyokuni S., *Free Radic. Biol. Med.* **20** (1996) 553-566.
Van Aelst L., Joneson T. and Bar-Sagi D., *EMBO J.* **15** (1996) 3778-3786.
Walsh TP., Clarke F.M. and Masters C.J., *Biochem. J.* **165** (1977) 165-167.
Westwick J.K., Lambert Q.T., Clark G.J., *et al.*, *Molec. Cell. Biol.* **17** (1997) 1324-1335.
Wong A.J., Pollard T.D. and Herman I.M., *Science* **219** (1982) 867-869.
Yin H.L. and Stossel T.P., *Nature* **281** (1979) 583-586.
Yun H.Y., Dawson V.L. and Dawson T.M., *Mol. Psychiatry* **2** (1997) 300-310.
Zweier J.L., Broderick R., Kuppusamy P., Thompson-Gorman S. and Lutty G.A., *J. Biol. Chem.* **269** (1994) 24156-24162.

LECTURE 6

Tensegrity and the Emergence of a Cellular Biophysics

D.E. Ingber

Departments of Pathology and Surgery,
Harvard Medical School and Children's Hospital,
Boston, MA 02115, USA

1. INTRODUCTION

Most physicists have kept their distance from living cells for good reason; they are just too complex. It is a long jump from a two body problem to cellular mechanics. However, the enormous advances made in polymer physics in recent years has greatly narrowed the distance between these two fields. As can be seen from the accompanying chapters in this volume, highly talented physicists are now working with biological polymers and molecular motors and are finding that the behavior of these complex systems are no longer beyond their reach. This is tremendously exciting because we will never fully understand life without defining its physical basis. At the same time, enormous advances have been made in the area of molecular cell biology. For example, the human genome project has led to the discovery of incredible numbers of new molecules and genes as well as many useful chemical reagents. However, we still do not understand how these molecules join together to form living cells that can change shape, grow, and move. This question of how life is structured is the ultimate problem in biology and solving this riddle will likely require the effort of physicists as well as biologists.

In their natural evolution, well established scientific disciplines occasionally make an abrupt change in focus and move forward in a different direction. This usually occurs when new information emerges that fills an existing void and creates a rudimentary vocabulary that can link previously isolated disciplines. As a result, scientific interbreeding can occur and natural selection can proceed. The goal of this chapter is to describe a new discipline that appears to be in the process of emerging which combines elements of Molecular Cell Biology, Polymer Physics, and Mechanical Engineering in an attempt to deal with questions of molecular regulation within the structural context of the whole living cell. I will review these developments in this new area of "Cellular Biophysics" from the perspective of a biologist who is interested in the relation between cell structure and function. In the process, I hope to change the average physicist's view of living cells and to convince them that much of the complexity of life already may be approachable using existing mathematical approaches and physical techniques; only the frame of reference has to be changed. The question of whether Cellular Biophysics has already begun to emerge can be best answered by reading the accompanying chapters in this volume which summarize work by leading investigators at the interface of these different fields presented at the recent meeting on "Dynamical Networks in Physics and Biology" at Les Houches, France in March of 1997.

2. CONVENTIONAL VIEW OF THE CELL

In the early part of this century, the cell was viewed as a small mass of protoplasm with a nucleus in the center. By the 1940's, biology textbooks described the cell as a membrane-lined sac filled with a viscous protoplasm. With the development of the electron microscope, it soon became clear that cells exhibited complex structure on the microscale. The initial emphasis in the 1960's and 70's was on the role of lipid membranes and membrane-lined intracellular organelles in cellular organization. In part, this focus was due to the electron microscopic technique itself which utilized chemical stains (uranyl acetate, lead citrate) that highlighted lipid bilayers as well as embedding media (plastic resins) that effectively masked much of the proteinaceous structures of the cytoplasm and nucleus. The use of ultrathin sections (400 – 800 Å thick) also provided a view of the cell that lost all feeling of its true three dimensional architecture.

Refinement of antibody preparation and immunofluorescence staining techniques in the 1970's, for the first time, permitted analysis of molecular structure within intact cells (Lazarides, 1976; Osborn *et al.*, 1978). These studies showed that, in addition to organelles, cells also contained a filamentous supporting framework or "cytoskeleton" that is comprised of three different major polymer systems (microfilaments, microtubules, and intermediate filaments) as well as an insoluble lattice within the nucleus, known as the nuclear matrix (Coffey and Berezney, 1975; Fey *et al.*, 1984). More recent work has revealed that these insoluble filaments do more than provide a structural backbone to the cell; they also provide a mechanism to regulate cellular biochemistry and gene expression. Specifically, much of the cell's metabolic machinery appears to function in an immobilized form or "solid-state". For example, the enzymes, substrates and chemical reactions that mediate protein synthesis, RNA processing, glycolysis, DNA synthesis, and signal transduction all channel along the surface of insoluble molecular scaffolds within the cytoplasm and nucleus (reviewed in Ingber, 1993a; Plopper *et al.*, 1995). This mechanism, in part, explains the incredible efficiency of biochemical reactions which we observe in living cells and which can not be mimicked in a test tube.

Although filamentous scaffolds have been demonstrated within intact cells since the 1970's and solid-state biochemistry is gaining increased attention, many molecular biologists and bioengineers still assume that cell structure is dominated by polymerization-depolymerization cycles and viscosity. For example, the leading engineering models for the mechanical behavior of living cells still consider the viscous fluid cytosol and surrounding tensed membrane to be the key load-bearing elements of the cell (Skalak *et al.*, 1981; Elson, 1988; Evans and Yeung, 1989). These models also assume that external forces are distributed evenly across the cell surface. Yet, work over the past fifteen years has revealed that living cells, like terrestrial animals, are point-loaded: the cytoskeleton anchors to the underlying extracellular matrix and to surrounding cells only at spot weld-like contacts at the cell base ("focal adhesions") and lateral cell borders (*e.g.*, desmosomes, adherens junctions) (Burridge *et al.*, 1988; Ingber, 1991; Yamada and Geiger, 1997). Furthermore, mechanical coupling between the cytoskeleton and the external supporting structures is mediated by specific molecular linkages that involve transmembrane receptors, such as the "integrins" in focal adhesions (Wang *et al.*, 1993) and cadherins in cell-cell junctions (Yamada and Geiger, 1997). Thus, the view of the cell as an elastic balloon filled with a molasses-like cytoplasm can not take us into future nor permit us to really understand how cell structure and function are regulated at the molecular level.

2.1 Molecular polymers of the cytoskeleton

Before addressing the question of how living cells are structured, it is necessary for the reader to have an understanding of the three major classes of molecular polymers that cells use to form their supporting cytoskeletal framework:

Microfilaments are 4-6 nanometer wide polymers of the protein, actin (Condeelis, 1993). They can form loose networks or stiffened bundles containing many cross-linked filaments aligned in parallel, within different locations in the same cell. Actin filaments are also often found associated with myosin filaments. These composite filament bundles are known as "contractile microfilaments" because they can generate mechanical tension via an actomyosin filament sliding mechanism similar to that utilized in skeletal muscle. Contractile filaments can be organized within loose networks, linear bundles ("stress fibers", Fig. 1), highly triangulated polygonal nets, or fully organized geodesic domes ("actin geodomes") in living cells (Lazarides, 1976; Osborn *et al.*, 1978; Heuser and Kirschner, 1980). Individual microfilaments are highly dynamic and, in certain localized regions, the entire cytoskeleton can form de novo as a result of new polymerization, such as in "lamellopodia" that extend forward in the leading edge of certain migrating cells (Stossel, 1993). However, the high flux of actin monomers observed in cells does not mean that all of the cytoskeleton is constantly depolymerizing and repolymerizing (*i.e.*, undergoing sol-gel transitions). For example, even spherical cells retain their insoluble cytoskeleton (Heuser and Kirschner, 1980; Mooney *et al.*, 1995). Furthermore, large microfilament bundles often can remain in place and bear mechanical loads for many hours, even though their outer coating of filaments exhibits a high rate of monomer flux (high on/off rates) on the order of minutes. Two simple examples are the stiffened actin bundles that act like struts to push out the surface membrane into finger-like extensions called "filopodia" in moving cells (Sheetz *et al.*, 1992) and the large bundles of contractile microfilaments (stress fibers) that actively pull against the cell's basal focal adhesion sites (Fig. 1, Kreis and Birchmeier, 1980; Harris *et al.*, 1980). It is also interesting to note that although microfilaments are highly flexible and exhibit a curved form when studied *in vitro*

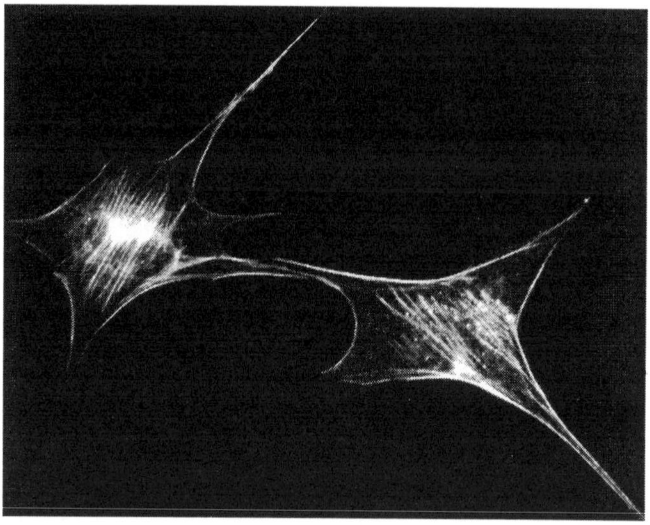

Fig. 1. Immunofluorescence micrograph showing linear actin microfilament bundles within the cytoplasm of living cells.

(Janmey, this volume), these polymers almost always appear highly extended and linear in form in living cells (Fig. 1), as if all of these filaments are under tension.

Microtubules are polymers of tubulin protein that exhibit a tubular form 24 nanometers in diameter (Gelfand and Bershadsky, 1991). They commonly initiate their polymerization from a single site in the cell, called the "centrosome", that positions itself in the perinuclear region of the cytoplasm near the Golgi apparatus. Microtubules are also very dynamic structures, however, the total content of tubulin polymer generally remains relatively constant in most cells (Mooney *et al.*, 1995): while one microtubule depolymerizes, another reassembles at a distant site. In other words, somehow all of the individual microtubules that are distributed throughout the entire cell are regulated as one. *In vitro* mechanical measurements also have revealed that microtubules are approximately ten times stiffer than microfilaments because of their increased second moment of inertia (hollow configuration; Mizushima-Sugano *et al.*, 1983) and, in fact, they almost always exhibit a highly linear form when isolated from the cell (Janmey, this volume). In contrast, microtubules characteristically exhibit a curved or buckled appearance in living cells (Osborn *et al.*, 1978).

Intermediate filaments are cytoskeletal polymers that are intermediate in size (10 nanometers) and which differ in composition depending on the cell type (Fuchs and Weber, 1994). For example, intermediate filaments in endothelial cells and fibroblasts are composed of polymers of the protein, vimentin whereas keratins, desmin, and neurofilaments form this lattice in epithelial cells, smooth muscle cells, and nerve cells, respectively. All intermediate filaments form very strong, entangled networks of protein polymers (Janmey *et al.*, 1991) that interweave and connect with the microfilament and microtubule networks (Osborn *et al.*, 1978). Intermediate filaments also appear to connect to discrete sites on the cell surface, including cell-matrix and cell-cell adhesion sites, as well as to the surface of the nucleus at the cell center (Fey *et al.*, 1984). In skin, the keratin intermediate filament lattice that remains after cell differentiation and death is largely responsible for the high tensile strength of the epidermis. Although these filaments always appear in a tightly "balled up" configuration *in vitro* (Janmey, this volume), they usually appear in a highly extended, crenulated form stretching through the cytoplasm of living cells.

Finally, it is important for the reader to realize that living cells can undergo enormous changes in shape (*e.g.*, spherical to almost completely flattened) over a period of minutes solely through the action of cytoskeletal tension. Furthermore, this dramatic cell shape transformation does not require a change in the total amount of cytoskeletal polymer (Mooney *et al.*, 1995). Thus, the structural framework of the cell must simultaneously exhibit both high mechanical strength and great flexibility. Viscous gels are flexible and they are excellent at bearing compression, however, they can not effectively withstand tension. We must therefore look for another model to explain how cells structure themselves to provide the dynamic plasticity and internal tone that is so characteristic of living materials.

3. AN ALTERNATIVE MODEL: CELLULAR TENSEGRITY

While the physicist comforts in theory, the biologist seeks data so that theory never has to be spoken. I know this well because I published a theory of cell structure and tissue organization early in my career (Ingber *et al.*, 1981; Ingber and Jamieson, 1985) and found that, for many years, few biologists would listen. However, since I was trained as an experimentalist, I learned that I could gain their attention by designing experiments that would test my hypothesis. Once I began to generate data that could explain results where existing models failed, obtaining their attention came much easier.

The theory I proposed assumed that living cells are literally "hard-wired" by a series of interconnected molecular struts and cables that physically interconnect cell surface matrix receptors (which we now know as integrins) to structural scaffolds within the nucleus (Ingber *et al.*, 1981; Ingber and Jamieson, 1985; Ingber, 1993b; Ingber, 1997). Most importantly, the model assumed that the entire supporting framework of the cell is under tension and that it is stabilized using a particular form of architecture that is known as "tensegrity" (Fuller, 1961). If cells use this form of tensile architecture, then pulling or pushing on integrins at the cell periphery would result in coordinated structural alterations in the cytoskeleton and nucleus and thereby, alter solid-state biochemistry and gene expression inside the cell.

3.1 Tensegrity architecture

Tensegrity structures gain their mechanical stability from continuous tension and local compression (tensional integrity). In contrast, most man-made structures depend on compressional continuity for their stability as observed, for example, in the construction of a pyramid which relies on the huge mass of each stone being compressed down upon the stone below for its shape stability. Simple tensegrity structures can be constructed by interconnecting a set of isolated compression struts with a continuous series of tension wires that hold the struts up against the force of gravity and stabilize them in an open array (Fig. 2). These sculptures clearly visualize the basic tensegrity mechanism of self-stabilization, however, structures do not have to contain isolated struts and cables to be tensegrity structures. Rather, it is how a framework distributes stresses to establish a force balance and stabilize itself against shape distortion that defines tensegrity. For example, fully triangulated structures, such as the tetrahedron (Fig. 2) and geodesic dome, are also tensegrity structures even though they can be composed entirely from stiff elements. Because each strut is stiff, it can resist either tension or compression at a particular location depending on the loading conditions. However, direct contact between all compression elements is never required for the stability of these structures; forces are always balanced by continuous tension and local compression (Fig. 2). The major difference between a fully geodesic structure and a tensegrity stick and "string" sculpture is that the joints are stabilized in different ways: through triangulation or by imposing a prestress (pre-existing tension), respectively. The structural efficiency of these networks can be further increased by building in a hierarchical manner (Lakes, 1993): individual supporting members can themselves be tensegrity structures composed of discrete networks on a smaller size scale that are then interlinked and stabilized by continous tension (Ingber, 1993b; Chen and Ingber, 1997; Ingber, 1998).

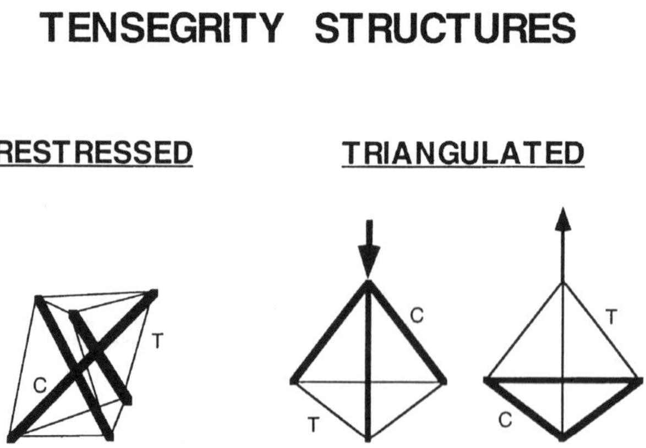

Fig. 2. Tensegrity Structures. (Left) Prestressed tensegrity models are constructed by interconnecting isolated compression struts by a continuous series of filamentous tensile elements. Each element is fully specialized to bear either tension (T) or compression (C). (Right) Fully triangulated tensegrity structures, such as the tetrahedron shown, also self-stabilize through use of continuous tension and local compression (tensegrity), however, depending on the loading conditions, each strut may bear either tension or compression. (Downward arrow, compressive load; Upward arrow, application of tension).

Although the stick and string tensegrity structure looks unusual, this form of architecture explains how our bodies are stabilized. If you ever see a skeleton in an anatomy laboratory, it is usually wired together and hung from a stand. This is because the bones of the living skeleton are pulled up against gravity and stabilized in place through interconnection with a series of tensile muscles, tendons, and ligaments. Furthermore, the stability (stiffness) of our arm or leg depends on the level of tension or "tone" in our muscles, just as the stiffness of a tensegrity structure depends on the prestress in its internal supporting network. In tensegrity structures, increasing prestress results in decreased movement without resistance (*i.e.*, decreased play), immediate mechanical responsiveness, proportional action at a distance, and decrease wear at points of interconnection. All of these features can be critical in biological systems; the knee joint provides a simple example.

3.2 Tensegrity in the living cytoskeleton

In the cellular tensegrity model, the three different biopolymers of the cytoskeleton join together to form a prestressed tensegrity structure at one level in the organic structural hierarchy. However, the same mechanism of shape stabilization is also used at both larger (tissue, organ) and smaller size scales (organelles, macromolecular assemblies, individual molecules) (Ingber and Jamieson, 1985; Ingber, 1993b; Chen and Ingber, 1997; Ingber, 1998). At the cell level (Fig. 3), contractile microfilaments create mechanical tension which distributes throughout the entire interconnected cytoskeletal-nuclear matrix lattice. The stiffer, hollow microtubules and cross-linked bundles of actin act as isolated compression elements to resist the inward-directed pull of the surrounding cytoskeleton and thus, to a create a prestress that stabilizes the entire structure. Another key compression element is the underlying extracellular matrix which separates individual focal adhesions and resists cytoskeletal tension that tends to pull these anchoring sites together (Fig. 3). Intermediate filaments function as guy wires or tensile stiffeners in this model which pull together the other filaments and link them to external support elements at the cell periphery (matrix and surrounding cells) as well as to the nucleus at the cell center. In certain microdomains (*e.g.*, small membrane protrusions), the fluid cytosol itself can act osmotically to resist compression locally, however, the discrete network that comprises the cytoskeleton forms the major tension-bearing element at the whole cell level.

The cellular tensegrity model therefore predicts that cell shape will depend on a balance of forces that are distributed between extracellular matrix and the different interacting polymers that comprise the cytoskeleton. Thus, all three cytoskeletal filaments and the level of prestress in the cytoskeleton should contribute significantly to the mechanical stiffness of the cell (*i.e.*, its ability to resist shape distortion). Furthermore, mechanical forces should not be transmitted equally at all points across the cell surface, as previously assumed in the field. Instead, cell surface adhesion receptors should provide a preferred molecular pathway for mechanical stress transfer across the plasma membrane.

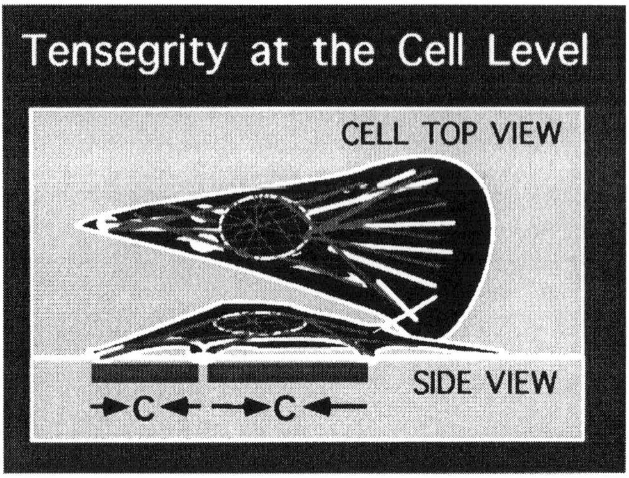

Fig. 3. Diagram of tensegrity at the whole cell level. Side and top views are shown to visualize hard-wiring from the cell surface to the nucleus as well as regions of the underlying extracellular matrix between the spot-like focal adhesion sites that resist compression (C) locally.

We recently showed all of these predictions to be true by developing a "magnetic twisting device" (Wang *et al.*, 1993; Wang and Ingber, 1994). Micron-sized ferromagnetic beads coated with different receptor ligands were bound to the surface of living cells. While still on the outside of the plasma membrane, controlled mechanical twisting stresses were applied to the bound surface receptors and interconnected cytoskeleton by first magnetizing the beads in one direction and then applying a weaker, but sustained magnetic field, in the orthogonal direction. Cell deformation was measured by quantitating the rate and degree of bead rotation. Using this approach, we showed that integrins, but not all transmembrane receptors, mechanically couple to the cytoskeleton and that all three cytoskeletal filament systems contribute to the response to stress (Wang *et al.*, 1993). More recently, using the same technique, cytoskeletal tone (prestress) was found to be a critical determinant of mechanical stiffness of living cells (Hubmayr *et al.*, 1996; Lee *et al.*, 1997) and microtubules were shown to act as internal struts that resist actin-based tension in the cytoplasm (Tagawa *et al.*, 1996). Other investigators also have demonstrated directly that microtubules resist compression locally within living cytoplasm using cells in which microtubules were labelled with green fluorescent protein (Kaech *et al.*, 1996).

Importantly, tensegrity can explain how whole living cells respond when mechanically distorted. Specifically, when analyzed using magnetic twisting cytometry, cells exhibited monotonic stiffening behavior: the mechanical stiffness of the cell increased in direct proportion as the level of applied stress was raised (Wang *et al.*, 1993). Many living tissues (*e.g.*, muscle, mesentery, skin, cartilage, etc.) also exhibit similar linear stiffening behavior, however, this fundamental property of living materials can not be explained starting from first principles (Fung, 1988; McMahon, 1984). We first showed that the novel mechanical behavior of tensegrity structures can explain this response using stick and string models (Wang *et al.*, 1993). More recently, working with Dimitrije Stamenovic (Boston U.), we developed a mathematical explanation for this behavior beginning from first principles (Stamenovic *et al.*, 1996). This mathematical approach revealed that there are two key features that determine the cellular (and tissue) response to mechanical stress: architecture and prestress. The prestress defines the initial stiffness of the system whereas the architecture determines how the structural stiffness changes during deformation. Those interested in the mathematical basis of tensegrity should refer to the most recent work by Stamenovic which utilizes non-extensible tension elements and buckleable compression struts and which more effectively predicts results obtained with living cells in multiple experimental studies (Coughlin and Stamenovic, 1997). Additional mathematical treatment of cellular tensegrity can be found in the chapter by Oddou and coworkers in this volume.

3.3 How robust is the model?

In the meeting at Les Houches, many physicists used the term "robust" to describe a model they proposed which could explain results from 2 or perhaps even 3 different experiments. In this context, tensegrity is probably the most robust model of cell architecture currently in existence. In addition to explaining the mechanical behavior of the cell measured using the magnetic twisting device, tensegrity also predicts the dependence of cell shape on extracellular matrix flexibility (Harris *et al.*, 1980; Ingber and Jamieson, 1985; Ingber and Folkman, 1989); the coordination between cell and nuclear spreading observed in adherent cells (Ingber *et al.*, 1987; Ingber, 1990); the establishment of cell polarity in response to cell anchorage (Ingber and Jamieson, 1986); and even how stress fibers and actin geodomes form at the molecular level within the cytoskeleton of living cells (Ingber, 1993b). It is also important to emphasize that the appearance of actin geodesic domes in cells, by definition, indicates the use of tensegrity architecture at a smaller size scale (actomyosin polymer lattice) in the organic structural hierarchy. In fact, tensegrity is seen at many size scales (*e.g.*, cathrin-coated pits, nuclear lamina, enzyme complexes, protein structure, *etc.*) (Ingber, 1998).

Finally, we also have been able to experimentally confirm that cells are indeed "hard-wired", such that pulling on integrin receptors on the surface of living cells using microneedles results in coordinated structural rearrangements throughout the cytoskeleton and nucleus (Maniotis *et al.* 1997a,b), again as predicted by the tensegrity model. Using this micromanipulation approach, we confirmed that intermediate filaments (and to a lesser degree, microfilaments) function as guy wires which mechanically suspend open and stiffen the nucleus of the cell, while microtubules enhance the ability of the cytoplasm and nucleus to resist lateral compression. Work from other labs also has confirmed that cross-linked actin bundles and individual microtubules are stiff structures that can bear compression locally in living cells (Kaech *et al.*, 1996; Sheetz *et al.*, 1992). In addition, we have demonstrated tensional continuity within interphase nuclei and between individual chromosomes (Maniotis *et al.*, 1997b). Taken together, these data provide direct support for the tensegrity model and demonstrate that a micromechanical approach involving discrete load-bearing networks is required to effectively deal with the biological reality of the living cell at the molecular level.

3.4 Tensegrity *versus* percolation theory

At the Les Houches meeting, there was much discussion about the differences and similarities between Tensegrity and Percolation Theory, another mathematical approach that has gained the attention of biophysicists interested in cytoskeletal mechanics (Forgacs, 1995; Forgacs, this volume; Frey *et al.*, this volume). Percolation theory is a mathematical method for analyzing the importance of phase-transitions and connectivity within networks (Forgacs, 1995). The mechanical properties of the network can be predicted if the material properties of each of the elements are known.

As a result of these discussions, it became clear that there is a major difference between tensegrity and percolation theory: tensegrity's reliance upon prestress and triangulation for mechanical stabilization. For example, Frey's mathematical modeling of cytoskeletal networks using percolation theory predicts that the mechanical stiffness of the whole network will be dominated by the bending (rather than buckling modulus) of its individual semiflexible support struts (Frey *et al.*, this volume). In contrast, the buckling (compressive) modulus dominates in tensegrity structures. Most importantly, data obtained with living cells from our laboratory and others are not consistent with the bending modulus dominating. In contrast, these experimental results are predicted by the tensegrity model (Wang *et al.*, 1993; Stamenovic *et al.*, 1996; Coughlin and Stamenovic, 1997; Oddou, this volume). Interestingly, the possibility that a percolation network may transform into a tensegrity network, if it becomes prestressed, also was raised in these discussions.

In general, percolation theory is excellent at predicting phase transitions, such as the concentration of filaments required for a polymer solution to gel and become a mechanically connected network. However, percolation theory can not explain how different types of cytoskeletal filaments can contribute uniquely to the overall mechanical behavior of the cell, as tensegrity can. In addition, percolation can not explain how specific three dimensional patterns (*e.g.*, geodesic domes) form within the molecular cytoskeleton of living cells or provide a mathematical basis to predict the material properties of living cells, independently of changes in cytoskeletal connections, as tensegrity can (Ingber, 1993b; Stamenovic *et al.*, 1996; Coughlin and Stamenovic, 1997). In summary, while tensegrity provides a mathematical basis for shape stability and a mechanistic explanation for pattern formation in living cells, percolation could offer a complementary approach to describe how the mechanical behavior of tensegrity-based networks may change in response to alterations in molecular connections.

4. IMPLICATIONS FOR CELLULAR BIOPHYSICS

My basic purpose in this chapter is to introduce physicists to the world of cell biology, with particular emphasis on the molecular and physical determinants of cell shape and function. I specifically wanted to clarify that oversimplified models of the cell that attribute most of the load-bearing capabilities of the cell to a viscous cytosol are incorrect. Even though mathematical descriptions of this type can be used to effectively describe certain behaviors of whole cells, this does not mean that this is how the cell is actually constructed. A simple analogy: hydrodynamic theory was used in early stages of the Manhattan Project to model the atom because, at that time, this was the state-of-the-art. This approach actually was incredibly effective, however, this does not mean that the atom is a fluid. The usefulness of viscous models of cell mechanics similarly does not mean that the cell is a waterballoon. Furthermore, the future of the field requires that we place mechanical and physical descriptions of the cell within a precise molecular context; only this approach will lead to the development of new molecular approaches for drug therapy or tissue repair.

In this context, the tensegrity model is important because it provides a larger frame of reference for the individual molecular behaviors that are already under study by many biophysicists. Defining the mathematical basis for the flexibility of a microtubule or the physics of its interactions with a binding molecule is extremely useful. However, tensegrity suggests that this type of reductionist approach will not completely explain how these molecular polymers behave in living cells nor expose their potential importance for cellular function. Instead, higher order structural interactions, mechanical stress distributions, and a hierarchical form of structural organization all must be taken into account. A good example of the impact of tensegrity in this area is a thermodynamic model for control of microtubule polymerization in neurites developed by Buxbaum and Heidemann (1988) which incorporates the pull of contractile microfilaments as well as the resistance of both the microtubules and the cell's extracellular matrix adhesions when calculating the entire cellular force balance.

Another excellent example of how tensegrity can help explain biological complexity comes in the area of mechanotransduction: how living cells sense and respond to mechanical signals. The old view is that mechanical stresses deform the cell membrane and somehow this is transduced into a change in chemistry by activating an illusive cell surface "mechanoreceptor". In contrast, tensegrity predicts that external forces are imposed on a preexisting force balance (cell pulling against its matrix adhesions), much like plucking on a violin string (Ingber, 1991; Ingber, 1997). Mechanical stresses will therefore be preferentially transmitted across anchoring receptors, such as integrins. Mechanoreception, in turn, will involve an integrated cellular response whereby forces are transmitted over long distances, causing structural rearrangments and mechanical changes throughout the interconnected cytoskeletal-nuclear lattice. At the molecular level, stress-dependent changes in molecular shape (deformation) would result in changes in chemical potential and hence, alterations in thermodynamics and kinetics (Ingber and Jamieson, 1985; Ingber, 1997) leading to a biochemical response. Furthemore, the same cell might produce an entirely different response depending on the tone in its cytoskeletal lattice or on the shape of its supporting cytoskeletal framework. Recent studies provide direct experimental support for this model (Wang *et al.*, 1993; Chen and Grinnell, 1995; also reviewed in Ingber, 1997). This type of integrated cellular mechanotransduction mechanism could never be explained by studying the biophysical or chemical properties of any single molecule or cytoskeletal polymer.

5. CONCLUSION

From this discussion, it should be clear that biologists and biophysicists must collectively develop a new frame of reference that takes into account the importance of mechanics and cellular architecture when considering how biochemistry and gene expression are regulated in living cells. The same can be said about understanding the processes of self-assembly and progressive remodeling that mediate cellular morphogenesis and tissue development. For example, we have shown that cells can be switched between complex behavioral gene programs, including growth, differentiation, and apoptosis (programmed cell death), simply by physically distorting the cell and changing cell shape (Ingber and Folkman, 1989; Ingber, 1990; Singhvi *et al.*, 1994; Chen *et al.*, 1997). In essence, given the same set of chemical inputs (*e.g.*, hormones, growth factors, extracellular matrix binding), the cell will produce a different functional output depending on the structural organization of its internal cytoskeletal scaffolds on which solid-state biochemistry proceeds. Local changes in physical parameters (*e.g.*, cell stretching at the tip of a growing gland) may therefore explain how the local cell growth differentials that drive pattern formation can be established within a chemical microenviroment (*e.g.*, growing tissue) that is likely saturated with soluble growth factors and hormones (Ingber and Jamieson, 1985). Thus, a new frame of reference that incorporates tensegrity architecture can help to explain how mechanics and chemistry interplay at many size scales in the biological systems hierarchy.

The future of cell regulation will likely involve a similar wholistic perspective whereby our defined chemical understanding of individual molecules is placed within the structural context of the cell, and eventually, of the entire living tissue. Describing a biophysical basis for a particular molecular polymer that incorporates its chemistry as well as its prestress and higher order architecture is what we are looking for: this is the first glimmer of the field of Cellular Biophysics I speak of. Hopefully, tensegrity will provide a conceptual and mathematical handle for additional physicists who are interesting in joining in our continuing effort to understand the physical basis of what we call "life". The potential is enormous.

ACKNOWLEDGEMENTS

This work was supported by grants from NIH and NASA.

REFERENCES

Berezney R. and Coffey D.S., *Science* **189** (1975) 291-293.

Burridge K., Fath K., Kelly T., Nuckolls G. and Turner C., *Ann. Rev. Cell. Biol.* **4** (1988) 487-525.

Buxbaum R.E. and Heidemann S.R., *J. Theor. Biol.* **134** (1988) 379-390.

Chen B.M. and Grinnell A.D., *Science* **269** (1995) 1578-1580.

Chen C.S., Mrksich M., Huang S.E., Whitesides G.W. and Ingber D.E., *Science* **276** (1997) 1425-28

Chen C. and Ingber D., Osteoarthritis and Articular Cartilage (1998) in press.

Condeelis, *J. Ann. Rev. Cell. Biol.* **9** (1993) 411-444.

Coughlin M.F. and Stamenovic D., *ASME J. Appl. Mech.* **64** (1997) 480-486.

Elson E.L., *Ann. Rev. Biophys. Biophys. Chem.* **7** (1988) 397-430.

Evans E. and Yeung A., *Biophys. J.* **56** (1989) 151-160.

Fey E.G., Wan K.M. and Penman S., *J. Cell. Biol.* **98** (1984) 1973-1984.

Frey E., Kroy K., Wilhelm J. and Sackmann E. (1998) this volume.

Forgacs G., *J. Cell. Sci.* **108** (1995) 2131-2143.

Forgacs G. (1998) this volume.

Fuchs E. and Weber K., *Ann. Rev. Biochem.* **63** (1994) 345-382.

Fuller R.B., *Portfolio Artnews Ann.* **4** (1961) 112-127.

Fung Y.C., Biomechanics: Mechanical Properties of Living Tissues (Springer-Verlag, New York, 1988).

Gelfand V.I. and Bershadsky A.D., *Ann. Rev. Cell. Biol.* **7** (1991) 93-116.

Harris A.K., Wild P. and Stopak D., *Science* **208** (1980) 177-80

Hubmayr R.D., Shore S., Fredberg J.J., Planus E., Panettieri Jr. R.A., Moller W., Heyder J. and Wang N., *Am. J. Physiol.* **40** (1996) C1660-C1668.

Heuser J.E. and Kirschner M.W., *J. Cell. Biol.* **86** (1980) 212-234.

Ingber D.E., *Proc. Natl. Acad. Sci. USA* **87** (1990) 3579-3583.

Ingber D.E., *Curr. Opin. Cell. Biol.* **3** (1991) 841-848.

Ingber D.E., *Cell.* **75** (1993a) 1249-1252.

Ingber D.E., *J. Cell. Sci.* **104** (1993b) 613-627.

Ingber D.E., *Ann. Rev. Physiol.* **59** (1997) 575-599.

Ingber D.E., *Scientific American* **278** (1998) 48-57.

Ingber D.E. and Folkman J., *J. Cell. Biol.* **109** (1989) 317-330.

Ingber D.E., Madri J.A. and Jamieson J.D., *Proc. Natl. Acad. Sci. USA* **78** (1981) 3901-3905.

Ingber D.E. and Jamieson J.D., "Cells as tensegrity structures: architectural regulation of histodifferentiation by physical forces tranduced over basement membrane", in Gene Expression During Normal and Malignant Differentiation, edited by L.C. Andersson, C.G. Gahmberg, P. Ekblom (Academic Press, Orlando, 1985) pp. 13-32.

Ingber D.E., Madri J.A. and Folkman J.D., *In vitro Cell. Dev. Biol.* **23** (1987) 387-394.

Ingber D.E., Madri J.A. and Jamieson J.D., *Am. J. Pathol.* **122** (1986) 129-139.

Janmey P.A., Euteneur U., Traub P. and Schliwa M., *J. Cell. Biol.* **113** (1991) 155-161.

Janmey P. (1998) this volume.

Kaech S., Ludin B. and Matus A., *Neuron.* **17** (1996) 1189-1199.
Kreis T.E. and Birchmeier W., *Cell.* **22** (1980) 555-561.
Lakes R., *Nature* **361** (1993) 511-515.
Lazarides E., *J. Cell. Biol.* **68** (1976) 202-219.
Lee K.-M., Tsai K., Wang N. and Ingber D.E., *Am. J. Physiol.* **274** (Heart Circ. Physiol. 43, 1998) H76-H82.
Maniotis A., Chen C.S. and Ingber D.E., *Proc. Nat. Acad. Sci. USA* **94** (1997) 849-854.
Maniotis A., Bojanowski K. and Ingber D.E., *J. Cellul. Biochem.* **65** (1997) 114-130.
McMahon T.A., Muscles, Reflexes, and Locomotion (Princeton U. Press, Princeton, 1984).
Mizushima-Sugano J., Maeda T. and Miki-Noumura T., *Biochim. Biophys. Acta* **755** (1983) 257-262.
Mooney D., Hansen L. and Ingber D.E., *J. Cell. Sci.* **108** (1995) 2311-20.
Oddou C. (1998) this volume.
Osborn M., Born T., Koitsch H.-J. and Weber K., *Cell.* **14** (1978) 477-488.
Plopper G., McNamee H., Dike L, Bojanowski K. and Ingber D.E., *Mol. Biol. Cell.* **6** (1995) 1349-1365.
Sheetz M.P., Wayne D.B. and Pearlman A.L., *Cell. Motil. Cytoskel.* **22** (1992) 160-69.
Sims J.R., Karp S. and Ingber D.E., *J. Cell. Sci.* **103** (1992) 1215-1222.
Singhvi R., Kumar A., Lopez G., Stephanopoulos G.N., Wang D.I.C., Whitesides G.M. and Ingber D.E., *Science* **264** (1994) 696-698.
Skalak R., Keller S.R. and Secomb T.W., *J. Biomech. Eng.* **103** (1981) 102-115.
Stamenovic D., Fredberg J., Wang N., Butler J. and Ingber D., *J. Theor. Biol.* **181** (1996) 125-36.
Stossel T.P., *J. Biol. Chem.* **264** (1989) 18261-18264.
Tagawa H., Wang N., Narishige T., Ingber D.E., Zile M.R. and Cooper IV G., *Circ. Res.* **80** (1997) 281-289.
Wang N., Butler J.P. and Ingber D.E., *Science* **260** (1993) 1124-1127.
Wang N. and Ingber D.E., *Biophys. J.* **66** (1994) 2181-2189.
Yamada K.M. and Geiger B., *Curr. Opin. Cell. Biol.* **9** (1997) 76-85.

LECTURE 7

The Leukocyte Actin Cytoskeleton

F. Richelme, A.-M. Benoliel and P. Bongrand

Laboratoire d'Immunologie, INSERM U387, BP. 29,
13274 Marseille Cedex 09, France

1. INTRODUCTION

The main task of leukocytes is to patrol throughout the body in order to detect and eliminate potentially harmful structures such as invading microorganisms or effete cells. The fulfilment of these functions often requires that cells undergo drastic morphological changes. These changes may be passive, *e.g.* when a spherical neutrophil flowing in blood acquires an elongated shape in order to pass through a small capillary vessel, or they may be active, when a resting cell sends forward a thin lamellipodium and starts migrating on a surface (Stossel, 1993). It is usually considered that cell shape is essentially controlled by a complex cytoplasmic network including at least three structural species: microfilaments, microtubules and intermediate filaments (see the review by Richelme *et al.*, 1996, for more information). Extensive interactions between these structures were described, and all of them may contribute cell morphological and mechanical properties. However, for the sake of simplicity, we shall focus on actin microfilaments. Indeed, actin represents more than 10% of leukocyte cytosolic proteins, and much experimental evidence demonstrated that microfilament alteration strongly influenced cell mechanical properties.

As depicted in Figure 1, cell actin organization is controlled by a variety of so-called actin binding proteins that may influence the availability of polymerizable actin, rate of actin association and dissociation on microfilament ends, or interaction between any microfilament and other microfilaments or cellular organelles. It would be of obvious interest to understand how cells regulate the functions of these actin-associated proteins. According to current biochemical knowledge, protein functions are usually regulated by phoshorylation/dephosphorylation reactions or interactions with small messenger molecules such as calcium and hydrogen ions, lipids or cyclic nucleotides. Thus, at first sight, it would seem an attractive prospect to determine the effect of each of these messengers on cytoskeletal organization. The problem with this approach is that any mode of cell stimulation usually triggers a complex cascade of interrelated intracellular reactions that are both localized and time-dependent. Thus, a rise of intracellular calcium may activate protein kinase C, which in turn will modulates both proton transport (with subsequent change of intracellular pH) and tyrosine protein kinases.

The aim of this paper is to summarize a series of experiments that were planned to assess the influence of intracellular calcium on cell mechanical properties. Calcium was considered as a suitable candidate for cytoskeletal control in view of the following data: i) many actin-associated proteins, particularly microfilament severing species such as gelsolin or villin, are calcium dependent. ii) living cells can exert a tight regulation on cell calcium. Indeed, the cytosol Ca^{2+} concentration is of order of 100 nM, which is about 10,000 fold lower than the calcium content of extracellular medium or intracellular stores. This resting concentration may be increased very rapidly following the opening of regulated calcium channels, and this rise may be regulated by efficient calcium pumps, as illustrated by the occurrence of calcium oscillations following appropriate cell stimulation. iii) As pointed out by Albritton *et al.* (1992), locally released intracellular calcium is buffered before it diffuses by more than a few micrometers, which makes it a suitable messenger for the control of localized events such as deformation. iv) Simple and efficient techniques are available to manipulate intracellular calcium by combining the use of ionophores (*i.e.* substances that make the membrane permeable to particular ions), calcium buffers and chelators. v) Finally, intracellular calcium distribution may be visualized by loading cells with calcium sensitive fluorescent probes and combining fluorescence microscopy and digital image processing (Tsien and Poenie, 1986).

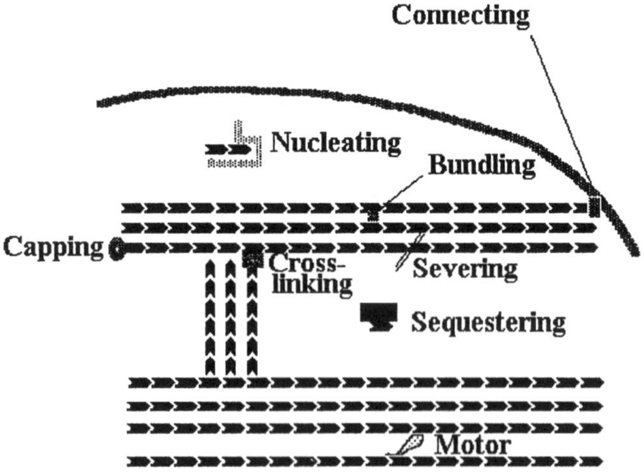

Fig. 1. General organization of the actin skeleton. About 50% of actin molecules are incorporated in microfilaments (F-actin). Actin arrangement is regulated by a variety of actin-binding proteins whose various functions are depicted. Note that a given molecule may perform different functions depending on surrounding conditions.

In view of aforementioned problems, we looked for a methodology allowing simultaneous control of intracellular calcium and cell mechanical properties with high temporal and spatial resolution. This consisted of aspirating myeloid cells into micropipettes with calibrated pressure on the stage of a fluorescence microscope. Cell deformation could be recorded together with pressure and possibly calcium distribution. After some experiments, conventional staining procedures allowed direct examination of cytoskeletal organization. We shall now describe some results obtained in our laboratory with this approach. First, we shall briefly summarize other data obtained with more conventional approaches.

2. A CONVENTIONAL STUDY OF THE ROLE OF CALCIUM IN THE SHAPE CONTROL OF HUMAN NEUTROPHILS

As shown in Figure 2A, a resting neutrophil is fairly spherical with uniform concentration of polymerized actin outside the nuclear area. However, if cells are exposed to chemotactic stimuli such as bacterial products or oligopeptides (such as N-formyl-methionyl-leucyl-phenylalanine) that are considered as bacterial analogs, they rapidly acquire an elongated shape with anisotropic actin localization (Fig. 2B). The importance of calcium in this process was studied (Lépidi *et al.*, 1992; Zaffran *et al.*, 1993a,b) leading to the following conclusions:

- A marked rise of intracellular calcium was detectable 5 seconds after stimulation.

- Chemotactic stimulation also triggered a marked increase of actin polymerization within 30 seconds. At this stage, topological microfilament reorganization could be detected in perfectly round cells, strongly suggesting that this reorganization preceded rather than followed deformation.

- After 5 minutes exposure to the chemoattractant, 50% of neutrophils displayed polarized shape.

It was thus tempting to ascribe a role of calcium change in cell deformation. However, when cells were pretreated with a calcium buffer (BAPTA/AM) in serum deprived medium, then subjected to chemotactic stimulation, the calcium rise was completely abolished whereas morphological changes were essentially unaffected by this treatment.

Taken at face value, these results would suggest that intracellular calcium has no effect on microfilament control. However, these data were felt difficult to reconcile with the aforementioned calcium sensitivity of many actin binding proteins. Thus, we hypothesized that cytoskeletal control involved highly redundant mechanisms, making difficult to analyze the importance of a single messenger with standard inhibition experiments. This concept is in line with the surprisingly modest consequences of inactivation of genes coding for seemingly important molecules such as interleukin 2 (Schorle *et al.*, 1991). Therefore, we reasoned that more sensitive methods were required to detect a possible role of calcium in microfilament regulation.

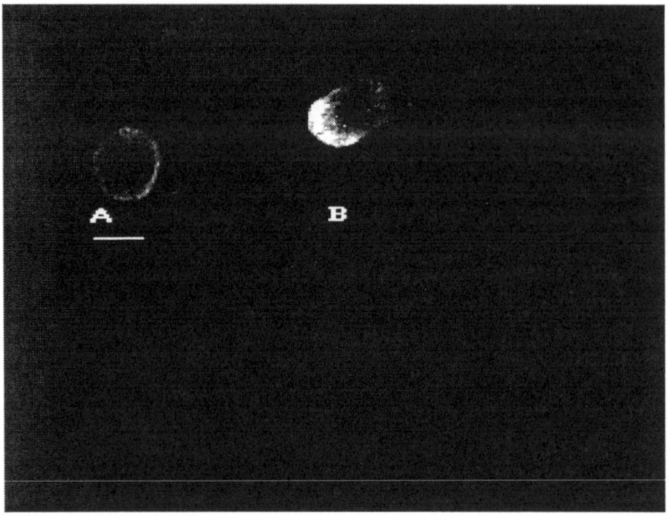

Fig. 2. Chemoattractant-induced neutrophil polarization. Human neutrophils were treated with a fluorescent microfilament probe (bodipy-phallacidin) without (A) or with (B) five minute exposure to a chemoattractant (100 nM formyl-met-leu-phe). They were then examined with a confocal microscope. Bar length is 10 μm.

3. USE OF THE MICROPIPETTE ASPIRATION TECHNIQUE TO STUDY THE INFLUENCE OF CALCIUM ON CYTOSKELETAL REGULATION

A widely used method for studying leukocyte mechanical properties consists of monitoring the deformation induced by aspiration into a micropipette of a few micrometer diameter (Evans and Kukan, 1984). This allows real time monitoring of individual cells. Association with fluorescence microscopy may yield simultaneous information on cell mechanical properties and cytosolic biochemical events. Thus, we thought that this was suitable for our purpose. First, we checked that the cell cytoskeleton was indeed involved in such deformation. This was demonstrated as follows (André *et al.*, 1990; Richelme *et al.*, 1997): cells from the murine phagocytic line P388D1 or the human phagocytic line THP-1 were aspirated in order to induce the formation of a protrusion. A few minutes later, they were fixed on the tip of the micropipette and treated with a fluorescent phalloidin derivative in order to visualize microfilaments (Barak *et al.*, 1980). An image analysis device was then used to compare the fraction of total fluorescence displayed by the protrusion to the relative volume of this protrusion (the methodology is exemplified in Fig. 3). No significant difference was found between both ratios, suggesting that the protrusion contained a relative amount of microfilaments comparable to that found in the whole cell, thus demonstrating that protrusions were not cytoskeleton-free blebs whose mechanical properties would be essentially contributed by the plasma membrane. Note that additional experiments would be required to rule out the possibility that the cytoskeleton deformation might start only a few tens of seconds after the onset of cell deformation (see Zhelev and Hochmuth, 1995). Interestingly, preliminary results suggested that the relative amount of microfilaments in the protrusion was decreased if i) cells were treated with cytochalasin B, in order to disorganize microfilament assembly, or ii) cells were treated with phallacidin before aspiration, which might impair cytoskeletal flexibility.

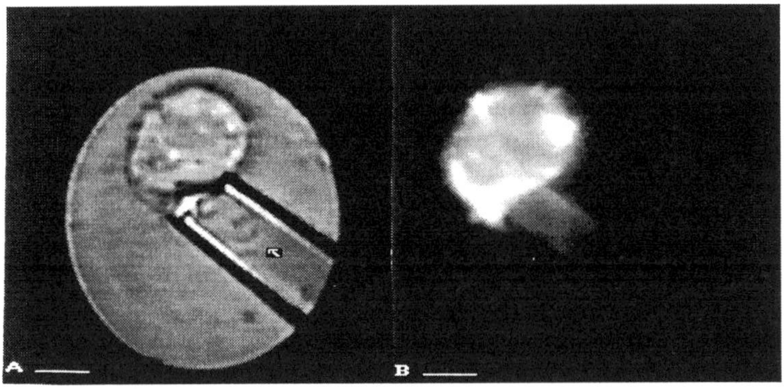

Fig. 3. Quantification of the microfilament content in mechanically induced cell protrusions. A murine P388D1 phagocytic cell was sucked into a micropipette for 2 minutes, then fixed and labeled with NBD-phallacidin to stain microfilaments. The visible (A) and fluorescent (B) images are shown. Modeling the protrusion as a cylinder and the cell as a sphere, the protrusion was estimated to represent 13.5% of total volume, and the fluorescence of the protrusion represented 11.5% of total fluorescence.

Then, we looked for a possible involvement of calcium in this process. Cells from the murine RBL line were loaded with fura-2, a calcium-sensitive fluorescent probe, and subjected to micropipette aspiration (Horoyan *et al.*, 1990). In most cells, aspiration induced a significant calcium rise (about fourfold increase) within a few tens of seconds, and basal values were recovered a few minutes later. This rise was generated by stimulation of intracellular calcium stores, since it remained detectable when experiments were performed in calcium-free medium. Further, when cells were labeled with a fluorescent phallacidin derivative after aspiration, a local concentration of polymerizated actin was often detected (mostly within the protrusion or near the pipette tip) whereas fluorescence was quite uniform in control cells. However, no correlation was found between the localization of calcium rise and microfilament redistribution in individual cells. More recent experiments performed on THP-1 cells labeled with fluo-3, another calcium probe, revealed a transient calcium increase in about 50% of treated cells. This was substantially lower than that found in RBL cells.

Since it was felt that a possibly transient effect of calcium on cytoskeletal organization might not be clearly revealed by *delayed* microfilament observation, we looked for a relationship between instantaneous intracellular calcium level and deformability. Thus, in another series of experiments, human neutrophils were labeled with calcium probes (fluo-3 or fura red) and subjected to micropipette aspiration on the stage of a confocal microscope. Fluorescence images were recorded every second and the fluorescence content was measured together with protrusion length (Fig. 4). In 6 cells out of 8 that displayed a clearcut calcium rise, this increase occurred within one second after the start of protrusion formation (Zaffran *et al.*, 1993a). However, the concomitance of cell deformation and calcium rise did not indicate whether calcium changes were a cause or a consequence of cell deformation.

The next step then consisted of determining whether a drastic modification of intracellular calcium concentration modified cell mechanical properties. This question was addressed by incubating cells for a few minutes in presence of a calcium ionophore in a medium containing millimolar calcium concentration, or in a medium where calcium depletion was achieved by adding a divalent cation chelator such as EGTA or EDTA. In three series of experiments performed with different populations of myeloid cells, calcium manipulation only resulted in moderate changes of mechanical properties. Indeed, the length of protrusions obtained after about 30 seconds aspiration was not significantly altered in human granulocytes (Zaffran *et al.*, 1993a), it was decreased in RBL cells (Horoyan *et al.*, 1990) and it was found increased in the human monocytic THP-1 line (see below for more details).

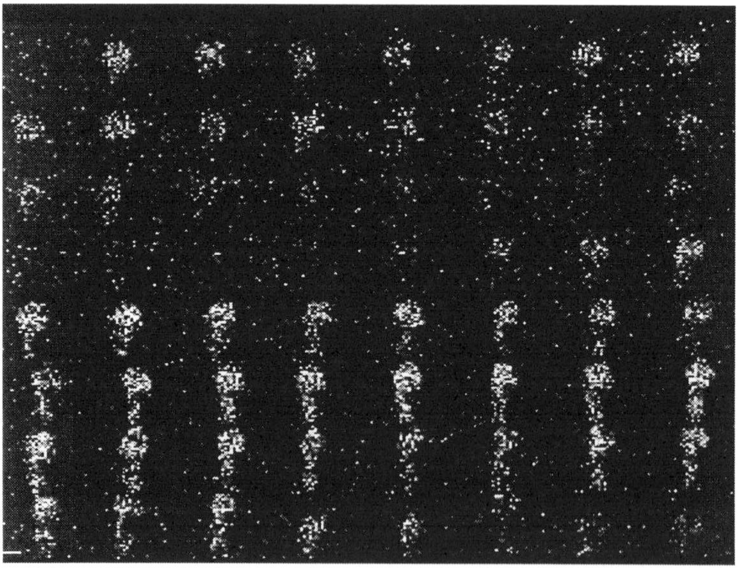

Fig. 4. Relationship between calcium changes and cell deformation. Human neutrophils were labeled with fluo-3, a molecule whose fluorescence is strongly increased in presence of high calcium concentrations. Cells were then subjected to mechanical aspiration with a micropipette mounted on the stage of a confocal microscope under continuous scanning. The rate of image formation is about one per second and sixty four sequential images are shown. Formation of a protrusion is clearly accompanied with a significant fluorescence increase. Bar length is 10 μm.

Three possible hypotheses might be suggested to account for our results. First, calcium might not be an important regulator of microfilament organization, and observed effects would be a consequence of secondary biochemical cascades depending on particular properties of a given cell population. Second, cells might be endowed with powerful homeostatic mechanisms allowing rapid restoration of mechanical properties after important calcium changes. Third, the length of protrusions generated after 30 seconds aspiration might not faithfully reflect cell mechanical properties. Several complementary approaches were used to address these possibilities.

First, in order to prevent a possible adaptation of treated cells to calcium changes, ionophores were added *during* micropipette aspiration with concomitant monitoring of the rate of protrusion increase: we found that in a proportion of treated cells, calcium increase resulted in a clearcut enhancement of the rate of elongation of the protrusion (Zaffran *et al.*, 1993b; Richelme *et al.*, 1996). This enhancement might be immediate (Zaffran *et al.*, 1993b) or delayed by a few seconds (Richelme *et al.*, 1996). This means that either the effect of calcium on cell deformability was indirect or an immediate change of mechanical properties might result in delayed alteration of elongation rate. It was thus found necessary to assess more carefully the mechanism of protrusion formation.

The kinetics of cell deformation was studied by taking advantage of an image analysis system allowing up to 20 milliseconds resolution in the monitoring of cell deformation. This is illustrated in Figure 5. The main conclusion is that cell deformation may be divided in 3 phases:

- A first step is a limited elastic deformation that occurs within less than 20 milliseconds (Fig. 5C), in accordance with previous reports (Schmid-Schonbein *et al.*, 1981).
- Then, cell deformation may be divided in two sequential phases with approximately constant elongation rate (Fig. 6). The duration of the first phase is of order of 60 seconds.

The interest of this discrimination is that these two phases were found to display differential sensitivity to metabolic inhibitors such as cytochalasin B or azide (Richelme *et al.*, 1997). Indeed, when cells were aspirated after calcium manipulation, the mean deformation rate of control cells was significant lower than that of cells with high (millimolar) or low calcium during the first minute of aspiration (mean elongation rates were respectively 0.21, 0.34 and 0.29 µm/s). On the contrary, during the following 90 seconds, elongation rates were respectively 0.062, 0.024 and 0.032 µm/s. These results would be consistent with the possibility that calcium might regulate differentially the short term cell mechanical resistance and adaptative response to exogenous forces.

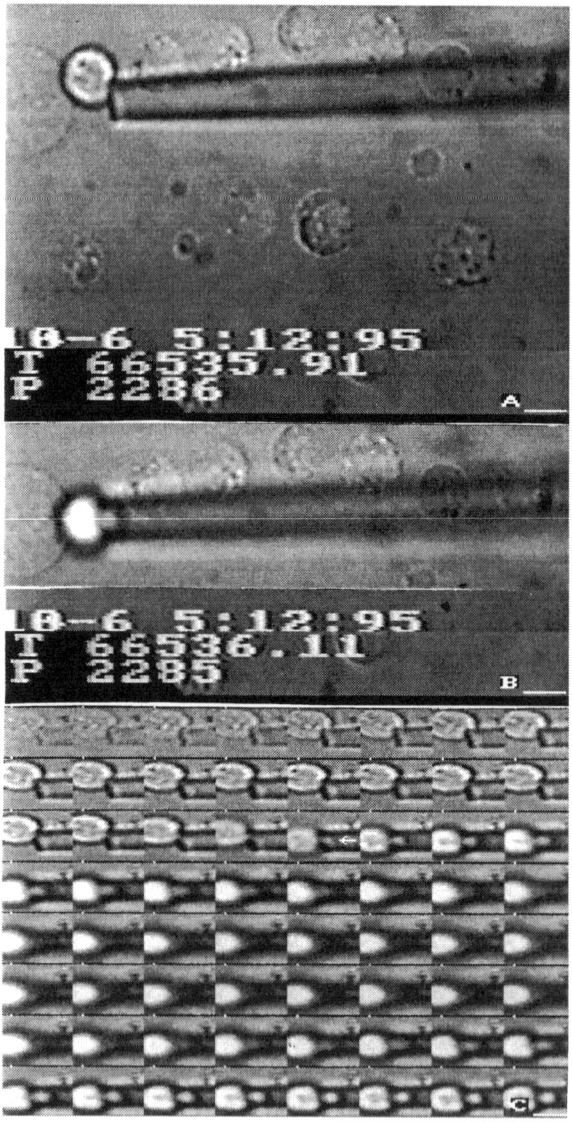

Fig. 5. Rapid monitoring of initial cell deformation in a micropipette. A micropipette with inside negative pressure of about 100 Pa was displaced towards a THP-1 cell under video monitoring. Time (T) is displayed in seconds, Pressure is displayed with arbitrary units of about 1 Pa. As shown in Figures 5 A and B, within 200 milliseconds, a clearcut protrusion was formed. The image of a small area surrounding the pipette tip was continuously digitized with 50 Hz frequency (odd and even lines of a given frame were processed sequentially, allowing twofold increase of time resolution with respect to the 25 Hz video rate, at the price of a 50% reduction of image height). As shown in Figure 5 C (arrow) the protrusion appeared within a 20 milliseconds interval.

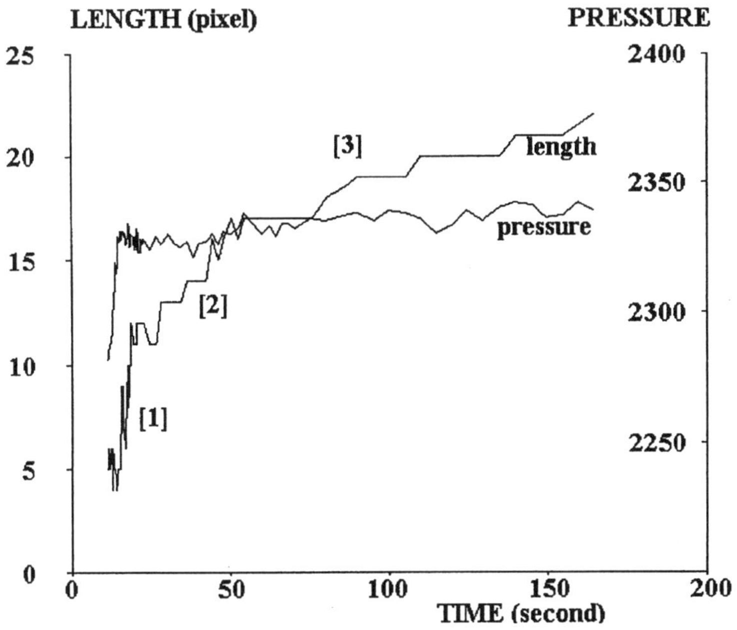

Fig. 6. Kinetics of protrusion formation within a micropipette. The time dependence of micropipette pressure and protrusion length are shown (1 pixel represents about 0.5 µm, a pressure unit is about 1 Pa, with arbitrary zero). Data are consistent with the idea that cell deformation might involve an elastic step [1] followed by two sequential phases [2] and [3] with differential sensitivity to metabolic inhibitors (Richelme *et al.*, 1997).

It was important to determine whether these phases were actually representative of different cell behavior (*e.g.* passive deformation followed by active response) or they might be artefactual consequences of our methodology. An important question concerns the behavior of the cell nucleus during deformation. Indeed, most studies on micropipette aspiration were performed with polymorphonuclear leukocytes whose lobulated nucleus is probably quite particular with respect to the capacity to enter a micropipette. It is therefore important to assess the influence of the nucleus of the cell deformation behavior. This point was addressed by labeling monocytic THP-1 cells with a vital nuclear probe before aspiration. As shown in Figure 7, The cell nucleus went close enough to the pipette tip to allow some perturbation of the cytoplasmic flow into the protrusion. This point must be considered to achieve a quantitative understanding of the deformation process.

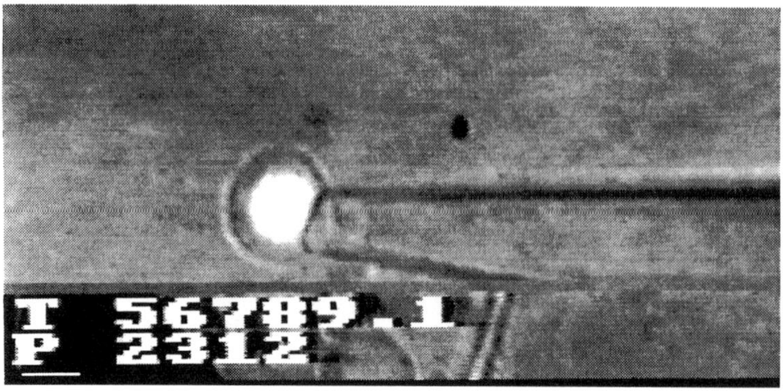

Fig. 7. A phagocytic THP-1 cell was labeled with Hoechst 33342 stain to make the nucleus more apparent. The cell was then aspirated into a micropipette with dual (visible and ultraviolet) illumination. The nucleus thus appeared as a bright disc that came close to the pipette tip.

4. CONCLUSION

The problem addressed in the present study was to determine whether calcium ions play a role in cytoskeletal regulation within intact cells. Due to the high number of microfilament-associated proteins and variety of regulatory mechanisms, it was not feasible to predict the effects of calcium changes on bulk cell organization by merely summing up the known effects of calcium on individual actin binding proteins. An experimental approach was therefore warranted.

Preliminary studies made at the *population level* on leukocyte response to chemoattractant stimulation revealed that a drastic impairment of cell calcium control did not perturb the complex morphological changes and cytoskeletal reorganization resulting in cell polarization. This would argue against an important role of calcium on cytoskeletal organization. Experiments performed at the *single cell level* with micropipette aspiration technique led to somewhat different conclusions:

When cell calcium was strongly increased or decreased *before* aspiration, higher deformation rates were observed during the first tens of seconds following the application of mechanical forces. Also, if cell calcium was increased *during* cell exposure to mechanical forces, the deformation rate was increased in about 50% of observed cells. This would suggest that a marked change of the cytosolic calcium content is sufficient to decrease mechanical strength within a few seconds..

Several tens of seconds after the onset of mechanical cell deformation, a spontaneous calcium rise is often detected, and this may be correlated to simultaneous deformation. Further, when calcium-clamped cells are subjected to the same manipulation, deformation rate is lower than that of controls after the first minute following force application. This would suggest that living cells may respond to prolonged mechanical forces by an increase of this deformability, and this response may involve a change of calcium concentration.

These results illustrate the multiplicity of mechanisms available in a living cell to regulate an important structures such as the actin cytoskeleton. Also, they illustrate the difficulty of a quantitative understanding of cell behaviour, despite the impressive amount of information gathered on the structure and function of cell components. It is suggested that unravelling these processes will require the development of methods allowing continuous monitoring of the behavior of individual cells with high spatial and temporal resolution.

ACKNOWLEDGMENTS

Part of this work was supported by a grant from the M.E.N.E.S.R. (ACC-SV n°11).

REFERENCES

Albritton N.L., Meyer T. and Stryer L., *Science* **258** (1992) 1812-1815.

André P., Capo C., Benoliel A.M., Buferne M. and Bongrand P., *Cell. Biophys.* **16** (1990) 13-34.

Barak L.S., Yocum R.R., Nothnagel E.A. and Webb W.W., *Proc. Natl. Acad. Sci. USA* **77** (1980) 980-984.

Evans E.A. and Kukan B., *Blood* **64** (1984) 1028-1035.

Horoyan M., Benoliel A.M., Capo C. and Bongrand P., *Cell. Biophys.* **17** (1990) 243-256.

Lépidi H., Benoliel A.M., Mège J.L., Bongrand P. and Capo C., *J. Cell. Sci.* **103** (1992) 145-156.

Richelme F., Benoliel A.M. and Bongrand P., *Bull. Inst. Pasteur* **94** (1996) 257-284.

Richelme F., Benoliel A.M. and Bongrand P., *Exp. Biol. Online* **2** (1997) 5.

Schmid-Schönbein G.W., Sung K.L.P., Tozeren H., Skalak R. and Chien S., *Biophys. J.* **36** (1981), 243-256.

Schorle H., Holtschke T., Hünig T., Schimpl A. and Horak I., *Nature* **352** (1991) 621-624.

Stossel T.P., *Science* **260** (1993) 1086-1094.

Tsien R.Y. and Poenie M., *T.I.B.S.* **11** (1986) 450-455.

Zaffran Y., Lépidi H., Benoliel A.M., Capo C. and Bongrand P., *Blood Cells* **19** (1993a) 115-131.

ZaffranY., Lépidi H., Bongrand P., Mège J.L. and Capo C., *J. Cell. Sci.* **175** (1993b) 675-684.

Zhelev D. and Hochmuth R.M., *Biophys. J.* **68** (1995) 2004-2014.

LECTURE 8

Branched Polymers and Gels

M. Daoud

Laboratoire Léon Brillouin, CE Saclay, 91191 Gif-sur-Yvette, France

1. INTRODUCTION

Polymers are giant molecules that are made of millions of atoms. They are the repetition of an elementary unit called monomer. They are used as plastics, paints, adhesives, rubbers, and in some future, as batteries. They have changed completely our everyday life by superseding natural materials such as wood, iron, etc. Our understanding of the structure of synthetic polymers has increased tremendously these last 20 years. They may be either linear or branched. In the former case, one makes react divalent monomers, that may interact only by two functional units, leading to a linear structure. In the latter case, multifunctional units react, leading to a structure that is randomly branched. A very interesting aspect of this type of reaction is that as time proceeds, one eventually goes from a viscous fluid called a *sol* to an elastic solid called a *gel.* The sol-gel transition was considered first in the forties in its mean field version by Flory and by Stockmayer and Zimm. It was improved recently by noting that it is directly related to the percolation transition in Physics (de Gennes, 1976; Stauffer, 1976). More recently, the fractal aspects of these branched polymers and gels were considered both theoretically and experimentally. Although the ideas developed for these synthetic polymers are probably too simplistic for direct use in biological problems, we believe that some of them might survive even when complications such as rigidity, presence of electrical charges, and eventually others are taken into account. In what follows, we will discuss the simplest possible case of sol-gel transition. We will consider electrically neutral solutions, and flexible polymers. Therefore, we do not consider the long range interactions that appear when the polymers are charged electrically. We also assume that each monomer reacts at random, and that the monomer solutions are sufficiently concentrated, so that we may neglect any diffusion motion before reaction. Even with these simplifying assumptions, we will see that the resulting structures are extremely rich. This richness comes from the fact that randomly branched polymers are extremely polydisperse. This means that the distribution of masses is very broad, and becomes more so as one approaches the sol-gel transition. As we will see, this has very important consequences. A first one is that measurements, that are averages over all molecular weights present, may lead to laws that are different from those that apply to a single mass. A second one is that there is also a very wide distribution in relaxation times. This leads to special, non exponential relaxations.

In what follows, we will first consider percolation, and its relation to the sol-gel transition. Then we will study the consequences of the distribution of molecular weights when one dilutes the solution, and consider the swelling effects. Finally, we will discuss the distribution of relaxation times and the rheological properties.

2. PERCOLATION AND SOL-GEL TRANSITION

As discussed by Forgacs in this conference, percolation is a purely geometrical phenomenon, that may be described in terms of phase transition although it is not directly a transition in the thermodynamic sense (Stauffer, 1985). In its simplest version, *bond* percolation is shown in Figure 1.

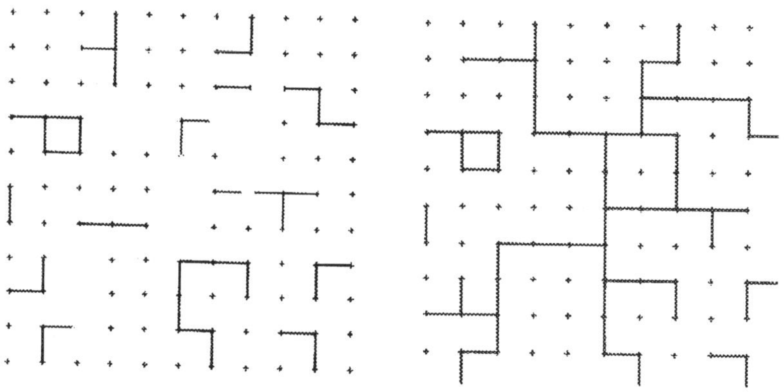

Fig. 1. Percolation. Bonds are thrown at random with probability p. When p is small (1a), only finite clusters are present. They model the sol, made of finite polymers. Above a threshold, in addition to the latter, an infinite cluster is present and models the gel (1b).

All sites of a lattice are present. One throws bonds at random between these sites with probability p. When p is small (Fig. 1a), only finite clusters are present, even though they may become very large. When p is large, in addition to the finite ones, an infinite cluster is present as shown in Figure 1b. The latter spans the whole lattice. If each bond were a resistor, the infinite cluster would allow a current to flow across the lattice. No current would flow if finite clusters only are present. There exists a threshold p_c between these two situations, when the infinite cluster appears. It corresponds to the percolation transition. Several exponents may be defined in the vicinity of the threshold, corresponding to various properties of the clusters, as discussed by Forgacs. In the following, we will use only two, that will be defined below. But let us consider the analogy between percolation and the sol-gel transition. The latter is a very special liquid-solid transformation that was found in polymers. It was modeled by Flory and by Zimm and Stockmayer in a special, loopless approximation, and improved by Gordon and collaborators. In this analogy, each of the sites models a multifunctional monomer, that may react by 4 functionalities in Figure 1. Bonds are assumed to be the chemical links between monomers. Therefore the clusters in percolation correspond to randomly branched polymers. The set of finite polymers below the threshold is called *sol.* At the threshold, an infinite molecule appears. This is called a *gel.* Because of its presence, an elastic modulus is present, whereas the sol has a viscous behavior. Therefore, the system goes from a viscous liquid state to an elastic solid state at the threshold. The latter corresponds to the sol-gel transition. Of course, the viscosity of the sol becomes large, and eventually diverges at the threshold, whereas the modulus is very small near but above the gel point. These facts correspond to the existence of very large polymers , inducing a very viscous behavior, and to a tenuous solid respectively below and above the transition. Again, we stress that no thermodynamic phase transition is present, but that we are merely discussing a connectivity transition where one goes from finite to infinite molecules in the system. A very important characteristics of the sol is that it involves a very broad distribution of molecular weights. This is called polydispersity. It was found that the probability $P(N,\varepsilon)$ of finding a polymer with mass N, at a distance $\varepsilon = p - p_c$ from the threshold varies as (de Gennes, 1979; Stauffer, 1985)

$$P(N,\varepsilon) \sim N^{-\tau} f(\varepsilon N^{s}) \qquad (1)$$

where τ and σ are exponents to be discussed below. The distribution is normalized below p_c: every monomer belongs to a polymer. Above the threshold, a finite part of the monomers belongs to the gel. For our purpose, we need to define a second exponent, namely the fractal dimension D_p of the macromolecules (Mandelbrot, 1977). This relates the mass N of a polymer to its radius R(N)

$$N \sim R(N)^{D_p} . \qquad (2)$$

The distribution may not be characterized by one but by two masses, namely the weight-average, and the z-averaged masses. The latter is the one that appears in relation (1):

$$N_z \approx \frac{\int N^3 P(N)\, dN}{\int N^2 P(N)\, dN} \approx N^{-1/\sigma} \qquad (3)$$

whereas the former involves another exponent, that will not be introduced here. It is found that both masses are related:

$$N_w \approx \frac{\int N^2 P(N)\, dN}{\int N P(N)\, dN} \approx N_z^{(2D_p - d)/D_p} \qquad (4)$$

where d is the dimension of space, that will be taken equal to 3 below. In the following, except in a few limited places, we will use only the fractal dimension for our discussion. This in principle describes the structure of the gel in the vicinity of the transition. We assume that all polymers have the same type of conformation in their reaction bath, so that the same relation is also valid for the description of any polymer in the distribution. This fractal dimension was calculated in many ways: by computer simulations in space dimension from one to 6, by renormalization group techniques, and is known with good accuracy. A generalization of the Flory theory for linear chains (Isaacson and Lubensky, 1981) leads to

$$N \sim R^{5/2}. \tag{5}$$

Finally, the exponent τ of the distribution is related to D: it was found that $\tau = 1 + 3/D_p = 11/5$. These results were checked experimentally by studying the distribution of masses by a combination of Gel Permeation Chromatography and light scattering (Patton *et al.*, 1989). The results are shown in Figure 2, which shows a log-log plot of the distribution as a function of a reduced mass M/M_Z. M_Z may be directly measured by Chromatography. One may check that the distribution decreases as a power law with exponent 2.3 ± 0.1, in good agreement with the previous discussion. The power law variation of the distribution, combined with the relation between the exponents τ and D_p may be interpreted in the following way (Cates, 1985): each of the polymers is a fractal. This implies that its structure contains holes with various sizes, including some of the size of the polymer itself. The distribution of masses is such that inside these holes, smaller polymers may be included. These also have a fractal behavior and contain holes, that in turn may also be filled with even smaller macromolecules, and so on.

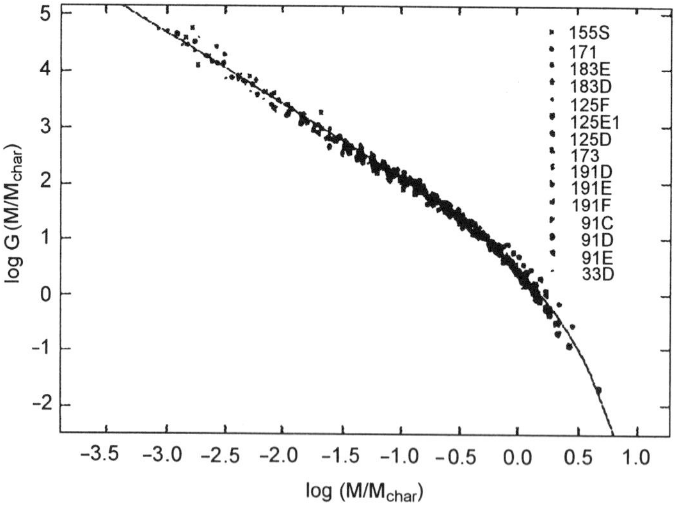

Fig. 2. Log-log plot of the distribution of masses as a function of the reduced mass. Various symbols correspond to different distances to gelation threshold. Linear part indicates a power law variation (from Patton *et al.*, 1989).

Thus the distribution is similar to Russian dolls, with holes at every scale that contain smaller polymers. From this point of view, it is also a fractal distribution, as one could have suspected because of its power law variation. In next section, we will see that this has the important consequence that measurements, that are averages over the distribution lead to results that are different from those that are made on single masses (Daoud *et al.*, 1984). This leads to effective exponents different from the actual ones.

3. EFFECTS OF DILUTION

Let us now consider a sol that has been synthesized, and where the distribution is frozen. This is realized by stopping the chemical reaction, for instance by working off stoechiometry. Synthesis is assumed to have been performed in the absence of solvent. Let us now add an excess of a good solvent to the sol. At the end of this stage, we have a dilute solution. We are going first to study the conformation of a single polymer, and then the average properties over the distribution.

3.1. The single polymer

It is possible to consider a single polymer in the Flory approximation. One finds:

$$N \sim R^2. \tag{6}$$

Thus the fractal dimension of a randomly branched polymer in a good solvent is identical to that of a random walk. This implies that the scattered intensity S(q) in a radiation scattering experiment is:

$$S_1(q) \sim q^{-2} \qquad (qR >> 1) \tag{7}$$

where q is the scattering vector. This may be understood as a generalization of (6) because $S_1(q)$ is basically the number of monomers in a sphere with radius 1/q. Relations (6) and (7) are what would be measured for a *monodisperse* solution, with only one mass. Because the diluted sol has a distribution of masses, what is measured is an average over the distribution:

$$S_t(q) \approx \frac{1}{N_w} \int N^2 \, P(N)\, S(q,N)\, dN\,. \tag{8a}$$

Using (8a) and (1), we get (Martin *et al.,* 1985)

$$S_t(q) \sim C\, N_W\, f(qR_Z) \tag{9}$$

with the observed z-average radius of gyration (Daoud *et al.,* 1984)

$$N_W \sim R_z^{2(3-\tau)} \tag{10}$$

and

$$S_t(q) \sim q^{-2(3-t)}\,. \tag{11}$$

Thus, any measurement leads to an effective exponent 2 (3 – τ), different from the exponent 2 that rules the behavior of every polymer in the solution. These relations are valid as long as τ is larger than 2. In the opposite case, both measurements lead to the same exponent. These results were used to get an independent estimate for τ.

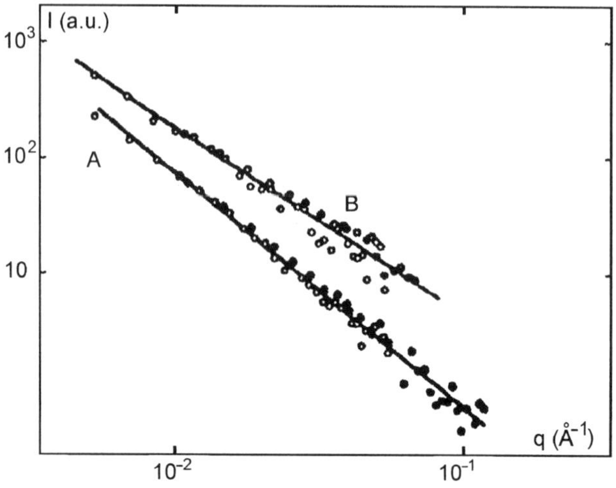

Fig. 3. Log log plot of the scattered intensity as a function of wavevector q for a fractionated (A) and polydisperse (B) samples. Difference in slopes shows the polydispersity effect. From these it is possible to get both dimensions of a polymer and of the distribution.

These are shown in Figure 3, where the scattered intensity by a polydisperse solution, and by a monodisperse one obtained by fractionation (Bouchaud *et al.*, 1986).

One can see clearly the difference in the slopes reflecting this polydispersity effect. These results are valid as long as the polymers are far from each other so that one may neglect the inter-polymer interactions. This implies low concentrations, and breaks down when the volume occupied by the macromolecules is on the same order as the volume of the container. Using equations (1) and (6), it is possible to estimate the volume occupied by the polymers and the cross-over concentration C*, above which the interactions become dominant. One finds $C^* \sim N_w^{-3/8}$.

3.2. The semi-dilute regime

For concentrations larger than C*, polymers start interacting strongly with each other (Daoud and Leibler, 1988). One gradually goes back to the situation in the reaction bath, where small polymers are located in the large ones. For any concentration larger than C*, this will holds in part: part of the small macromolecules enter the large ones and screen the interactions. This is reflected partly by the existence of a screening length, that characterizes the effect. This may be calculated by using a scaling approach. In the latter, we assume that concentration is not the relevant parameter. What is important is to know whether the polymers are below or above C*. Therefore, the adequate parameter is the ratio C/C*. Then we assume that one may describe the characteristic length, for any concentration in the simple following form

$$R_z \sim N_w^{5/8} f(CN_w^{3/8}) \tag{12}$$

where the function f(x) is regular for small x, and behaves as a power law for large x: $f(x) \sim x^a$, where the exponent a is determined by an adequate condition. The latter depends on the variable we are considering. For the screening length ξ for instance, this is the fact that screening is a local property. This implies that does ξ not depend on the mass, but only on concentration. This leads to

$$\xi \sim C^{-5/3}. \tag{13a}$$

For the average radius, the condition is that its dependence should be similar to that in the reaction bath, as long as fractal dimension is concerned. This implies

$$R_Z \sim N_w^{1/2} C^{-1/3} \tag{13b}$$

where we used the fact that an average has to be made, and it is not the percolation dimension that is observed, but an effective dimension as above. It is also possible to give an interpretation of these results by saying that one has to divide the distribution into two parts. Small masses, smaller than a value g to be determined, are swollen and behave as in a dilute solution. Large ones, larger than g, are screened and behave as in the reaction bath if one takes g as a unit mass. This is called the *blob* model. A blob is a part of a polymer, with g monomers and radius ξ. Thus equation (13b) may be recovered by assuming that

$$R_Z \sim [N_W/g_W]^{1/2}\, \xi$$

and

$$\xi \sim g_W^{5/8} \sim C^{-5/3}.$$

Relation (13a) was checked experimentally by light and neutron scattering by Munch *et al.* (1992). It is possible to apply the same considerations to the swelling of gels, but we will not discuss this here. We will rather turn to the dynamics and to the distribution of relaxation times.

4. RHEOLOGY

So far, we considered only static properties. As we mentioned in the introduction however, the sol-gel transition is mainly characterized by the divergence of the viscosity η and the appearance of a modulus G. These were more precisely defined by two exponents:

$$\eta \sim \varepsilon^{-s} \qquad (p < p_C) \tag{14}$$

and

$$G \sim \varepsilon^{\mu} \qquad (p > p_C)\,. \tag{15}$$

These relations hold for zero frequency. For non zero frequency, it is possible to introduce a complex response function

$$\overline{G}(\omega) = G(\omega) + j\,\omega\,\eta\,(\omega) = j\,\overline{\eta}(\omega) \tag{16}$$

where j = (–1), and to assume, following Efros and Schklowskii (1976), a scaling form for the frequency dependence of the complex modulus $\overline{G}(\omega)$:

$$\overline{G}(\omega) \sim \varepsilon^{\mu} f_{\pm}(j\omega T) \tag{17a}$$

where the function $f_{\pm}(x)$ is *a priori* not the same above and below the threshold, and T is a characteristic time that depends only on the distance ε to the threshold.

$$T \sim \varepsilon^{a} . \tag{17b}$$

The exponent a in relation (17b) is determined by expanding below the threshold, and identifying to the viscosity, relation (12), the term in the expansion that is proportional to frequency. This leads to

$$T \sim \varepsilon^{-s-\mu} . \tag{18}$$

Knowing the scaling behavior of the complex modulus, it is possible to consider the frequency dependence of f(x). For small frequencies, $\omega T << 1$, f(x) is analytic and may be expanded. For large frequencies, in the opposite limit, it behaves as a power law. The exponent of the latter may be determined by considering the fact that for high frequencies, one is probing finite regions of space. Therefore, the modulus should be independent of ε. This implies:

$$\overline{G}(\omega) \sim \{j\omega\}^{-\mu/(s+\mu)} . \tag{19a}$$

This has the important consequence that both real and complex part of the modulus are proportional to each other:

$$G(\omega) \sim \omega\, \eta(\omega) \sim \omega^{-\mu/(s+\mu)} . \tag{19b}$$

Note that also implies that the loss angle δ is constant, $\delta \sim \pi\, \mu/2(s + \mu)$. Relation (19b) was checked experimentally over several decades of frequencies (Durand *et al.,* 1987). Finally, the knowledge of the scaled form of the complex viscosity $\overline{\eta}(\omega)$ also allows us to determine the distribution of relaxation times $H(\tau)$ (Daoud, 1988). The latter is related to $\overline{\eta}(\omega)$ by

$$\overline{\eta}(\omega) \approx \int \frac{H(\tau)}{1 + j\omega\tau} d\tau . \tag{20}$$

Using relations (17) and (20), we get:

$$H(\tau) \sim \tau^{-\mu/(s+\mu)} g(\omega T) \tag{21}$$

where g(x) is a cut off function. Thus the distribution of relaxation times also decreases as a power law. We cannot reduce it to a single characteristic time, but to two of them. These may be defined as moments of the distribution, but we may see them directly on relations (20) and (21): the longest time is the time T that we found in the modulus, relation (18). We can also see that the distribution is not normalized, and that there is a time that is proportional to the viscosity: $T_1 \sim \varepsilon^{-s}$. Actually, the distribution is normalized, and T_1 is its first moment if one considers a logarithmic scale for times instead of a linear one. Such scale may be justified if one insists on having a dimensionless time scale, which is provided by a logarithmic scale. An important consequence of this special distribution is that one gets non exponential relaxations that may be either power laws or stretched exponentials depending on whether one is considering a time inside or outside the distribution (Martin, 1987).

5. CONCLUSION

We considered the structure of randomly branched polymers, and the sol-gel transition. These may be described by percolation. The most important property is that there is a very broad distribution of masses. This implies that although each of the polymers, and the gel, is a fractal, any measurement does not lead to the actual fractal dimension, but to an effective one that takes into account polydispersity. This was shown on a special case, namely when the sol is diluted in an excess of a good solvent. Then measurements made with fractionated and unfractionated samples give different q dependences for the scattered intensity for instance, or for the fractal dimension of the polymers. Because polydispersity is so important, the effective dimensions also depend on the exponent τ of the mass distribution. Another consequence is that there is a very broad distribution of relaxation times. The latter decreases as a power law, and may not be reduced to a single characteristic time. This implies that relaxations are not simple exponentials, but may have either power law or stretched exponential variations with time.

In this study, we considered the simplest case, and we neglected complications such as rigidity or electric charges for instance. These may be included, leading to a more complicated phase diagram. Still, the polydispersity effects that were discussed above will be present, and have a strong influence on measurements.

ACKNOWLEDGMENTS

The author is much indebted to M. Adam, M. Delsanti, P.G. de Gennes, A. Lapp, L. Leibler, and J.E. Martin for interesting discussions.

REFERENCES

Bouchaud E., Delsanti M., Adam M., Daoud M. and Durand D., *J. Phys. Lett.* **47** (1986) 1273.

Cates M.E., *J. Phys. Lett.* **38** (1985) 2957.

Daoud M., Family F. and Jannink G., *J. Phys. Lett.* **45** (1984) 119.

Daoud M. and Leibler L., *Macromol.* **41** (1988) 1497.

Daoud M., *J. Phys. A* **21** (1988) L973.

Durand D, Delsanti M., Adam M. and Luck J.M., *Europhys. Lett.* **3** (1987) 297.

Devreux F., Boilot J.P., Chaput F., Malier L. and Axelos M.A.V., *Phys. Rev. A* **47** (1993) 2689.

Efros A.L. and Schklovskii B.I., *Phys. Status Solidi* **B76** (1976) 475.

Flory P.J., Principles of Polymer Chemistry (Cornell University Press, Ithaca, 1953).

Gennes P.G. de, *J. Phys.* **37** (1976) 1445.

Gennes P.G. de, Scaling Concepts in Polymer Physics (Cornell Univ. Press, Ithaca, 1979).
Gordon M. and Ross-Murphy S.B., *Pure Appl. Chem.* **43** (1975) 1.
Isaacson J. and Lubensky T.C., *J. Phys.* **42** (1981) 175.
Mandelbrot B.B., The Fractal Geometry of Nature (Freeman, San Francisco, 1977).
Martin J.E. and Ackerson B.J., *Phys. Rev. A* **31** (1985) 1180.
Martin J.E., In Time dependent effects in disordered systems, edited by R. Pynn and T. Riste, *N.A.T.O. A.S.I. B* **167** (Plenum Press, 1987) p. 425.
Munch J.P., Delsanti M. and Durand D., *Europhys. Lett.* **18** (1992) 557.
Patton E.V., Wesson J.A., Rubinstein M., Wilson J.C. and Oppenheimer L.E., *Macromol.* **22** (1989) 1946.
Stauffer D., *J. Chem. Soc. Trans. II* **72** (1976) 1354.
Stauffer D., *Phys. Rep. 54* (1979) 1; Introduction to Percolation Theory (Taylor and Francis, London, 1985).
Zimm B.H. and Stockmayer W.H., *J. Chem. Phys.* **17** (1949) 1301.

LECTURE 9

Statistical Mechanics of Semiflexible Polymers: Theory and Experiment

E. Frey, K. Kroy, J. Wilhelm and E. Sackmann

Institut für Theoretische Physik und Institut für Biophysik, Physik-Department der Technischen Universität München, James-Franck-Straße, D-85747 Garching, Germany

1. INTRODUCTION AND OVERVIEW

Living cells are soft bodies of a characteristic form, but endowed with a capacity for a steady turnover of their structures. Both of these material properties, *i.e.* recovery of the shape after an external stress has been imposed and dynamic structural reorganization, are essential for many cellular phenomena. Examples are active intracellular transport, cell growth and division, and directed movement of cells. The structural element responsible for the extraordinary mechanical and dynamical properties of eukaryotic cells is a three-dimensional assembly of protein fibers, the *cytoskeleton*. A major contribution to its mechanical properties is due to actin filaments and proteins that crosslink them. Numerous experiments *in vivo* [5] and *in vitro* [14, 13] have shown that the mechanical properties of cells are largely determined by the cytoskeletal network.

The cytoskeletal polymers (actin filaments, microtubules and intermediate filaments) which build up this network are at the relevant length-scales (a few microns at most) all *semiflexible polymers*. In contrast to flexible polymers the *persistence length* of these polymers is of the same order of magnitude as their total contour length. Hence, the statistical mechanics of such macromolecules can not be understood from the conformational entropy alone but depends crucially on the bending stiffness of the filament. The nontrivial elastic response and distribution function of the end-to-end distance of a single semiflexible polymer illustrate instructively this interplay of energetic and entropic contributions. Figure 1 shows the extension (in units of fL^2/k_BT) of a semiflexible polymer of fixed length L grafted at one end when a weak force f is applied at the other end for different persistence lengths ℓ_p [21]. This is one of the few *exact* results for the wormlike chain model. Note that the largest response is obtained for a polymer with a persistence length of the order of its contour length. More flexible molecules contract due to the larger entropy of crumpled conformations, stiffer molecules straighten out to keep their bending energy low. In the flexible case the response is isotropic and proportional to $1/k_BT$, *i.e.*, the Hookian force coefficient is proportional to the temperature and we recover ordinary rubber elasticity. On the other hand, when the persistence length is longer than the contour length, the response becomes increasingly *anisotropic*. Transverse forces give rise to mechanical bending of the filaments and the transverse spring coefficient in the stiff limit is proportional to the bending modulus κ. The effective longitudinal spring coefficient turns out to be proportional to κ^2/T, indicating the breakdown of linear response at low temperatures $T \to 0$ (or $\ell_p \to \infty$). This is a consequence of the well known Euler buckling instability illustrated in Figure 1 (top). The anisotropy of the elastic response considerably complicates the construction of network models.

Energetic effects also have a strong influence on the probability distribution function $G(R;L)$ for the end-to-end distance R of a semiflexible polymer of length L. When the polymer is not much longer than its persistence length, $G(R;L)$ deviates strongly from the Gaussian shape found for flexible polymers (see Fig. 2) with the main weight being shifted towards full extension with increasing stiffness [34]. This is also reflected by the nonlinear force-extension relation, whose overall shape is determined by the Euler instability. Measurements of the radial distribution function as well as of the response of single polymers to applied forces can be used to measure the bending stiffness of biopolymers.

A useful experimental technique for investigating the short time dynamics of semiflexible polymers is dynamic light scattering [6, 9, 22]. We have calculated the dynamic structure factor for solutions of semiflexible polymers in the weakly bending rod limit. The result allows characteristic parameters of the polymers, such as their persistence length ℓ_p, their lateral diameter a and also the mesh size ξ_m of the network to be determined from light scattering measurements. The most important result of this work is that the decay of the dynamic structure factor shows two regimes, an initial simple exponential decay determined by the hydrodynamics of the solvent and a stretched exponential decay due to structural relaxation driven by the bending energy. For actin and intermediate filaments, where both regimes can be resolved, the method is well suited to investigate also more complex questions concerning for example the interactions with actin binding proteins which among others can induce cross-linking or bundling.

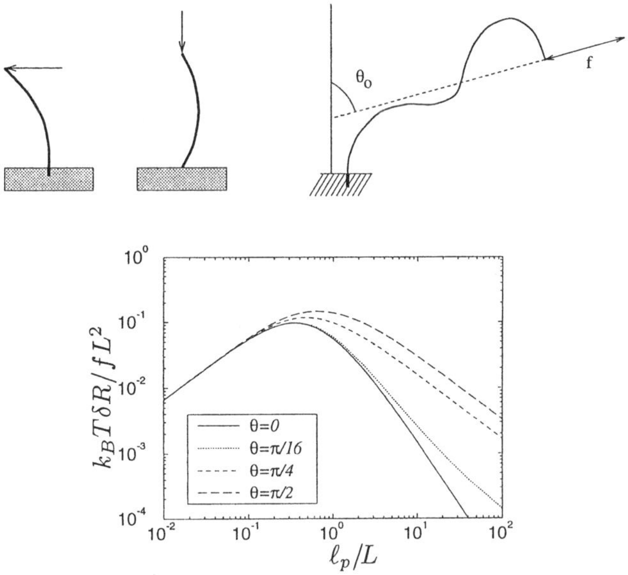

Fig. 1. — *Top*: The elastic response of a stiff rod is extremely anisotropic due to the Euler instability. *Bottom*: Response of a filament clamped at one end with a fixed initial orientation to a small external force at the other end. The normalized extension (inverse force coefficient) is plotted versus the persistence length in units of the total length of the polymer. The curves are parameterized by the relative angle θ_0 of the force **f** to the initial orientation.

In order to describe the material properties of the cytoskeleton, one has to understand how semiflexible polymers built up statistical networks and how the macroscopic stresses and strains are mediated to the single filaments in such a "rigid polymer network". This is a quite rapidly growing field in polymer physics with many questions still open [1].

If the polymers are not crosslinked, the response will depend on how fast one pulls. Roughly speaking, the solution will either show elastic behavior and obey Hook's law with a linear relation $\sigma = G\gamma$ between stress σ and strain γ or rather show viscous behavior and obey Newton's law $\sigma = \eta\dot{\gamma}$, where the stress is proportional to the strain rate $\dot{\gamma}$. The task of a theoretical description is to find out how the material parameters like the shear modulus G^0 and the viscosity η depend on the elastic and dynamic properties of the polymers and the architecture of the network. The mechanical and dynamical properties of the various crosslinking proteins are also expected to influence the viscoelasticity of the network [32].

Whereas it is known that on short time scales the effect of mutual steric hindrance ("entanglement") in conventional polymeric materials is very similar to the effect of permanent chemical crosslinks, this is an open problem for semiflexible polymer networks. Since forces between neighboring polymers can only be transmitted transverse to the polymer axis and there is no restoring force for sliding of one filament past another, a steric contact is acting completely different from a chemical crosslink on a microscopic scale.

But even the most simple of these problems, the static elasticity of a network of permanently crosslinked semiflexible polymers, is quite complex. Because of the strongly anisotropic behavior of the single elements, the predicted macroscopic properties of the network vary greatly with the explicit or implicit assumptions made about network geometry and stress propagation by recent theoretical treatments of both entangled solutions and crosslinked networks [24, 12, 30, 21]. We hope to represent a key aspect of the geometrical structure of both cellular and artificial stiff polymer networks by looking at *disordered* networks. Specifically, we use a crosslinked network of sticks randomly placed in a plane as a toy model for studying the origin of macroscopic elasticity in a stiff polymer network. Although quantitative predictions about the behavior of existing (three-dimensional) networks of semiflexible polymers are not attempted at this stage, this model is expected to reflect the salient features of the full problem and to promote its understanding by allowing the detailed discussion of questions like "What modes of deformation contribute most to the network elasticity?", "How many filaments do actually carry stress, how many remain mostly unstressed?", "What kind of effective description of the complicated microscopic network geometry should be used?" This approach connects the theory of cytoskeletal elasticity to the very active fields of transport in random media and elastic percolation. In Section 3.4 we will describe the model in more detail and discuss the question of the dominant deformation mode.

Previous fluorescence microscopic observations [18] suggest that the tube picture is a useful concept for understanding the viscoelastic properties of *entangled* semiflexible polymer networks. Starting from this model, where the matrix surrounding a single test polymer is represented by a tube-like cage, we have developed a phenomenological description that seems to be able to account for some of the observed elastic and dynamic properties. The predictions for the entanglement transition, the concentration dependence of the plateau modulus, and the terminal relaxation time are in good quantitative agreement with experimental data.

2. SINGLE-CHAIN PROPERTIES

2.1. The wormlike chain model

The theoretical understanding of the mechanical properties of a *single* semiflexible macromolecule in isolation is already a nontrivial statistical mechanics problem with quite a number of recent developments 50 years after it was first formulated [20]. The model usually adopted for a theoretical description of semiflexible chains like actin filaments is the *wormlike chain model.* The filament is represented by an inextensible space curve $\mathbf{r}(s)$ of total length L parameterized in terms of the arc length s. The statistical properties of the wormlike chain are determined by a free energy functional which measures the total elastic energy of a particular conformation.

$$\mathcal{H} = \int_0^L ds \, \frac{\kappa}{2} \left(\frac{\partial \mathbf{t}}{\partial s} \right)^2 ; \quad |\mathbf{t}| = 1 , \tag{1}$$

where $\mathbf{t}(s) = \partial \mathbf{r}(s)/\partial s$ is the tangent vector. The energy functional is quadratic in the local curvature with κ being the bending stiffness of the chain. The inextensibility of the chain is expressed by the local constraint, $|\mathbf{t}(s)| = 1$, which leads to non-Gaussian path integrals. Since there would be high energetic costs for a chain to fold back onto itself one may safely neglect self-avoidance effects for sufficiently stiff chains.

Despite the mathematical difficulty of the model some quantities can be calculated exactly. Among these is the tangent-tangent correlation function which decays exponentially, $\langle \mathbf{t}(s) \cdot \mathbf{t}(s') \rangle = \exp[-(s-s')/\ell_p]$, with the persistence length $\ell_p = \kappa/k_B T$ (in three dimensional space). Another example is the mean-square end-to-end distance

$$\begin{aligned} \mathcal{R}^2 := \langle R^2 \rangle &= 2\ell_p^2 (e^{-L/\ell_p} - 1 + L/\ell_p) \\ &= \begin{cases} L^2 & \text{for } L/\ell_p \to 0 \text{ (rigid rod)} \\ 2\ell_p L & \text{for } L/\ell_p \to \infty \text{ (random coil),} \end{cases} \end{aligned} \tag{2}$$

which reduces to the appropriate limits of a rigid rod and a random coil (with Kuhn length $2\ell_p$) as the ratio of L to ℓ_p tends to zero or infinity, respectively. The calculation of higher moments quickly gets very troublesome [10].

2.2. Linear force-extension relation

Another useful property of the model, which can be computed exactly is the linear force extension relation [21]. Adding a term $-\mathbf{f} \cdot \mathbf{R}$ to the Hamiltonian in equation 1, the extension of a wormlike chain with a weak force applied between its ends is computed as

$$\delta R = (\langle \mathbf{R} \rangle_f - \langle \mathbf{R} \rangle_0) \, \mathbf{f}/f = f(\mathcal{R}^2 - \langle |\mathbf{R}| \rangle^2)/k_b T = f/k \tag{3}$$

with k the force coefficient. The problem is the calculation of the moment of the the end-to-end distance $\langle|\mathbf{R}|\rangle$. This has not been achieved so far, but an expansion in the stiff limit is possible. For example, we can write $\langle|\mathbf{R}|\rangle = \mathcal{R}\langle\sqrt{1+(R^2-\mathcal{R}^2)/\mathcal{R}^2}\rangle$ and expand the square root to obtain to leading order for the force coefficient

$$4k_BT\frac{\mathcal{R}^2}{\langle R^4\rangle - \mathcal{R}^4} \xrightarrow{L/\ell_p \ll 1} \frac{90\kappa^2}{k_BTL^4}.$$

This is exact in the stiff limit, $L/\ell_p \ll 1$, and qualitatively correct over the whole range of stiffness. However, for the special boundary condition of a grafted chain we can even get an *exact result for arbitrary stiffness.* Consider a chain with one end clamped at a fixed orientation and let us apply a force at the other end. Then the linear response of the chain may be characterized in terms of an effective Hookian spring constant which depends on the orientation θ_0 of the force with respect to the tangent vector at the clamped end. In the appropriate generalization of the second part of equation 3 the force coefficient k is replaced by an angle dependent function, $k^{-1} \to L^2\mathcal{F}_{\theta_0}(\ell_p/L)/k_BT$. Noting that the conformational statistics of the wormlike chain is equivalent to the diffusion on the unit sphere [29] the function $\mathcal{F}$ can be calculated [21] (see Fig. 1). In the flexible limit, where the chain becomes an isotropic random coil, all curves coincide and reproduce entropy elasticity, $\mathcal{F}_{\theta_0}(x) \sim x$. But for stiff chains the force-extension relation depends strongly on the value of the angle θ_0 between the force and the grafted end. Transverse forces give rise to ordinary mechanical bending characterized by the bending modulus κ ($\mathcal{F}_{\pi/2} \sim 1/x$), whereas longitudinal deformations are resisted by a larger force coefficient κ^2/T ($\mathcal{F}_0 \sim 1/x^2$). The breakdown of linear response in the limit $T \to 0$ (or $\ell_p \to \infty$) is a consequence of the well known Euler buckling instability, as we already mentioned in the introduction. The linear response for longitudinal forces is due to the presence of thermal undulations, which tilt parts of the polymer contour with respect to the force direction.

2.3. Nonlinear response and radial distribution function

In viscoelastic measurements on *in vitro* actin networks one observes strain hardening [15], *i.e.* the system stiffens with increasing strain. This may either result from collective nonlinear effects or from the nonlinear response of the individual filaments. In the preceding section we have seen that the force coefficient obtained in linear response analysis for longitudinal deformation diverges in the limit of vanishing thermal fluctuations indicating that the regime of validity for linear response shrinks with increasing stiffness. Since the nonlinear response of a single filament may be obtained from the radial distribution function by integration, we discuss the latter first.

A central quantity for characterizing the conformations of polymers is the radial distribution function $G(\mathbf{R};L)$ of the end-to-end vector $\mathbf{R}$. For a freely jointed phantom chain (flexible polymer) it is known exactly and for many purposes well approximated by a simple Gaussian distribution. While rather flexible polymers can be described by corrections to the Gaussian behavior [2], the distribution function of polymers which are shorter or comparable to their persistence length shows very different behavior. It is in good approximation given by

$$\begin{aligned} G(\mathbf{R};L) &\approx \frac{\ell_p}{\mathcal{N}L^2} f\Big(\frac{\ell_p}{L}(1-R/L)\Big), \\ \text{where } f(x) &= \begin{cases} \frac{\pi}{2}\exp[-\pi^2 x] & \text{for } x > 0.2 \\ \frac{1/x-2}{8\pi^{3/2}x^{3/2}}\exp\left[-\frac{1}{4x}\right] & \text{for } x \le 0.2 \end{cases} \end{aligned} \tag{4}$$

and $\mathcal{N}$ is a normalization factor close to 1 [34]. This result is valid for $L \lessapprox 2\ell_p$, $x \lessapprox 0.5$ and $d = 3$ where d is the dimension of space. A similar expression exists for $d = 2$. As can be seen in Figure 2, the maximum weight of the distribution shifts towards full stretching as the stiffness of the chain is increased to finally approach the δ-distribution like shape required for the rigid rod.

The radial distribution function is a quantity directly accessible to experiment since fluorescence microscopy has made it possible to observe the configurations of thermally fluctuating biopolymers [7, 17, 27]. Comparing the observed distribution functions with the theoretical prediction for $d = 2$ is both a test of the validity of the wormlike chain model for actual biopolymers as well as a sensitive method to determine the persistence length which is the only fit parameter. It should be noted here that the determination of persistence length *e.g.* of actin is still an actively discussed subject [4, 33].

A very interesting possibility would be to attach two or more markers (*e.g.*, small fluorescent beads) permanently to single strands of polymers and to observe the distribution function of the marker separation. This would eliminate all the experimental difficulties associated with the determination of the polymer contour. Note that in contrast to existing methods of analysis it is not necessary to know the length of polymer between two markers; it can be extracted from the observed distribution functions along with ℓ_p by introducing L as a second fit parameter.

The nonlinear response of the polymer to extending or compressing forces can be obtained from the radial distribution function by integration. The result (Fig. 2) is in agreement with and provides the transition between the previously known limits of linear response and very strong extending forces (*e.g.*, [25]). For compressional forces, a pronounced decrease of differential stiffness around the classical critical force $f_c = \kappa\pi^2/L^2$ can be understood as a remnant of the Euler instability. For filaments slightly shorter than their persistence length the influence of this instability region extends up to and beyond the point of zero force corresponding to the maximum in the linear response coefficient for $\ell_p \approx L$ (see Fig. 1). For large compressions beyond the instability, the force-extension-relation calculated from the distribution function is only in qualitative agreement with numerical results because of the restricted validity of equation 4 for $x \to 1$.

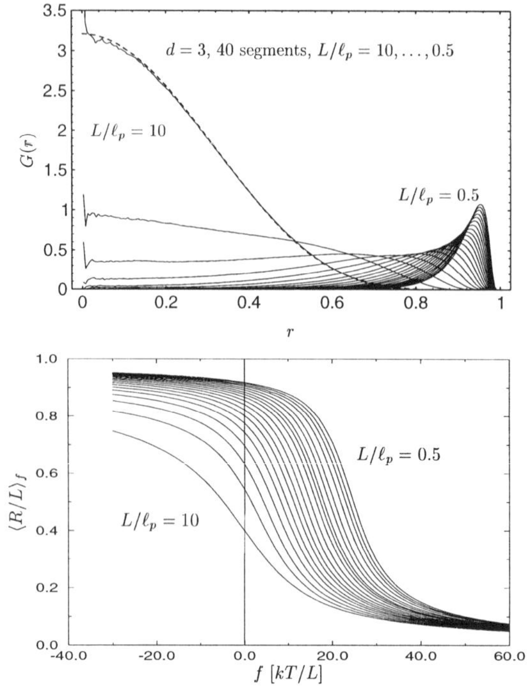

Fig. 2. — *Top*: End-to-end distribution function of a semiflexible polymer (numerical results). Note that with increasing stiffness of the polymer there is a pronounced crossover from a Gaussian to a completely non-Gaussian from with the weight of the distribution shifting towards full stretching. *Bottom*: The mean end-to-end distance R as a function of a force applied between the ends ($\mathbf{f} = -f\mathbf{R}/|\mathbf{R}|$). The step at positive (*i.e.* compressive) forces can be viewed as a remnant of the Euler instability.

2.4. Dynamic structure factor

The short time dynamics of polymers is most effectively measured by scattering techniques. The *dynamic structure factor* of a semiflexible polymers can be derived analytically in the limit of strong length scale separation $a \ll \lambda \ll \ell_p$, L ("weakly bending rod" approximation) and $\lambda \leq \xi_m$. Here we have introduced the symbols a, λ and ξ_m for the lateral diameter of the polymer, the scattering wavelength and the mesh size of the network, respectively. The length scale separation guarantees that the decay of the structure factor is due to the *internal dynamics of single filaments.* For $\lambda \approx \xi_m$ the structure factor decays in two steps due to the fast internal and to the slower collective modes, respectively. For the analysis we will restrict ourselves to rather dilute solutions and refer

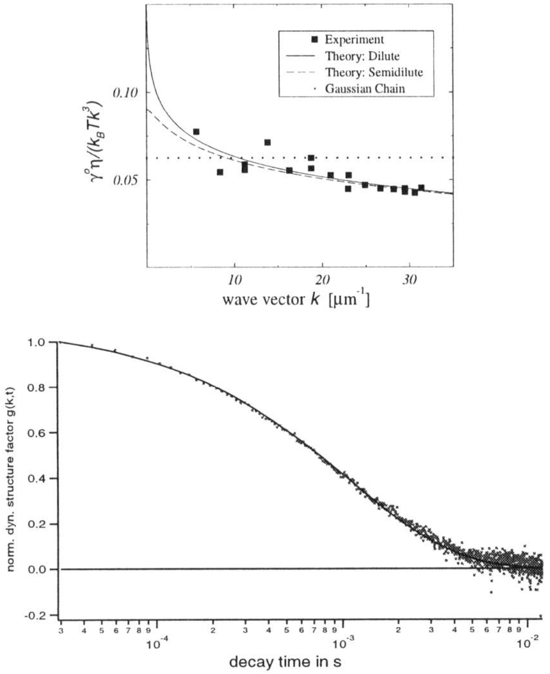

Fig. 3. — Results from dynamic light scattering experiments. *Top*: Correction to the classical prediction $\gamma_k^{(0)} \sim k^3$ for the initial decay rate of the dynamic structure factor. The theoretical predictions for dilute solutions (solid line) and semidilute solutions (dashed line) are compared with experimental data [31]. Also included is the prediction for Gaussian chains from Ref. [3]. *Bottom*: Fit of theoretical dynamic structure factor to experimental data for $k = 24.2\,\mu\text{m}^{-1}$ [8].

the reader for a more complete treatment to the literature [22].

The most important result of the calculations is that the time decay of the dynamic structure factor shows two regimes: the simple exponential initial decay regime for $t \ll \tau = \tilde{\zeta}_\perp/\kappa k^4$ with a decay rate

$$\gamma_k^{(0)} = \frac{k_B T}{6\pi^2\eta} k^3 \left(\frac{5}{6} - \ln ka\right), \tag{5}$$

and the stretched exponential decay

$$g(\mathbf{k}, t) \propto \exp\left(-\frac{\Gamma(\frac{1}{4})}{3\pi}(\gamma_k t)^{\frac{3}{4}}\right), \qquad \gamma_k = k_B T k^{\frac{8}{3}} / \ell_p^{1/3} \tilde{\zeta}_\perp \tag{6}$$

for $t \gg \tau$. The effective transversal friction coefficient per length, $\tilde{\zeta}_\perp$, accounts for the hydrodynamic damping of the undulations on length scales comparable to the scattering wavelength. For polymers with $\ell_p \approx \lambda$ mainly the first regime, equation 5, is observed. The microscopic lateral diameter a of the polymer enters the calculation as a microscopic cutoff and can be determined from a measurement of the *initial decay* of the dynamic structure factor. The method is easy to apply and has been shown to provide reasonably accurate results (see Fig. 3). For polymers with $\ell_p \gg \lambda$ mainly the stretched exponential decay regime, equation 6 is observed. The decay rate is determined by the persistence length of the molecule, which is also readily extracted from experimental data. Some applications of the methods to dynamic light scattering with actin are shown in Figure 3.

Our analytical results for the initial decay rate and the dynamic exponent for semiflexible polymers suggest that known deviations of the dynamic exponent for more flexible polymers from its classical value $z = 3$ are most likely due to the local semiflexible structure of these molecules, and that the above analysis is therefore of some relevance also to scattering from flexible polymers. The reason is that the singular ($\propto 1/r$) hydrodynamic interaction favors short distances. The possibility to determine the microscopic lateral diameter a of the polymer from the initial decay becomes important when the effects of bundling and side-binding are investigated, which can give rise to an effective thickening of the molecules.

3. MANY-CHAIN PROPERTIES

As we have mentioned in the introduction, we are still far from understanding on a microscopic basis the macroscopic viscoelastic properties of solutions and gels of semiflexible polymers. In the following we will review some idealized concepts we have developed to model certain aspects of this complicated behavior and compare our predictions with some recent experiments.

3.1. Entanglement transition

Upon increasing the polymer concentration c at fixed polymer length there is a critical concentration c^* above which a plateau develops in the frequency dependence of the storage modulus and the modulus increases steeply. The same phenomenon is observed at a critical length when the polymer length is varied at fixed concentration. It is called the entanglement transition. An experimental observation of the transition for short actin filaments ($L \approx 1.5\ \mu$m) as a function of concentration is shown in Figure 4 (top). Adapting concepts which have been developed for flexible polymers [19] we have proposed a mean field description for the entanglement transition in semiflexible polymer solutions [21]. The basic idea is that the ends of a polymer are less efficient in confining other polymers than internal parts of the polymer. As a consequence the critical concentration c^* is related to the overlap concentration

$\bar{c}$ by some universal number C which may be interpreted as an effective coordination number and is a measure of the mean number of neighboring polymers necessary to confine the lateral motion of the test polymer. The prediction for an ideal gas of rods, $G^0 \propto k_B T/\xi_{\text{eff}}^2 L$, is also shown in Figure 4. From the fit we have determined the coordination number C, which compares very well with the result for flexible polymer solutions. Away from the transition the effective mesh size reduces to the ordinary mesh size $\xi_{\text{eff}} \propto \xi_m$. With $\xi_m \propto c^{-1/2}$ [31] this implies $G^0 \propto c$, which is indeed observed for a regime of concentrations within the semidilute phase. However, at higher concentrations or lengths the slope of $G^0(c)$ increases, as can be seen from the bottom plot in Figure 4, which shows the measured plateau modulus G^0 for pure actin and actin with a small amount of gelsolin ($r_{\text{AG}} = 6000 : 1$ corresponding to an average actin filament length of 16 μm). The description of the polymers as rods, which neglects internal modes of the individual polymers, is no longer appropriate.
Note: Since the time of writing this manuscript, new experimental and theoretical investigations [11] have suggested another, probably superior, interpretation of the entanglement transition, which does not involve the concept of the coordination number.

3.2. Plateau modulus

To explain the observed elastic modulus on the basis of the elastic properties of the individual chains and the architecture of the network is in general a quite difficult task. Some of the theoretical work is based on a tube picture as depicted schematically in Figure 5. The effect of the network surrounding an arbitrary test polymer is represented by a cylindrical cage of diameter d. The main response of the polymer to various deformations of this tube is assumed to arise from distortions of undulations of wavelength L_e, with $L_e^3 \propto d^2 \ell_p$ [26]. It is far from obvious, whether the macroscopic elasticity can be explained solely in terms of the compressibility of the tubes [26, 12], or whether contributions from filament bending [21, 30] or buckling [24] are also important. The theoretical approaches differ quite significantly in their assumptions on how forces are transmitted through the network and which type of deformation of a single polymer gives the dominant contribution to the network elasticity. Due to those differences in the basic assumptions the predictions for the shear modulus of the network also differ. Our present data are reasonably consistent with entropy arguments assuming $G^0 \propto k_B T/\xi_m^2 L_e$, which leads to $G^0 \propto c^{4/3}$ for $d \propto \xi_m$ and $G^0 \propto c^{7/5}$ for $d \propto \xi_m^{5/6}/\ell_p^{1/5}$ [12], respectively. Above 0.4 mg/ml our data also agree with a scaling $G^0 \propto c^{5/3}$, which has been derived by a scaling argument based in Figure 5 for the case that the macroscopic elastic response is mainly due to filament bending [21]. It should also be noted that these concentrations are close to the critical volume fraction $3a/L$ ($a \approx 10$ nm is the lateral diameter of an actin filament) for the nematic transition of rigid rods.

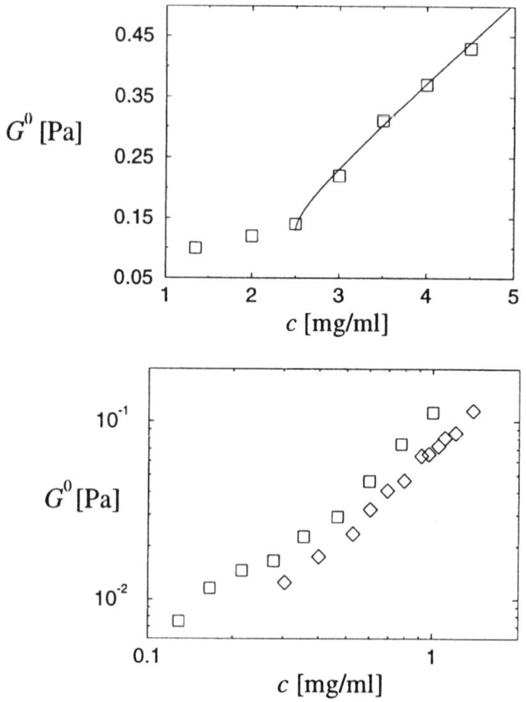

Fig. 4. — *Top*: The entanglement transition for short rod-like actin filaments ($L \approx 1.5\ \mu$m) of varying concentration. *Bottom*: Concentration dependence of the plateau modulus for pure actin (□) and actin with a small amount of gelsolin (◇) [11].

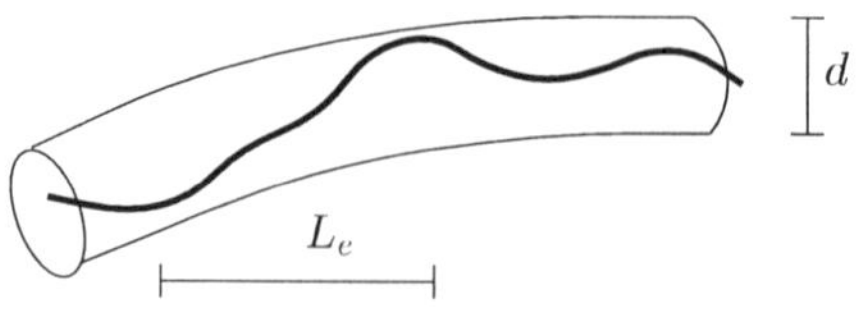

Fig. 5. — Some of the theoretical work is based on a tube picture as depicted schematically here. The effect of the network surrounding an arbitrary test polymer is represented by a cylindrical cage of diameter d. Different scaling laws for d have been proposed.

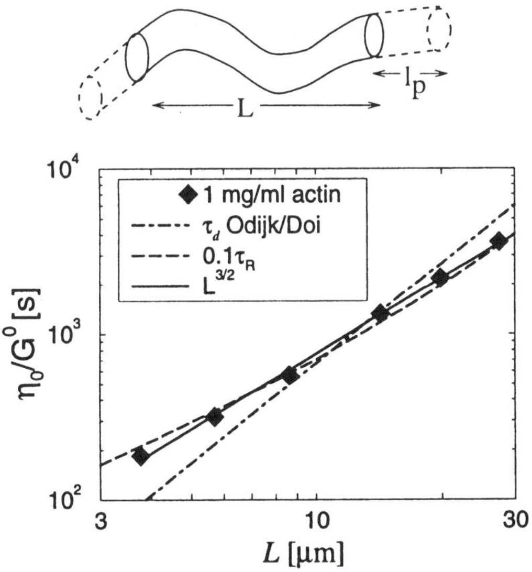

Fig. 6. — *Top*: Illustration of the generalized reptation picture for semiflexible polymer solutions. *Bottom*: Terminal relaxation time above the entanglement transition [11].

3.3. Terminal relaxation

At frequencies below the plateau regime the elastic response decreases and the polymer solution starts to flow. The corresponding time scale is called the terminal relaxation time. It can be determined from the measured plateau modulus and the zero shear rate viscosity using the relation $\eta_0 \sim G^0 \tau_r$ or directly from $G'(\omega)$. Intuitively, the mechanism for the terminal relaxation is obvious from the tube picture (see Fig. 6) described above: viscous relaxation only occurs, when the polymers have time to leave their tube-like cages by reptation, *i.e.*, by Brownian motion along their axis. The reptation model, which was originally formulated for flexible polymers, was adapted to the semiflexible case [26].

Odijk [26] estimated the disengagement time τ_d for a semiflexible chain diffusing out of its tube. However, the data for τ_r presented in Figure 6 (bottom) are not in accord with his result for τ_d. The dependence of the observed terminal relaxation time τ_r on polymer length L is substantially weaker than predicted for τ_d, even in the stiff limit where $\tau_d = \ell_p L / 4D_\parallel \propto L^2$ (dot-dashed line in Fig. 6), $D_\parallel = k_B T / 2\pi\eta L$ being the longitudinal diffusion coefficient of the chain in the free draining approximation. Instead, the solid line in Figure 6 corresponds to the scaling law $\tau_r \propto L^{3/2}$.

A tentative interpretation of the data can be given in terms of a semiflexible polymer diffusing along a strictly one dimensional path; *i.e.*, not being allowed to choose between infinitely many new directions at its ends. This situation is schematically depicted in the upper part of Figure 6. The characteristic decay time for self-correlations of the end-to-end vector $\langle \mathbf{R}(t)\mathbf{R}\rangle$ is then given by [11]

$$\tau_R = L^4 \ell_p^2 / D_{\|} \langle R^2 \rangle^2 \approx (L + 2\ell_p)^2 / 4D_{\|}. \tag{7}$$

This presents an upper bound for the terminal relaxation time within the tube model. As seen from the dashed line in Figure 6, τ_R (for $\ell_p = 17\ \mu$m [7, 27]) is in fact by a factor of ten too large compared to the data but describes fairly well the length dependence of τ_r. The restriction to one path implies a very slow decay of conformational correlations. An unusually slow decay of stress (the frequency dependence of $G'(\omega)$ is still less than linear in the measured frequency range) is indeed observed, but this might also in part be due to the broad length distribution of actin [16]. Clearly, further investigations are necessary to come to a better understanding of the terminal regime.

3.4. Stochastic network models

As motivated in the introduction, we use a two dimensional toy model of crosslinked semiflexible polymer networks to study some of the fundamental problems in semiflexible network theory [35]. Here we address the question whether longitudinal or transversal deformation dominate the network response and find an interesting crossover between the two extreme cases. Since such questions are already nontrivial for networks at $T = 0$ (*i.e.* no fluctuations), we consider this most simple case first: The linear elastic response of a network of classical compressible and bendable rods. A more sophisticated treatment will include contour fluctuations of the filaments (but no steric interactions) by replacing the rods with elements having the appropriate linear or nonlinear force extension relations described above.

Rods of unit length are randomly placed on a square piece of the plane. Wherever two sticks intersect, they are connected by a crosslink. Periodic boundary conditions are applied to the left and right sides of the square. A shearing horizontal displacement is enforced on all rods intersecting the upper or lower boundary of the square at the intersection points. The (linearized) elastic equations of the system are solved numerically for displacements and forces at all crosslink points. The shear modulus is obtained from the force needed to impose the displacements of the upper and lower boundary. Crosslinks are not allowed to stretch but do not fix the angle between the intersecting rods. The results reported here change quantitatively but not qualitatively when crosslinks do fix the intersection angles.

The parameters of this system are the density ρ of rods per unit area, the compressional and the bending stiffness of the rods. We choose to express the latter by the linear response force constants k_{comp} and k_{bend} for a rod of unit

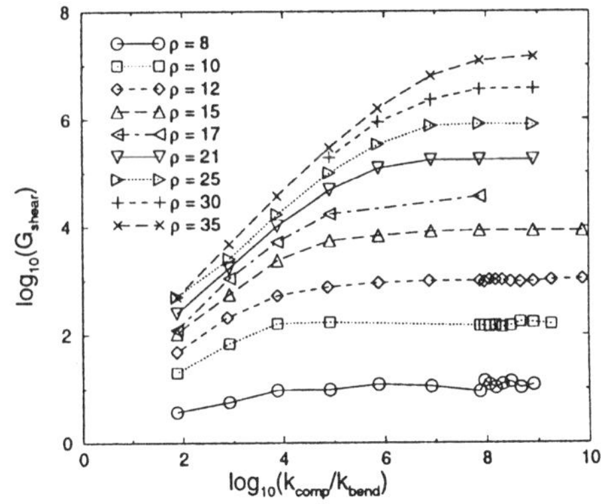

Fig. 7. — Shear modulus of a crosslinked two dimensional random network of rods of unit length for different densities ρ and different aspect ratios α plotted over the relative size $k_{\mathrm{comp}}/k_{\mathrm{bend}}$ of compressional and bending force constants. Units were chosen in such a way that the bending stiffness κ and hence k_{bend} were constant ($\kappa = 1$).

length. Geometrically, the relative size of the two force constants is controlled by the aspect ratio $\alpha = a/L$ of the rods where a is the rod radius.

While the system can show no elastic response (in the limit of infinite size) for densities below the geometrical percolation threshold $\rho_c^g \approx 5.72$ [28], the shear modulus G grows very rapidly with increasing ρ for $\rho > \rho_c$. For sufficiently slender rods G displays unusual power law behavior in ρ even for densities far above the percolation threshold.

To address the question whether the elasticity of a random stiff polymer network is dominated by transverse or by longitudinal deformations of the filaments, one can study the dependence of the shear modulus on the two force constants. We keep k_{bend} fixed and increase k_{comp} from values corresponding to $\alpha = 0.15$ (short, thick rod) to values corresponding to $\alpha = 1 \times 10^{-5}$ (long slender rod) for different system densities (see Fig. 7). We observe that beyond a certain point the shear modulus ceases to depend on k_{comp}, indicating that the elasticity is dominated by bending modes for thin rods. Since the relevant scale for the system elasticity at higher densities is set not by the rod length but by the mesh size, the point of onset for this behavior shifts upwards with density. The dominance of bending modes in this region is confirmed by the observation that almost all of the energy stored in the deformed network is accounted for by the transverse deformation of the rods. Even in the region were the shear modulus does depend on k_{comp}, a substantial part of the elastic energy is stored in bending deformations.

ACKNOWLEDGMENT

Our work has been supported by the Deutsche Forschungsgemeinschaft (DFG) under Contract No. SFB 266 and No. Fr 850/2.

REFERENCES

[1] Aharoni S.M. and S.F. Edwards, *Rigid Polymer Networks*, Vol. 118 of *Advances in Polymer Science* (Springer, Berlin, 1994).
[2] Daniels H.E., *Proc. Roy. Soc. Edinburgh* **63A** (1952) 290.
[3] Doi M. and Edwards S.F., *The Theory of Polymer Dynamics* (Clarendon Press, Oxford, 1986).
[4] Dupuis D.E., Guiford W.H. and Warshaw D.M., *Biophys. J.* **70**A (1996) 268.
[5] Eichinger L., Köppel B., Noegel A., Schleicher M., Schliwa M., Weijer K., Witke W. and Janmey P., *Biophys. J.* **70** (1996) 1054.
[6] Farge E. and Maggs A.C., *Macromol.* **26** (1993) 5041.
[7] Gittes F., Mickey B., Nettleton J. and Howard J., *J. Cell. Biol.* **120** (1993) 923.
[8] Götter R., Kroy K., Frey E., Bärmann M. and Sackmann E., *Macromol.* **29** (1996) 30.
[9] Harnau L., Winkler R.G. and Reineker P., *J. Chem. Phys.* **140** (1996) 6355.
[10] Hermans J.J. and Ullman R., *Physica* **44** (1952) 2595.
[11] Hinner B., Tempel M., Sackmann E., Kroy K. and Frey E., 1997, Entanglement, elasticity and viscous relaxation of actin solutions, to be published.
[12] Isambert H. and Maggs A.C., *Macromol.* **29** (1996) 1036.
[13] Janmey P., *Cell Membranes and the Cytoskeleton*, Vol. 1A of *Hanbook of Biologocal Physics*, Chap. 17 (North Holland, Amsterdam, 1995) p. 805.
[14] Janmey P.A., *Curr. Op. Cell. Biol.* **2** (1991) 4.
[15] Janmey P.A., Hvidt S., George F., Lamb J. and Stossel T., *Nature* **347** (1990) 95.
[16] Janmey P.A., Peetermans J., Zaner K.S., Stossel T.P. and Tanaka T.,*J. Biol. Chem.* **261** (1986) 8357.
[17] Käs J., Strey H., Bärmann M. and Sackmann E., *Europhys. Lett.* **21** (1993) 865.
[18] Käs J., Strey H. and Sackmann E., *Nature* **368** (1994) 226.
[19] Kavassalis T.A. and Noolandi J., *Macromol.* **22** (1989) 2709.
[20] Kratky O. and Porod G., *Rec. Trav. Chim.* **68** (1949) 1106.
[21] Kroy K. and Frey E., *Phys. Rev. Lett.* **77** (1996) 306.
[22] Kroy K. and Frey E., *Phys. Rev. E* **55** (1997) 3092.
[23] Kroy K. and Frey E., Viscoelasticity of gels and solutions of semiflexible polymers, 1998, to be published.
[24] MacKintosh F., Käs J. and Janmey P., *Phys. Rev. Lett.* **75** (1995) 4425.

[25] Marko J.F. and Siggia E.D., *Macromol.* **28** (1995) 8759.
[26] Odijk T., *Macromol.* **16** (1983) 1340.
[27] Ott A., Magnasco M., Simon A. and Libchaber A., *Phys. Rev. E* **48** (1993) R1642.
[28] Pike G.E. and Seager C.H., *Phys. Rev. B* **10** (1974) 1421.
[29] Saitô N., Takahashi K. and Yunoki Y., *J. Phys. Soc. Jpn* **22** (1967) 219.
[30] Satcher Jr. R.L. and Dewey Jr. C.F., *Biophys. J.* **71** (1996) 109.
[31] Schmidt C., Bärmann M., Isenberg G. and Sackmann E., *Macromol.* **22** (1989) 3638.
[32] Wachsstock D., Schwarz W. and Pollard T., *Biophys. J.* **66** (1994) 801.
[33] Wiggins C.H., Riveline D.X., Ott A. and Goldstein R.E., Trapping and wiggling: Elastohydrodynamics of driven microfilaments. cond-mat/9703244 (1997).
[34] Wilhelm J. and Frey E., *Phys. Rev. Lett.* **77** (1996) 2581.
[35] Wilhelm J. and Frey E., Elasticity of stiff polymer networks (1998) to be published.

LECTURE 10

Scale Effects, Anisotropy and Non-Linearity of Tensegrity Structures: Applications to Cell Mechanical Behavior

S. Wendling, E. Planus[1], D. Isabey[1] and C. Oddou

Laboratoire de Mécanique Physique, Université Paris12-Val-de-Marne, France
[1]INSERM U492, Physiopathologie et Thérapeutique Respiratoires, Hôpital Henry Mondor, 94010 Créteil Cedex, France

1. INTRODUCTION

Changes in cells shape and function are due to the rearrangement of individual molecular components that join together in space to form structural cytoskeletal frameworks [1]. Cells grow, survive, migrate, differentiate and remodel their architecture according to the physico-chemical processes which closely depend upon their mechanical environment [2]. However, the mechanotransduction phenomena and more generally mechanical properties of cells are yet poorly understood and it becomes of central interest to measure and analyze cells rheological behavior. Models of cytoskeletal networks and related cells mechanical responses have been recently proposed with, either, a physical viewpoint based upon the mechanics of continuous media [3, 4] or an analysis based upon the mechanics of discrete element structures. These recent models have taken into account the fact that the fibrous microscopic nature of the cytoskeletal components resembles the polymeric foam material [5] or have introduced the concept that living cells use tensegrity architecture for their organization [6]. The purpose of this study is to furthermore investigate these current modeling aspects with emphasis upon the structure mechanical approach: scale effect, anisotropy, non linearity of stress-strain relationships.

2. MECHANICAL MODELS OF THE CYTOSKELETON

The cellular solids theory [7] considers the steady mechanical behavior of an "open lattice" formed by interconnected filaments and states that the overall properties are derivable from the material properties of elements and network geometry. Among the constitutive materials of the living cell, this theoretical model solely considers the distributed network of actin filaments. The deformation of such cellular solid essentially occurs through bending of the fibers as illustrated on the planar view of an open-lattice unit (Fig. 1a). Any of those units is characterized by the same length L and radius a of the constituting filaments considered to behave like beams whose characteristic cross-sectional area inertia momentum is expressed by:

$$I = \frac{\pi a^4}{4} \quad . \tag{1}$$

Under uni-axial loading of the network the action of the transmitted force F, as illustrated in Figure 1a, results in a small deflection δ identical for all the beams and given by:

$$\delta = \frac{FL^3}{EI} \tag{2}$$

where E stands for the Young's modulus of the actin materials.

Under small applied load and related deformation of such a structure, the overall resulting strain $\varepsilon = \delta/L$ can then be related to the mean compressing or stretching equivalent stress $\varepsilon = F/L^2$ by the definition of a reduced effective Young's modulus whose expression can be approximated by:

$$E^* = \frac{\sigma/\varepsilon}{E} = \frac{\pi}{4}\left(L^*\right)^{-4} \tag{3}$$

where a reduced microscopic length L* represents:

$$L^* = \frac{L}{a} \quad . \tag{4}$$

Such an idealized foam model made of elementary units with identical properties is essentially isotropic and thus cannot necessarily describe the anisotropy resulting from the structural heterogeneity of living cells.

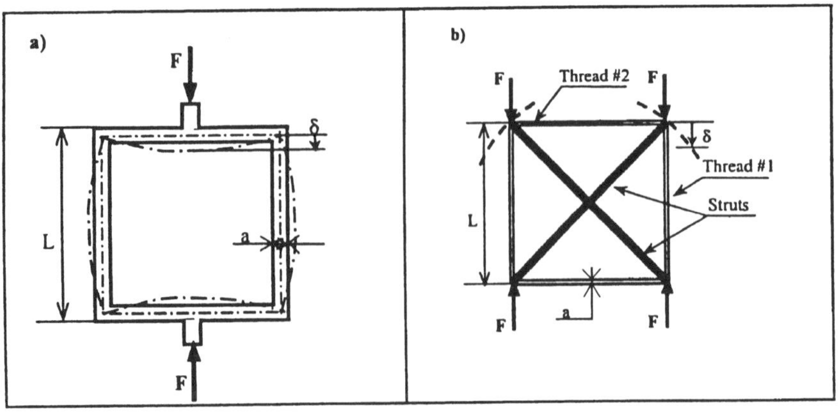

Fig. 1. a) Planar view of an open-lattice unit and b) illustration of a pseudo-tensegrity system.

By contrast, the tensegrity model considers effects of internal tension within elements and differences in properties of the cytoskeletal filaments. The geometrical arrangement and related properties of tensegrity structures result from the mechanical interactions between a discrete set of disconnected compressive struts and a continuous spatial network of tensile threads. As applied to the cellular level, such systems constitute potential knowledge-based models of the cytoskeleton mechanical behavior in which rheological properties of the individual constitutive elements -such as bundles of actin filaments or stress fibers on one part, and microtubules and intermediate filaments on the other part- can be modified to mimic specific biological effects.

As a simplified model, the planar "pseudo-tensegrity system" is composed of two rigid struts and four threads (Fig. 1b) whose rigidity k under extension, is given by:

$$k = \frac{\pi E a^2}{L_0} \tag{5}$$

where L_0 is the resting length and a the radius of the threads.

Under compression, the rigidity k of the threads equals zero because they are assumed to become slack. Moreover at reference state (*i.e.* no external force applied), the initial pre-stretching forces T in the threads are equal and expressed as:

$$T = kL_0(\lambda - 1) \tag{6}$$

where $\lambda = L/L_0$ is the initial elongation in each thread and L the length of the threads at reference state.

Considering the force equilibrium when an external and vertical force F is applied at the extremity of the strut (see Fig. 1b), the equivalent form of the equation (foam) becomes:

$$F = kL_0(\tan\theta - 1) \tag{7}$$

with

$$\theta = \mathrm{ASIN}\left[\frac{1}{\sqrt{2}}\left(1 + \frac{\delta}{\lambda L_0}\right)\right]. \tag{8}$$

This equation demonstrates that the force-elongation ($F - \delta$) relationship cannot be linear both in extension and in compression.

Under conditions where one of the two threads becomes slack, the previous equation has to be modified as follows:

$$F = -kL_0\left[\frac{2\lambda}{\sqrt{2}}\cos\theta - 1\right]\tan\theta \tag{9}$$

if thread #1 (in the direction of the applied force F) is slack, or:

$$F = kL_0\left[\frac{2\lambda}{\sqrt{2}}\sin\theta - 1\right] \tag{10}$$

if thread #2 (in the direction perpendicular to F) is slack. These equations indicate that the overall behavior of the system differs depending on (i) the type of loading and (ii) the pre-stretching force (internal tension at the reference state).

3. RESULTS AND DISCUSSION

The force displacement relationship of the overall system, *i.e.*, the solution of equations (7-10) above, can be normalized using the properties of the thread, *e.g.*, $F^* = F/kL_0$ and $\varepsilon = \delta/L$. The results, in Figure 2, expressed in terms of $(F^* - \varepsilon)$ relationship reveal (i) the non-linear behavior of the structure, (ii) the differences in this non linear behavior between uni-axial extension and compression and (iii) a discontinuity in this relationship associated to a slackening state in one of the two threads. Moreover, a stiffening response is observed during extension while a softening response is observed during compression. Interestingly enough, the $(F^* - \varepsilon)$ relationship becomes more or less linear in extension beyond the discontinuity which depends on $T^* = T/kL_0 = \lambda - 1$.

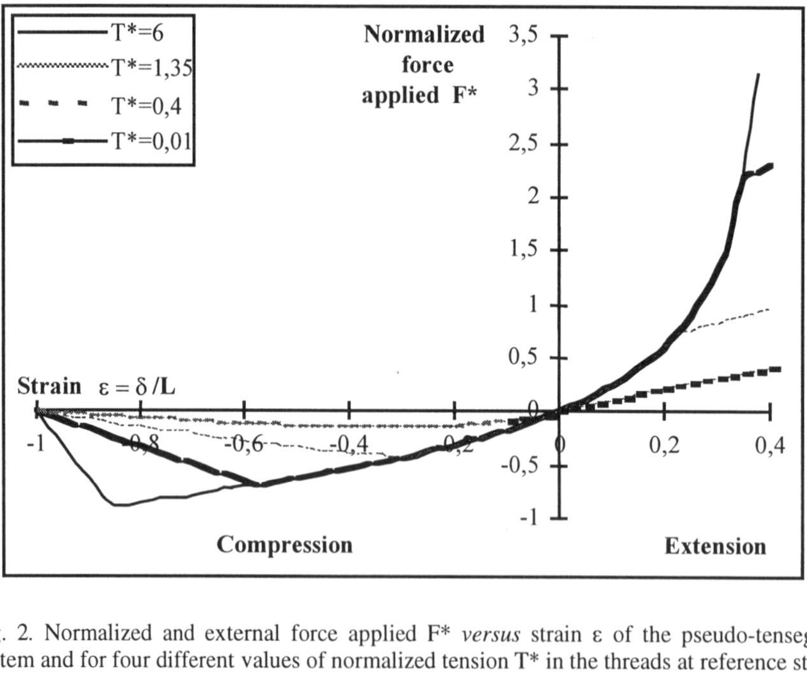

Fig. 2. Normalized and external force applied F* *versus* strain ε of the pseudo-tensegrity system and for four different values of normalized tension T* in the threads at reference state.

In order to characterize the mechanical behavior of this tensegrity structure in terms of rheological properties, we defined the apparent elastic modulus of the structure considering an equivalent rectangular volume whose one of the two struts forms one of the diagonals. The initial cross-sectional area of this volume is a rectangle whose thread #2 forms the diagonal. The elastic modulus E of the system was estimated from the ratio between stress and strain at values (σ′, ε′) corresponding to the curve discontinuity under extension. Actually, this discontinuity depends also on T*. Thus, this modulus E was normalized by the Young modulus of the threads and then expressed versus the normalized length of the thread L* (= L/a with a, thread radius). The following expression was then found for the normalized elastic modulus:

$$E^* = \frac{\sigma' / \varepsilon'}{E} = 2\pi L^{*-2} \left[\frac{(\tan\theta' - 1)}{(2/\sqrt{2})\sin\theta' - 1} \right] \tag{11}$$

where θ′ is a function of the displacement δ′ corresponding to the discontinuity.

This expression clearly demonstrates the L^{*-2}-dependence of E*. By contrast, the dependence on T* is less obvious but occurs through the T*-dependence of θ (Eq. 8). The L^{*-2}-dependence of E* is shown by the slope (–2) characterizing the (E* – L*) relationship when quantities E* and L* are plotted in a double logarithmic scale (Fig. 3). The results in Figure 3 confirm a much weaker T*-dependence of E*.

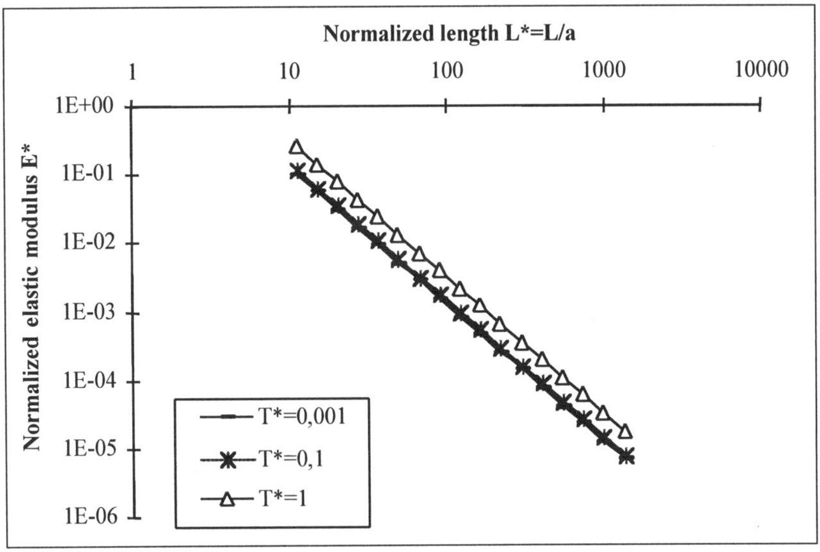

Fig. 3. Normalized elastic modulus E* *versus* normalized length L* of the thread plotted in log-log scale for three values of normalized tension T*.

The normalized apparent modulus E* given for the pseudo-tensegrity model (Eq. 11), can be compared to the equivalent expression given for foams (Eq. 3). This comparison reveals a fundamental difference between the behavior of foams and the pseudo-tensegrity models which concerns the normalized length dependence of $E^* \propto (L^{*-4})$ for foams and $E^* \propto (L^{*-2})$ for the tensegrity structure. This difference traduces the fundamental discrepancy between the underlying mechanisms of the deformation in each model as explained above. While the transition from macro to micro scales was obvious in the foam model, *i.e.*, the characteristic parameter is the relative density, this passage was not obvious for the tensegrity models. The normalization procedure used to analyze the results of the pseudo tensegrity model allows such a transition to be possible. Nevertheless, the results obtained in the present planar model cannot really be used to describe the mechanical properties of the living cell, due to the three-dimensional character of the cytoskeletal structure.

Thus, we analyzed in a separate study [8], the mechanical behavior of a spatial tensegrity structure constituted of thirty-elements, *i.e.*, 6 rigid struts maintained in equilibrium with 24 tensile elastic threads, already viewed as an elementary spatial model of the cytoskeletal complex architecture [9]. The L^{*-2}-dependence of E* of the spatial model was found to be similar to the planar model. However, the normalized elastic modulus E* was two orders of magnitude reduced.

The results obtained for the thirty-element tensegrity structure were used to describe the relationship between the assumed length of actin filaments and the elastic modulus of cytoskeleton for possible values of internal tension. The value of elastic modulus so predicted was compared to biological measured values. One of them were issued from magnetocytometry measurements which consist to apply mechanical stresses directly to cell surface receptors (*via* microbeads) with a magnetic twisting device. Using a length of actin filaments corresponding to the microbead diameter (∅ = 5 μm), and an internal tension of 10^{-12} Newton, we found values of elastic modulus in a range of values: [1 – 10 Pascal] similar to the one founded by Wang *et al.* during magnetocytometry experiments on endothelial cells in culture [10]. Satcher and Dewey [5] have predicted elastic modulus of the actin filaments network in endothelial cells using the foam model and considering a concentration of actin filaments: 10 mg/ml. They found elastic modulus values in a range of 10^3 – 10^4 Pascal which are not in agreement with the results obtained by magnetocytometry, likely because the theoretical models are different.

4. CONCLUSION

The present results suggest that the tensegrity models might be useful to describe quantitatively the cell mechanical behavior on the basis of the mechanical and geometrical properties of the individual filaments as well as internal tension at reference state. Further studies are necessary to confirm these results in a variety of attachment conditions, modes of loading and types of living cells.

REFERENCES

[1] Bereiter-Hahn J., Anderson O.R. and Reif W.E., Cytomechanics: the mechanical basis of cell form and structure (Springer Verlag, 1987).

[2] Elson E.L., Cellular mechanism as an indicator of CSK structure and function, *Ann. Rev. Biophys. Biophys. Chem.* **17** (1988) 397-430.

[3] Theret M.J., Levesque, Sato M., Nerem R.M. and Wheeler L.T., The Application of a Homogeneous Half-Space Model in the analysis of Endothelial Cell Micropipette Measurements, *J. Biomech. Eng.* **110** (1988) 191-199.

[4] Fung Y.C. and Liu S.Q., Elementary Mechanics of the Endothelium of Blood Vessels, *J. Biomech. Eng.* **115** (1993) 1-12.

[5] Satcher Jr. R.L. and Dewey C.F., Theoretical Estimates of Mechanical Properties of the Endothelial Cell Cytoskeleton, *Biophys. J.* **71** (1996) 109-118.

[6] Ingber D.E and Jamieson J.D., Cells as tensegrity structures: architectural regulation of histodifferentiation by physical forces transduced over basement membrane, in Gene expression during normal and malignant differentiation, edited by L. Anderson, C. Gahmberg and P. Ekblom (1985).

[7] Gibson L.J. and Ashby M.F., Cellular Solids, Structure and Properties (Pergamon Press, 1988).

[8] Wendling S., Oddou C. and Isabey D., Loi de comportement d'une structure de tenségrité élémentaire; application en mécanique cellulaire, *Arch. Physiol. Biochem.* **104** (1996) 598.

[9] Stamenovic D., Fredberg J.J., Wang N., Butler J.P. and Ingber D.E., A Microstructural Approach to Cytosqueletal Mechanics based on Tensegrity, *J. Theoret. Biol.* **181** (1996) 125-136.

[10] Wang N., Butler J.P. and Ingber D.E., Mechanotransduction across the cell surface and through the cytoskeleton, *Science* **260** (1993) 1124-1127.

Cellular and Extracellular Networks

LECTURE 11

Networks of Extracellular Matrix and Adhesion Proteins

J. Engel

Biozentrum, Dept. Biophysical Chemistry, Klingelbergstrasse 70,
4056 Basel, Switzerland

1. INTRODUCTION

The extracellular matrix is not just the glue between cells or a kind of an intercellular cement as it was assumed before details of its molecular architecture were elucidated. Together with the adhesion receptors located at the cell surface it serves highly sophisticated functions including specific cell-cell recognition, cell communication, growth regulation, differentiation guidance of cell migration and maintenance of tissues.

The complexity of extracellular matrix (ECM) components and adhesion proteins is very high both in a quantitative and qualitative sense. All these proteins are composed of many different building blocks which are called modules because they occur in many different proteins and are frequently repeated in the same protein. Each module in a protein can have a different function giving rise to the multifunctionality of extracellular matrix and proteins. In most cases modules are autonomously folding protein domains of 40 to 200 amino acid residues. Functions of modules in a given protein are not independent of each other because of their physical linkage in the same space element and also because of binding interactions between modules of the same molecule. Binding of parts of the large ECM molecule or adhesion receptor to other ligands may influence the functional activity of other domains in the same protein. A well known example is the initiation of signalling cascades inside cells by changes of the activity of the cytosolic domain induced by binding of a specific ligand to the extracellular part of a receptor. In a way each of the large modular proteins may therefore be looked at as a small network.

The size of the network is tremendously increased by the multiple interactions between proteins. Linkage is frequently mediated by oligomerization units of which α-helical coiled-coil structures are the most frequent ones. These mediate hetero- and homoassociations of two to five subunits (Kammerer, 1997). Other interactions may lead to fibrillar arrangements or to structures in which many different proteins are interlinked (Timpl, 1996). Of importance for the establishment of interactions and high local concentrations is also membrane attachment. For example E-cadherin in adhesion belts may assume a very high surface concentration in which individual E-cadherin proteins contact each other. As an example for the importance of network interactions I shall briefly summarize our recent data on the structure and function of a novel five-stranded coiled-coil module. This module was fused to the C-terminus of the extracellular portion of E-cadherin rendering this initially single stranded molecule pentameric with a potential for multivalent interaction similar to those of membrane bound E-cadherin.

2. THE FIVE STRANDED COILED-COIL DOMAIN OF CARTILAGE OLIGOMERIC MATRIX PROTEIN COMP

COMP is an ECM protein which is predominantly located in cartilage. It belongs to the family of thrombospondins whose modular organization is shown in Table I.

Table I. Modular organization of thrombospondins and COMP. Modules are designated as follows: TN, thrombospondin N-terminal; TC, thrombospondin C-terminal; T1, thrombosponding type 1/properdine; EG, epidermal growth factor-like, thrombospondin type 1; T3, thrombospondin type 3. CC, α-helical coiled-coil region.

thrombospondin 1 and 2	$[TN\ CC\ VC\ (T1)_3\ (EG)_3\ (T3)_7\ TC]_3$
thrombospondin 3 and 4	$[TN\ CC\ (EG)_4\ (T3)_7\ TC]_5$
cartilage oligomeric	
matrix protein (COMP)	$[CC\ (EG)_4\ (T3)_7\ TC]_5$

It plays an important role in cartilaginous tissues but also in the formation of bones. This is exemplified by the short stature disease pseudoachondrodysplasia which is caused by mutations in the calcium binding T3 modules of COMP. Probably for the interaction with cells (chondrocytes) by its TC modules multivalent binding and therefore the pentameric state of COMP is substantial (see Fig. 1). It may be assumed (but this is not experimentally proven yet) that hypothetical monomeric COMP proteins would not bind sufficiently strongly to a single cellular receptor but that interaction with a cluster of receptors is required. Interestingly two members of the thrombospondin family are also pentameric whereas thrombospondins 1 and 2 are rendered trimeric by different coiled-coil modules than those found in thrombospondins 3, 4 and COMP. The functional reasons for these variations in subunit numbers are unknown. Heteroassociation of different thrombospondin subunits has also been proposed and this may offer an additional functional advantage.

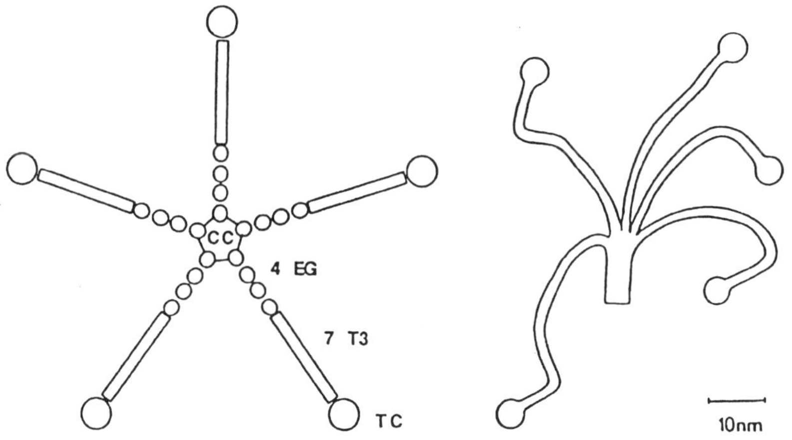

Fig. 1. Schematic representations of cartilage oligomeric matrix protein COMP in a top view (left) and side view (right). Each of the 5 polypeptide chains consisting of 4 epidermal growth factor like module EG, 7 thrombospondine 3/ properdine modules (B) and a C-terminal thrombospondin (TC) module is connected to a five-stranded α-helical coiled-coil domain (CC). The right figure represents the electron microscopically observed shape (Mörgelin *et al.*, 1992) demonstrating the flexible shape of the about 30 nm long arms.

The coiled-coil module of COMP was the first domain with five strands to be discovered. Two-stranded and three-stranded coiled-coil structures are much more common (Kammerer, 1997). The COMP domain was recombinantly expressed in E. coli bacteria (Efimov *et al.*, 1994) in sufficient yields to solve the crystal structure (Malashkevich *et al.*, 1996). As all other coiled-coil structures it is predominantly stabilized by complementary hydrophobic interaction between neighboring α-helices, but salt bridges and hydrogen bonds also play a role (Kammerer, 1997). The balance of these interactions determine the unusual number of 5 strands in the case of the COMP module. It was shown in other cases that modifications of the hydrophobic interactions or of salt bridges may alter the number of strands in a coiled-coil domain (Harbury *et al.*, 1993).

The COMP domain is also stabilized by a ring of five disulfide bonds between chains. With all disulfide bonds intact it preserves its native structure at the boiling temperature of water. Even after reductive cleavage of the disulfide bonds a high stability is preserved and the transition temperature is 60 °C. The COMP coiled-coil domain folds into the pentameric structure reversibly with a high rate. The high stability and the spontaneous folding are important for its oligomerization function and also qualify the COMP domain as tool for artificial crosslinking of proteins.

3. IS THE COILED-COIL DOMAIN OF COMP A PROTOTYPE ION CHANNEL?

This question was raised when looking at the structure of the COMP domain which displays an essentially hydrophobic channel of 2 to 6 Å diameter (Fig. 2). This channel can accommodate hundreds of water molecules and about 13 of these were visualized in the X-ray structure (Fig. 2). The visible water molecules are located in preferred positions, but may still retain a very high exchange rate with water from the bulk phase. Very interestingly the COMP domain acted as an ion trap for Cl^- which was found to be bound to a ring of 5 glutamine groups inside the channel.

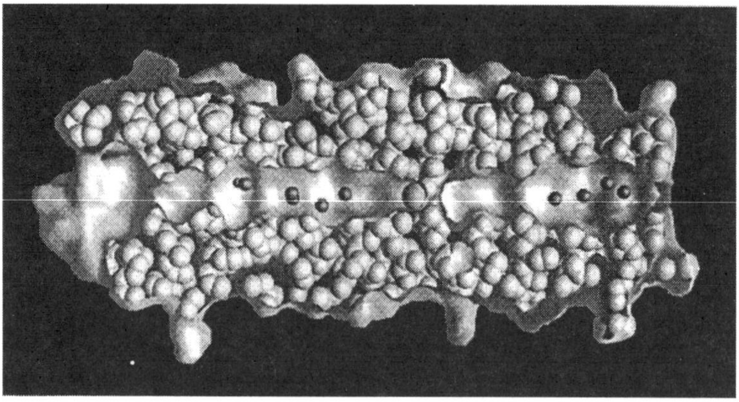

Fig. 2. Cross section of the five-stranded coiled-coil domain of COMP as resolved from the crystal structure at 2 Å resolution (Malashkevich *et al.*, 1996). Five chains, each with 46 residues form a parallel coiled-coil structure with a length of 73 Å and a pore which varies in diameter between 2 and 6 Å. The interior of the pore is hydrophobic with the exception of a ring of 5 glutamine groups to which a chloride ion was found to be bound (big sphere in the pore). The pore is filled with water and some water molecules were visualized in the structure (small spheres).

The COMP domain cannot function as a transmembrane channel because of the hydrophilicity of its outer surface. It is, however, very likely that the three-dimensional structure of the COMP domain closely resembles the structure of transmembrane ion channels of the α-helical bundle type (Montal, 1996) for example in phospholamban or acetylcholine receptor. The structure of none of these channels has been solved yet, but it is known that they consist like the COMP domain of five parallel α-helices. Furthermore structures rather similar to that of the COMP domain have been proposed on the basis of site directed mutagenesis and model building for phospholamban (Adam *et al.*, 1995; Simmerman *et al.*, 1996). The structural similarity of the COMP domain and the ion channels underlines a feature which is widely observed: Modules of the same structure which originate from common phylogenetic ancestors are used in biology for rather different functions, in our example oligomerization of five protein subunits of COMP and for ion transport.

4. ARTIFICIAL PENTAMERIZATION OF THE ADHESION PROTEIN E-CADHERIN

E-cadherin consists of five repeating modules EC1 to EC5 which are linearly arranged in its about 22 nm long extracellular part (Fig. 3). These are followed by a membrane spanning region TM and a cytosolic domain. At cell-cell contact sites for example in adhesion belts of epithelial cells E-cadherin molecules occur at high concentrations at the outer cell membrane. The extracellular parts of the adhesion receptor of two adjacent cells bind to each other by homophilic interactions.

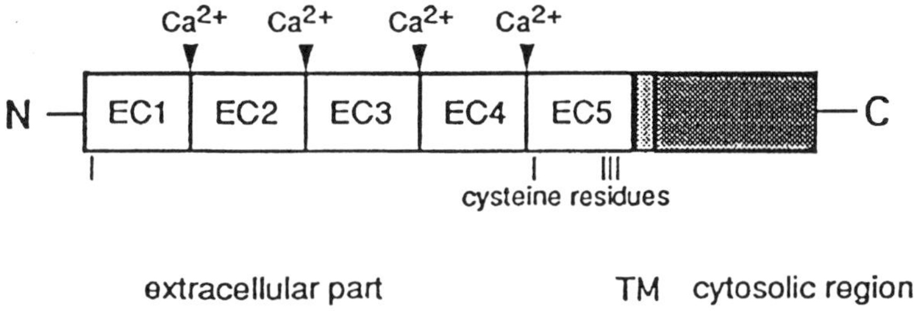

Fig. 3. Modular organization of the adhesion molecule E-cadherin.

The binding strength between two E-cadherin molecules is however very low. In solution phase only at very high concentrations dimer formation is observed and a binding constant of K = 12500 ± 2500 M^{-1} at 20 °C was determined (Koch *et al.*, 1997). This corresponds to a standard free enthalpy of binding of ΔG^{o} (20 °C) = – (5.5 ± 0.1) kcal/mol. Cadherin activity is strongly calcium dependent and the binding constant is unmeasurably small in its absence. This dependence is explained by the binding of three Ca^{2+} ions between domains EC1 and EC2 (Koch *et al.*, 1997) and probably also between other EC domains. The Ca^{2+} ions are needed for the stabilization of the extracellular structure (Pokutta *et al.*, 1994) and have been visualized by X-ray crystallography (Nagar *et al.*, 1996).

In cell-cell contacts large clusters of closely spaced E-cadherins linked to the two opposing cell membranes bind simultaneously. Cadherin layers were seen by electron microscopy in cell-cell contacts but the resolution of this technique was not sufficient to decide which of the EC domains of E-cadherin are involved in the homophilic interaction. Mutagenesis studies indicated that domains EC1 and EC2 are involved and responsible for the high specificity of cadherin recognition (Takeichi, 1990). X-ray crystallography was so far only performed with domains EC1 of N-cadherin (Shapiro *et al.*, 1995) and the domain pair EC12 of E-cadherin (Nagar *et al.*, 1996). These data were translated to the arrangement of cadherins between cells. A *dimer* contact between equally orientated domains and an *adhesion contact* between domains of opposing orientations were proposed. Unfortunately in the crystals both contacts were rather different for the two systems inspite of the high similarity of E- and N-cadherin. The question remains whether the contacts seen in the crystals between isolated domains reflect the true binding sites between cells. The physiological but rather weak binding interactions may be easily compensated by crystal forces.

In order to mimic the linking effect of membranes we decided to cross-link the extracellular parts of five E-cadherins by the five-stranded coiled-coil domains of COMP. For this purpose a chimeric protein ECAD-COMP consisting of modules EC1 to EC5, a flexible linker sequence and the COMP coiled-coil region at the C-terminus was recombinantly expressed in human kidney 293 cells (Tomschy *et al.*, 1996). In electron micrographs only about 25% of the molecules had the appearance of five-armed stars (Fig. 4A) in which the five 22 nm long arms emerged from the small and not visible COMP domain. In most cases the arms associated with their end to rings (Fig. 4B) which merged to pairs of rings (Fig. 4C). Frequently two rings per pentamer were observed. In all cases contacts between the arms and contacts between the two rings of two molecules occurred at sites at which EC1 and EC2 are located and in contrast to proposals for other authors (Shapiro *et al.*, 1995) no interactions between other domains were observed.

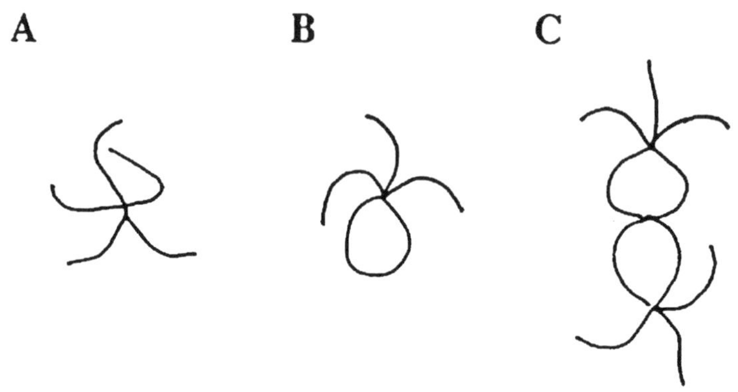

Fig. 4. Drawings of three species with abundant appearance in electron micrographs of ECAD-COMP. A: pentamer with unassociated cadherin arms, B: pentamer in which two of the five arms associated at their N-terminal region, C: two pentamers associated by interaction of the N-terminal regions of two rings.

As a control monomeric extracellular regions EC1 to EC5 were also recombinantly prepared and investigated by electron microscopy. At the highest possible concentrations of about 0.1 μM essentially no dimers were observed. An about 10 000 fold higher internal concentration of N-terminal domains in the pentamers may be estimated by dividing their number (five) by the volume of the outer 4.5 nm thick shell of a sphere with the radius of the length of the arms (22 nm). At this high concentration and with the binding constant determined by analytical ultracentrifugation of $K = 12500\ M^{-1}$ about 55% of the N-terminal domains should be associated. This is in agreement with the qualitative electron microscopic results (Tomschy *et al.*, 1996).

5. OUTLOOK

The findings demonstrate that multiple interactions in networks are of importance and provide some examples of how they can be created. For the cadherin system the physiological network is by far larger than described so far. In Figure 5 the interactions of the cytosolic domains of cadherins by the actin filaments are included. This interaction is mediated by catenin (symbolized by α, β, γ in Fig. 5) and regulated by phosphorylation (Stappert and Kemler, 1994) and in this way the surface concentration of E-cadherins may be controlled. Due to the weak interactions of isolated E-cadherins no adhesion will take place at low surface concentrations. At high concentrations the model discussed in the preceding section predicts self association of EC domains 1 and 2 of neighboring E-cadherins to a highly reactive complex which can associate with a similar complex of a neighboring cell. Note that intrinsically weak interactions between cadherins are substantial for the fast and reversible induction and release of adhesion.

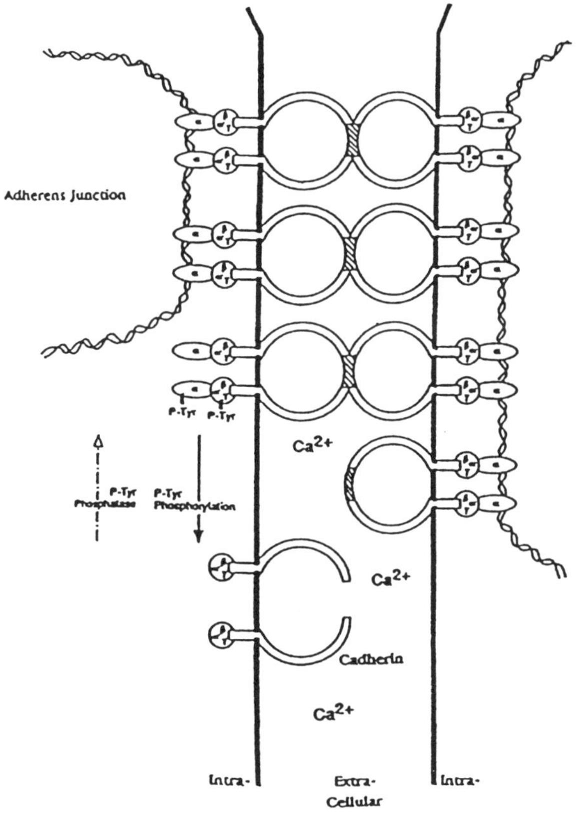

Fig. 5. Model for the cell-cell adhesion of cadherins and its regulation by the catenin (symbolized by α, β, γ) mediated interaction to the actin network (symbolized by filaments).

It was recently shown that the cadherin catenin complex not only provides stable cell-cell contacts but also serves as an organizing center for many signalling pathways. Release of β-catenin during cell dissociation may lead to gene expression by direct interaction with a transciption factor (Behrens *et al.*, 1996; Gumbiner, 1995). This provides a link between release of cell adhesion during periods of cell growth. It is part of a complex network of events which we are just starting to understand.

ACKNOWLEDGMENTS

We gratefully acknowledge financial support of the Swiss National Science Foundation.

REFERENCES

Adams P.D., Arkin I.T., Engelman D.M.and Brünger A.T., *Nature Struct. Biol.* **2** (1995) 154.

Behrens J., von Kries J.P., Kühl M., Bruhn L., Wedlich D., Grosschede R. and Birchmeier W., *Nature* **382** (1996) 683-642.

Efimov V.P., Lustig A. and Engel J., *FEBS Lett.* **341** (1994) 54-58.

Gumbiner B.M., *Curr. Opin. Cell. Biol.* **7** (1995) 634-640.

Harbury P.B., Uhang T., Kim P.S. and Alber T., *Science* **262** (1994) 1401-1407.

Kammerer R.A., *Matrix Biol.* **15** (1997) 555-565.

Koch A.W., Pokutta S., Lustig A. and Engel J., *Biochemistry* **36** (1997) 7697-7705.

Malashkevich V.N., Kammerer R.A., Efimov V.P., Schulthess T. and Engel J., *Science* **274** (1996) 761-765.

Mörgelin M., Heinegard D., Engel J. and Paulsson M., *J. Biol. Chem.* **267** (1992) 6137-6141.

Montal M., *Curr. Opinion Cell. Biol.* **6** (1996) 499-510.

Nagar B., Overduin M., Ikura M. and Rini J.M., *Nature* **380** (1996) 360-364.

Pokutta S., Herrenknecht K., Kemler R. and Engel J., *Eur. J. Biochem.* **223** (1994) 1019-1026.

Shapiro L., Fannon A.M., Kwong P.D., Thompson A., Lehmann M.S. Grübel G., Legrand J.-F., Als-Nielsen J., Colman D.R. and Hendrickson W.A., *Nature* **374** (1995) 327-337.

Simmerman H.K., Kobayashi Y.M., Autry J.M. and Jones L.R., *J. Biol. Chem.* **271** (1996) 5941.

Stappert J. and Kemler R. *Cell. Adhes. Commun.* **2** (1994) 319-327.

Takeichi M., *Ann. Rev. Biochem.* **59** (1990) 237-252.

Timpl R., *Curr. Opinion Cell Biol.* **8** (1996) 618-624.

Tomschy A., Fauser C., Landwehr R. and Engel J., *EMBO J.* **15** (1996) 3507-3514.

LECTURE 12

Networks of Extracellular Fibers and the Generation of Morphogenetic Forces

S.A. Newman

Department of Cell Biology and Anatomy, New York Medical College, Valhalla, NY 10595, USA

1. INTRODUCTION

During the development of multicellular organisms the earliest tissues to appear are *epithelioid*, consisting of cells in direct contact with one another. In contrast, the *connective tissues*, which appear somewhat later, are composed of cells surrounded by complex microenvironments known as extracellular matrices (ECM). ECMs consist of proteins, polysaccharides, hybrid molecules known as proteoglycans, and sometimes mineral, all in a highly hydrated state (see Comper, 1996 for reviews). When all categories are taken into consideration, the physical state of living tissues can range from liquid (blood), to elastic sheet (skin), to solid (bone), but mostly fall into the category of semi-solid condensed materials which de Gennes has termed "soft matter" (de Gennes, 1992).

Because living tissues contain cells that are capable of self-propulsion and of performing a variety of other non-random functions, it is not always evident to what extent changes in tissue form and shape can be attributed to the "thermodynamic" processes usually invoked to account for the behavior of non-living condensed materials. Such changes can be quite spectacular: in the course of embryonic development, regeneration, wound healing, and various pathological processes, tissues may disperse into individual cells, lengthen or shorten, or develop one or more internal boundaries across which cell mixing is selective or prohibited. As part of this suite of processes, collectively known as "morphogenesis", a tissue may engulf, or be engulfed, by a neighboring tissue across their common boundary.

For epithelioid tissues in embryos, and in other morphogenetically active settings, it has been persuasively argued that the capacity of cells to readily make and break contacts, slip past one another, and perambulate with no preferred direction, qualifies them to be considered as elasticoviscous liquids, with such generic properties as surface tension and immiscibility (Steinberg, 1978; Glazier and Graner, 1993; see Newman, 1998 for review). Indeed, many morphogenetic properties of epithelioid tissues can be accounted for in these terms (Armstrong,

1989; Foty *et al.*, 1995; 1996). The physical determinants of connective tissue morphogenesis, however, are less straightforward. In particular, although connective tissue cells are the synthetic source of the organic components of the ECM, they contribute only fractionally to the mass of the formed tissue. The physics of the ECM then, rather than the collective properties and behaviors of cells, must be considered the dominant factor in connective tissue morphogenesis (Newman and Tomasck, 1996).

Because the relevant secreted macromolecules spontaneously assemble into ECMs when new tissues and tissue interfaces are established, it is reasonable to ask what forces come into play as a consequence, and whether such effects can plausibly contribute to the morphogenesis of these tissues. Natural ECMs *in vivo* are highly complex, multicomponent systems, so as a first approximation we have studied an *in vitro* model ECM system that exhibits a number of features that appear relevant to morphogenesis in the living state.

2. MATRIX-DRIVEN TRANSLOCATION

The most abundant organic component of connective tissue ECMs is type I collagen, a rod-like protein that assembles into macromolecular fibrils and subsequently into macroscopic fibers and fiber bundles. As type I collagen fibers grow in width and length, they are capable of becoming organized into extensive parallel arrays, as in tendons and ligaments. Alternatively, they may form regular sheets, in which fiber orientation changes abruptly from layer to layer, as in bone, or most dramatically, in the transparent cornea of the eye, which contains orthogonal, paracrystalline arrays of collagen fibers. More typically however, type I collagen fibers become randomly arranged, constituting an isotropic structural network in the so-called "irregular" connective tissues.

Heat-denatured type I collagen is referred to as gelatin. When gelatin is dissolved in water and cooled down, the denatured collagen molecules fail to form fibrils and fibers; rather, they become entangled and form a resilient gel. In contrast, type I collagen that is extracted from tissues under mild conditions will, if gradually returned to physiological pH and ionic strength, assemble first into microscopic (~1 μm long) fibrils, which, in turn, aggregate into macroscopic ($10 - 10^2$ μm long) fibers. These *in vitro*-generated fibrils and fibers have the size distribution and cross-banded ultrastructural appearance typical of type I collagen *in vivo* (Elsdale and Bard, 1972). The assembly of this native collagen also results in a gel, but one which is more delicate and easily collapsed than that formed by gelatin. Living cells intermixed with native collagen during the *in vitro* assembly process will attach to and thrive within this microenvironment, taking on the appearance and physiological functions of tissue cells (Tomasek *et al.*, 1982), in contrast to cells grown on planar, artificial substrates. The normal response of cells to the *in vitro*-generated collagen matrices is a further indication of their relevance as a model for the material properties of connective tissues. Using such *in vitro*-produced collagen matrices, we have devised an experimental system which has allowed us to explore some of the physical determinants of connective tissue morphogenesis (Newman *et al.*, 1985). In this preparation, a droplet of soluble type I collagen at the beginning of the assembly process is deposited adjacent to a second droplet, which differs from the first by being populated with a small, but critical number of living cells, or cell-sized polystyrene beads. Surprisingly, an interface forms between the two droplets, indicating the presence of an interfacial tension, despite the fact that the composition of the two droplets is the same with the exception of the presence of particles in one of them. Over the next few minutes the droplet containing particles spreads over and partially engulfs the droplet lacking particles, extending over several mm as a consequence (Newman *et al.*, 1985; Forgacs *et al.*, 1989) (Fig. 1). While depicted in Figure 1 as occurring with droplets bounded above by air, MDT also occurs when the preparation is bounded above and below by polystyrene plates (Forgacs *et al.*, 1991).

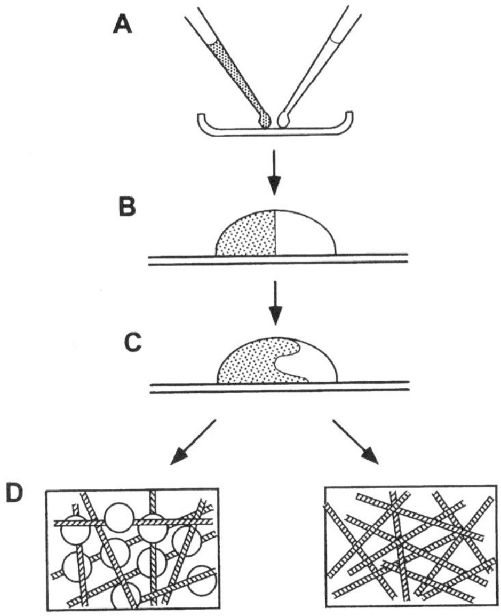

Fig. 1. Diagrammatic representation of matrix-driven translocation (MDT) assay. *A*, two droplets of type I collagen, one containing and one lacking cells, or cell-sized polystyrene latex beads, are deposited on the surface of a petri dish. *B*, a higher magnification representation of a side view of the two droplets shortly after their fusion. *C*, The translocation effect — a reconfiguration of the apparent interface between the two droplets during the collagen assembly process. *D*, a higher magnification view of the growing collagen fibrils shown in *C*, disrupted by beads or cells on the left, and forming a macroscopic network on the right. A particle-to-fibril ratio greater than actually used is depicted for representational purposes.

The phenomenon described occurs within a narrow range of initial collagen concentrations — typically $1.5 < c < 1.7$ mg/ml. At lower collagen concentrations there is no apparent phase separation, and at or around 1.7 mg/ml, although an interface forms, there is no relative movement of the two phases. Movement is restored at the higher collagen concentration if μg/ml amounts of the extracellular matrix protein fibronectin, or its amino-terminal heparin binding domain, is present in the droplet that lacks particles (Newman *et al.*, 1985; 1987).

There is also a critical dependence on the number density of particles (cells or beads) present. If this number is below ~ 2×10^6 per ml of collagen, a sharp interface does not form between the droplets, and translocation fails to occur (Forgacs *et al.*, 1989).

We have proposed that this set of phenomena, referred to as "matrix-driven translocation" (MDT), can be interpreted as follows (Forgacs *et al.*, 1989; 1991; 1994; Newman *et al.*, 1997): when the assembling collagen fibrils in the collagen solution reach a critical length, they can "percolate", forming a pervasive macroscopic network (Stauffer, 1985), but when their interaction is disrupted by the presence of cells or polystyrene beads, the network fails to form. Collagen droplets containing and lacking beads would thereby constitute separate "phases", with the bead-lacking drop being more cohesive than the bead-containing drop. Then, if the relative configurations of the droplets when fused with one another are dictated by the principles of thermodynamic equilibrium, they would behave like oil and water or any two immiscible fluids with different surface tensions: the less cohesive phase would envelop the more cohesive one — hence the translocation effect. At the higher collagen concentrations the interface is suggested to become "pinned" in a metastable state (de Gennes, 1985), and only able to attain its equilibrium configuration when the interaction of the particle surfaces with fibronectin on the opposite side of the interface lowers an energy of activation.

We have suggested that the establishment of domains of immiscibility within actual connective tissues, as well as the topographic rearrangement of such domains, and the regulation of this rearrangement by cell surface-ligand interactions, could all potentially be accounted for by physical interactions among ECM and cell surface components encompassed in the MDT model system (Forgacs and Newman, 1994). Although the congruences of the model with living systems are difficult to evaluate directly, an important step in this direction would be to explore the validity of the physical explanation we have provided for MDT, and the extent to which the principles involved may be applicable to living tissues.

3. PERCOLATION OF COLLAGEN FIBRILS

The assembly of collagen *in vitro* is a biphasic process: a growth phase in which properties such as turbidity and viscosity increase exponentially with time (until a plateau is reached and the solution has become a gel), is generally preceded (depending on initial concentration and temperature) by a lag phase, during which these properties do not measurably change, but nuclei or assembly intermediates form (Veis and George, 1994).

The gelation of collagen is due to entanglement of the long fibrils and fibers that are present at the end of the exponential phase of the assembly process (Veis and George, 1994). Indeed, scanning electron micrographs of the interior of the collagen gels that have formed after MDT has occurred indicate the presence of a dense network of long collagen fibers about 1.4 μm in diameter, each bearing numerous side branches (Forgacs *et al.*, 1991). However, MDT takes place early during the exponential phase, well before gelation, when the collagen solutions are still fluid. Is there any basis upon which to assume that network formation can occur among the fibrils present at the end of, or just after the lag phase?

Percolation theory is an analytic tool to describe network formation (Stauffer, 1985). It has been used to describe phenomena like gelation, magnetization, galaxy formation, continent formation, and the spread of fires and infectious diseases. A system of elements is said to percolate when a "macroscopic cluster", or network, has formed, and a pathway of linked or contacting elements can be found which spans the system's entire spatial domain. At the percolation transition the system's properties (*e.g.*, surface tension, viscosity) can change dramatically.

Bug, Safran and Webman (1985) have investigated the conditions under which a system of rigid rods will undergo a percolation transition. They derived a relationship between the volume fraction v, and the aspect ratio (length/diameter) x, of the rods that needs to be satisfied for percolation to occur:

$$v \geq 1/x \, , \, x >> 1. \quad (1)$$

Since collagen has a specific volume of 0.66 cm^3/g (Brokaw *et al.*, 1985), and the minimum mass density at which the interface forms in the MDT experiment is 1.5 mg/ml, percolation could provide an explanation for phase differences across the interface only if x is at least 1000. Molecular collagen, before any fibril assembly has taken place, has a length of 300 nm and a diameter of 1.4 nm. With an aspect ratio of 214, such molecular rods would not percolate. However, electron microscopic examination of samples of collagen directly after the lag phase showed it to be organized into networks of long, thin microfibrils 2 to 4 nm in diameter (Gelman *et al.*, 1979). Although the lengths of these filaments are difficult to measure, they were evidently much longer than individual collagen molecules, and growing primarily in length during this period (Gelman *et al.*, 1979). The assumption that 1 – 2 mg/ml collagen solutions, at the early stages of fibrillogenesis, when the solutions still had liquid properties, would contain a percolating population of microfibrils, thus appears well-justified. Such percolating networks would, of course, be dynamic, with individual contacts being made and broken with great frequency, as has been described for hydrogen bond networks in water (Stanley *et al.*, 1981).

4. EFFECT OF PARTICLES ON COLLAGEN SURFACE TENSION

In the MDT experiment the collagen solution containing cells or polystyrene beads appears to engulf the solution lacking particles (Fig. 1). By our assumption, MDT depends on a driving force resulting from an imbalance among various interfacial tensions. In particular, if σ_{w1} and σ_{w2} represent the effective interfacial tensions between the particle-containing and particle-lacking phases with their bounding substrata (polystyrene or air, but assumed here, for simplicity, to be the same), and σ_{12} represents the interfacial tension between the two phases, then the spreading coefficient γ (de Gennes, 1985) can be defined as:

$$\gamma = \sigma_{w2} - \sigma_{w1} - \sigma_{12} . \quad (2)$$

If L is the length of the contact line between the two phases, the driving force F for MDT would be:

$$F = \gamma L . \quad (3)$$

For practical reasons the interfacial tension σ_{12} is very difficult to measure. However, the surface tensions at the air-collagen interface can be measured with high accuracy using the pendant drop method (Cheng *et al.*, 1990). We have performed such measurements on the collagen solutions used in the MDT experiment, and have found that solutions containing particles at or above the number density required for the MDT effect (~ 2×10^6 per ml) consistently exhibited surface tensions that were 3 – 4 dyn/cm lower than those of the pure collagen solutions (Forgacs *et al.*, 1994). This difference was apparent at the earliest stages of fibrillogenesis, and persisted during the entire period relevant to MDT. Although in the absence of a value for σ_{12} this value for $\sigma_{w2} - \sigma_{w1}$ does not uniquely define the spreading coefficient γ, the result is consistent with our expectation that the particle-containing droplet is less cohesive that the particle-lacking one. The effect of the beads on collagen surface tension is somewhat surprising from a mechanistic point of view, however. At a number density of 2×10^6 per ml, the 6 μm polystyrene beads used in these experiments only constitute 0.1% of the droplet volume. Clearly they are having an effect on the collagen well beyond what would be expected from their numbers.

5. EFFECT OF PARTICLES ON COLLAGEN VISCOSITY

We reasoned that if the particles were affecting the collagen surface tension by altering the organization of percolating microfibrils, this might show up in certain bulk properties of the solutions, such as viscosity. We used a Couette viscometer to measure the low shear viscosity of assembling collagen solutions with and without particles present (Newman *et al.*, 1997). This device consists of a rotating cylinder, separated from a larger one concentric to it by a small gap, into which the test solution is poured. On the basis of earlier work by Einstein (see Jeffrey and Acrivos, 1976) and others (Thomas, 1965; Campbell and Forgacs, 1990) we expected that the presence of particles (even at a small volume fraction) would increase the viscosity of the solution, and this indeed turned out to be the case, but the results were more complicated than we anticipated. The viscosity of collagen increases exponentially once the lag period ends (Comper and Veis, 1977a,b), and this held true for both bead-containing and bead-lacking preparations. (See Newman *et al.*, 1997 for a discussion of the connection between this exponential rate law and the power law dependency expected from the percolation-based theory of gelation).

Typically, once the viscosity rise began, the rate of the rise was substantially greater if beads were present. This was reflected in an exponential rate constant K that was 1.3 – 2-fold greater for the bead-containing than the bead-lacking preparation (Newman *et al.*, 1997). While previous theory suggested that the viscosity of the particle-containing solution should be greater than that of the particle-lacking one at comparable points during the assembly process, the increased *rate* of viscosity rise in the presence of beads was not predicted by existing theory. We speculated that the beads, which contain charged surface moieties (Jozefonvicz and Jozefowicz, 1990) with known affinity for collagen (McPherson *et al.*, 1988), might be nucleating the assembly of collagen on their surfaces, thus increasing the effective diameter of the beads at the expense of the surrounding network (Newman *et al.*, 1997). A phenomenological equation derived by Thomas (1965) relates the relative viscosity of particle-containing and particle-lacking solutions to the effective bead volume fraction. When we plotted our results on relative viscosity against effective bead volume fraction using Thomas's expression, we indeed found an apparent increase in bead radius. This increase varied as $t^{1/3}$, a growth law identical to that observed in many diffusion-induced growth processes (Perrot *et al.*, 1994).

The interpretation that collagen preferentially polymerizes on the bead surfaces also has the advantage of providing a possible explanation for the reduced surface tension of the bead-containing collagen droplets relative to that of droplets that lack beads (Forgacs *et al.*, 1994, and discussion above). If the tenuous network of collagen microfibrils present at the end of the lag phase is just at its percolation threshold, any process that withdraws collagen from the medium (to generate numerous short "bristles" on the bead surfaces, for example) will tend to drop their number below the critical value. This would have the effect of causing the bead-containing and bead-lacking regions to constitute distinct phases, and would also imply that the bead-containing phase should be the less cohesive of the two (Fig. 2).

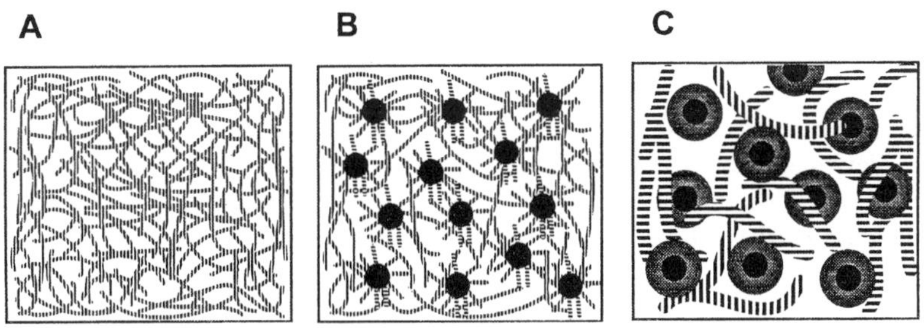

Fig. 2. Schematic representation of model for alteration of collagen microfibril network by particles or cells capable of binding collagen. In A, a percolating network of collagen microfibrils is present after the lag phase. In B, the presence of polystyrene latex particles or cells during the lag phase has caused microfibril assembly to nucleate on their surfaces at the expense of the network, causing the solution to be more viscous but less cohesive. In C, the assembly of microfibrils into fibrils during the exponential growth phase leads to solutions of increasing viscosity. The growth in the effective radii of the particles by the accretion of collagen leads to an acceleration in the viscosity rise.

The viscosity measurements have also provided additional information that bears on our proposed mechanism for MDT. The spreading coefficient γ, defined in equation (2), above can be thought of as providing the motive force for dragging a viscous liquid (*e.g.*, the bead-containing collagen solution) across several mm over several minutes. Thus, MDT can be driven by interfacial forces only if

$$\gamma = \sigma_{w2} - \sigma_{w1} - \sigma_{12} \geq \eta v \,, \quad (4)$$

where η is the viscosity, and v is the velocity of translocation. As noted above, $\sigma_{w2} - \sigma_{w1}$ is 3 – 4 dyn/cm (Forgacs *et al.*, 1994), and σ_{12} would be expected to have a small value, not too different from zero. The velocity of translocation v is $\sim 10^{-3}$ cm/s (Newman *et al.*, 1985). This puts an upper limit on η of 10^3 Poise, for the inequality in (4) to hold. The viscosity of both bead-containing and bead-lacking collagen reaches this value at about 3 min after the lag phase ends (Newman *et al.*, 1997), a value entirely consistent with the duration of MDT (Newman *et al.*, 1985).

6. CONCLUSIONS

The foregoing discussion has shown that the MDT effect can be accounted for by several fundamental physical properties of collagen matrices. These are: (i) the constitution of a cohesive, network-based material in the early stages of collagen fibrillogenesis, by the percolation of collagen microfibrils; (ii) the formation of a separate phase with lower cohesivity from the same material, by the presence of particles or cells which, by recruiting collagen to their surfaces, destroy the tenuous microfibril network; and (iii) the generation of unbalanced interfacial forces in the bounded, two-fluid system which are sufficient to promote the rearrangement of phases shown in Figure 1, so long as the ECM viscosities do not exceed that of, say, ethylene glycol. In principle, all of these properties may exist *in vivo*, although the composition of the matrices in living systems, and therefore the contributions to network formation, interfacial tensions, and viscosities, will be much more complex than in our simple model system. Moreover, such interfacially-driven morphogenetic forces could be held in abeyance *in vivo*, as they can be in the model system (Newman *et al.*, 1985; Forgacs *et al.*, 1989), to be released when required by interaction of additional components, fibronectin, for example (Newman *et al.*, 1987; Jaikaria *et al.*, 1991; Kishore *et al.*, 1997) with cell or bead surfaces.

The coordination of ECM assembly with invasion of loosely associated populations of cells into new regions during gastrulation (Critchley *et al.*, 1979; Wakely and England, 1979), neural crest dispersion (Newgreen and Erickson, 1986), and formation of the corneal stroma (Hay and Revel, 1969; Hay, 1980), to give only a few examples, suggests the opportunity for such forces to be mobilized during embryonic development. Given this opportunity, it would be unusual for evolution not to have made use of it.

ACKNOWLEDGMENTS

The model for MDT presented here was developed with the participation of Gabor Forgacs, Michel Cloître, and Daniel Beysens. Support from the National Science Foundation (IBN-9305628) is gratefully acknowledged.

REFERENCES

Armstrong P.B., *Crit. Rev. Biochem. Mol. Biol.* **24** (1989) 119-149.

Brokaw J.L., Doillon C.J., Hahn R.A., Birk D.E., Berg R.A. and Silver F.H., *Int. J. Biol. Macromol.* **7** (1985) 135-140.

Bug A., Safran S. and Webman I., *Phys. Rev. Lett.* **54** (1985) 1412-1415.

Campbell G.A. and Forgacs G., *Phys. Rev. A* **41** (1990) 4570-4573.

Cheng P., Li D., Boruvka L., Rotenberg Y. and Neumann A.W., *Colloids Surf.* **43** (1990) 151-167.

Comper W.D., Extracellular Matrix, Vols. 1 and 2 (Harwood Academic Publishers, Amsterdam, 1996).

Comper W.D. and Veis A., *Biopolymers* **16** (1977a) 2113-2131.

Comper W.D. and Veis A., *Biopolymers* **16** (1977b) 2133-2142.

Critchley D.R., England M.A., Wakely J. and Hynes R.O., *Nature* **280** (1979) 498-499.

de Gennes P.G., *Rev. Mod. Phys.* **57** (1985) 827-863.

de Gennes P.G., *Science* **256** (1992) 495-497.

Elsdale T. and Bard J., *J. Cell. Biol.* **54** (1972) 626-637.

Forgacs G., Jaikaria N.S., Frisch H.L. and Newman S.A., *J. Theor. Biol.* **140** (1989) 417-430.

Forgacs G. and Newman S.A., *Int. Rev. Cytol.* **150** (1994) 139-148.

Forgacs G., Newman S.A., Obukhov S.P. and Birk D.E., *Phys. Rev. Lett.* **67** (1991) 2399-2402.

Forgacs G., Newman S.A., Polikova Z. and Neumann A.W., *Colloids Surf.* **3** (1994) 139-146.

Foty R.A., Forgacs G., Pfleger C.M. and Steinberg M.S., *Phys. Rev. Lett.* **72** (1994) 2298-2301.

Foty R.A., Pfleger C.M., Forgacs G. and Steinberg M.S., *Development* **122** (1996) 1611-1620.

Gelman R.A., Williams B.R. and Piez K.A., *J. Biol. Chem.* **254** (1979) 180-186.

Glazier J.A. and Graner F., *Phys. Rev.* **47** (1993) 2128-2154.

Hay E.D., *Int. Rev. Cytol.* **63** (1980) 263-322.

Jaikaria N.S., Rosenfeld L., Khan M.Y., Danishefsky I. and Newman S.A., *Biochemistry* **30** (1991) 1538-1544.

Jeffrey D.J. and Acrivos A., *Amer. Inst. Chem. Eng. J.* **22** (1976) 417-432.

Jozefonvicz J. and Jozefowicz M., *J. Biomater. Sci. Polym. Ed.* **1** (1990) 147-165.

Kishore R., Samuel M., Khan M.Y., Hand J., Frenz D.A. and Newman S.A., *J. Biol. Chem.* **272** (1997) 17078-17085.

McPherson J.M., Sawamura S.J., Condell R.A., Rhee W. and Wallace D.G., *Coll. Relat. Res.* **8** (1988) 65-82.

Newgreen D.F. and Erickson C.A., *Int. Rev. Cytol.* **103** (1986) 89-145.

Newman S.A., "Epithelial Morphogenesis: A Physico-Evolutionary Interpretation", in Molecular Basis of Epithelial Appendage Morphogenesis, edited by C.-M. Chuong (R.G. Landes Co., Austin, 1998) pp. 341-358.

Newman S., Cloître M., Allain C., Forgacs G. and Beysens D., *Biopolymers* **41** (1997) 337-347.

Newman S.A., Frenz D.A., Hasegawa E. and Akiyama S.K., *Proc. Natl. Acad. Sci. USA* **84** (1987) 4791-4795.

Newman S.A., Frenz D.A., Tomasek J.J. and Rabuzzi D.D., *Science* **228** (1985) 885-889.

Newman S.A. and Tomasek J.J., "Morphogenesis of Connective Tissues", in Extracellular Matrix, Vol. 2: Molecular Components and Interactions, edited by W.D. Comper (Harwood Academic Publishers, Amsterdam, 1996) pp. 335-369.

Perrot F., Guenoun P., Baumberger T., Beysens D., Garrabos Y., and Le Neindre B., *Phys. Rev. Lett.* **73** (1994) 688-691.

Stanley H.E., Texeira J., Geiger A. and Blumberg R.L. *Physica* **106A** (1981) 260-277.

Stauffer D., Introduction to Percolation Theory (Taylor and Francis, London, 1985)

Steinberg M.S., "Specific Cell Ligands and the Differential Adhesion Hypothesis: How Do They Fit Together?", in Specificity of Embryological Interactions, edited by D.R. Garrod (London: Chapman and Hall, 1978) pp. 97-130.

Thomas D.G., *J. Colloid Sci.* **20** (1965) 267-277.

Tomasek J.J., Hay E.D. and Fujiwara K., *Dev. Biol.* **92** (1982) 107-122.

Veis A. and George A., "Fundamentals of Interstitial Collagen Self-Assembly", in Extracellular Matrix Assembly and Structure, edited by P.D. Yurchenco, D.E. Birk and R.P. Mecham (Academic Press, San Diego, 1994) pp. 15-45.

Wakely J. and England M.A., *Proc. R. Soc. Lond. B* **206** (1979) 329-352.

LECTURE 13

First Steps Towards a Comprehensive Model of Tissues, or: A Physicist Looks at Development

J.A. Glazier and A. Upadhyaya

Department of Physics, University of Notre Dame, Notre Dame, IN 46556-5670, USA

1. INTRODUCTION

To justify an intrusion into biology, physicists must bring humility and novel approaches. Our approaches must yield results meaningful to biologists or we reenact the old joke of the spherical cow. Naively, we can look at embryological development as a problem in pattern formation. How does a fertilized (or even unfertilized) egg, give rise to the complex structure of an animal? Clearly, the question is arrogant and too hard: animals are much more complex spatially and temporally than even the most complex hydrodynamics. On the other hand, we can recognize, at least locally and over short times, processes that resemble pure physical or chemical phenomena.

Biologists traditionally describe pathways, enumerating reactants and the interactions among them. This focus provides qualitative explanations, *e.g.* the expression of particular homeotic genes gives a body segment its identity. The ultimate understanding of organisms will require this level of detail, but we still cannot model all of the reaction kinetics in even fairly simple processes because of the large number of rate constants, each of which represents a major experimental effort. (Though Arkin's paper in this volume shows that we are making rapid progress in this direction). More seriously, if even one layer of a detailed reaction kinetics pathway is missing, quantitative modeling becomes hopeless. Studies of cAMP reception pathways in *Dictyostelium discoideum* combined with those on the role of capping proteins in creating leading edges of actin polymerization and hence pseudopod formation (Hug *et al.*, 1995) may form a nearly closed pathway for chemotaxis in *Dictyostelium*. If so, we will soon be able to develop true molecular kinetics models of cell motion. For the moment our goal is to develop a general framework which allows us to include reaction kinetics but does not require it.

Physicists look for general patterns of behavior that can be captured in universal, maximally simplified, few parameter models which are independent of smaller scale processes. If we have a model that indistinguishably reproduces cell motion in response to chemical gradients, we don't ask about internal cellular processes. Our experiments and models try to separate out individual mechanisms. When we have understood all processes individually, we recombine them and look for novel effects arising from interactions.

We can conveniently describe cell motion and differentiation by fields and what a physicist would call an **effective energy,** E. Some of the fields are material, *e.g.* the local concentration of a diffusant; others, like an orientation vector field, are not. Similarly, the effective energy will be a mixture of true energies, like cell-cell adhesion, and terms that mimic energies, *e.g.* the response of a cell to a chemotactic gradient. The names are merely a shorthand for the mathematical structure. Given an effective energy we can calculate the resulting cell motion, since differences in energy produce forces, $\vec{F} = \vec{\nabla} E$. Since we are in an extremely viscous regime, the velocity, not acceleration, is proportional to force, with mass replaced by an effective cell mobility, $\vec{v} = m\vec{F}$:

$$\vec{v} = -m\vec{\nabla} E. \quad (1)$$

Equation (1) implies that cells move to minimize the total effective energy. Our cells generalize as well; we can treat additional materials, *e.g.* medium, substrate and extracellular matrix (**ECM**) simply as cells with special properties.

We can begin our subdivision as shown in Figure 1. Development results from the interaction of four main processes, cell division, cell migration, cell differentiation, and the formation of extracellular structures. While cell division is inherently complex, the final result, two daughter cells, can be simulated easily if we neglect cell polarity. The key is to define a rate of cell growth. The secretion of ECM can be expressed naturally in our formalism. Differentiation involves two processes, temporal changes in cell properties and spatial changes leading to anisotropy in cell behavior, *e.g.* cell elongation. Both of these involve large numbers of experimental parameters, and we will initially restrict ourselves to cases in which cells are effectively **isotropic** and **static**, though we can relax the latter within the formalism. The information determining differentiation can be internal (*e.g.* cascades of gene expression or information carried in asymmetric components of cytoplasm during cell division, *e.g.* selection of body axis after fertilization) or external (*e.g.* contact signaling or external diffusible morphogens). Cell migration is the easiest process to model. Its causes subdivide into short and long range signaling mechanisms, the former including differential adhesion, contact inhibition and haptotaxis, the later including chemotaxis, galvanotaxis and gravity.

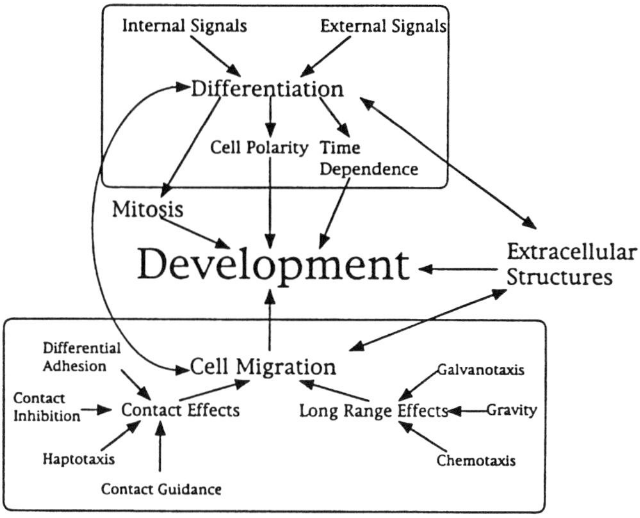

Fig. 1. — Flow diagram depicting the interaction of four processes leading to development.

2. MODEL

Our goal is to develop a set of standard terms describing individual processes, which can then be combined as necessary. This brief overview cannot show the details of implementing each mechanism. However, the references in each section point to full information. Similarly we will not discuss the many other approaches to tissue modeling (Graner, 1993; Agarwal, 1995).

The minimal definition of a cell: Label each cell under consideration by a unique index σ. At any time, t, a cell, of type, τ, has a volume $v(\sigma, t)$. Since we know that cell volume is not absolutely fixed (*e.g.*, due to changes in osmotic pressure) we describe the cell volume in terms of an effective pressure or membrane elasticity, λ_σ, and a target volume $v_{target}(\sigma, t)$:

$$E_{volume} = \sum_{all\ cells} \lambda_\sigma \big(v(\sigma, t) - v_{target}(\sigma, t)\big)^2. \tag{2}$$

Equation (2) has the form of an elastic volume constraint with membrane elasticity, λ_σ. If we neglect cell polarity and assume that growth of cells is simple and deterministic, both strong assumptions, we can model cell growth by setting:

$$v_{target}(\sigma, t) = v_o(\sigma) + g \int \big(\text{nutrient supply} * \text{cell cycle factors}\big) dt. \tag{3}$$

To reduce the number of parameters we will usually assume that all the cells of a given type have the same λ and v_0, and that the growth rate of a cell is constant. Where cells are not growing or dividing, we simply take $v_{target}(t) = v(\tau)$. These simplifications are reasonable in homogeneous tissues and in plants, but are far from true in early stage embryos.

Mitosis: Mitosis can occur either when the cell reaches a fixed type dependent volume, or when the ratio between cell surface area and cell volume reaches a critical value (since cell nutrient absorption is roughly proportional to surface area and metabolism is roughly proportional to volume). In yeast and in other simple organisms as well as in cancer cells, the growth rate depends mainly on the supply of nutrients. Cell division is triggered only after the cell reaches a critical size (Hayles *et al.*, 1986). As the cell grows, the concentration of a diffusible chemical falls and the decrease of its concentration below a threshold flips the molecular switch that triggers cell division. In the model, we create two daughter cells at this critical stage, with the plane of separation along the cell's longest axis and new target areas, $v_{target}/2$.

Contact between cells: Cells express surface adhesion molecules on their membranes, *e.g.* integrins, cadherins, N-CAMS, some highly specific, others nonspecific. If we bring two cells together, pulling them apart requires work. Since they are only now being measured (Engel, this volume), we neglect changes in binding strength due to diffusion of adhesion molecules into the contact area, local reorganization of adhesion molecules, *e.g.* into adhesive plaques, and changes in the number or adhesivity of the binding molecules induced by the adhesion itself. We can describe the net interaction between two cell membranes by an effective binding energy per unit area, $J_{\tau,\tau'}$, which depends on the types of the cells. We can measure $J_{\tau,\tau'}$ with laser tweezers (Sato-Maeda *et al.*, 1994), a shaker, a compression apparatus (Foty *et al.*, 1996) or a Taylor-Couette shear cylinder. The effective surface energy is:

$$E_{contact} = \oint_{cell\ surface} J_{\tau,\tau'} ds, \tag{4}$$

where ds is a unit of surface area. Based on this Differential Adhesion Hypothesis (DAH), Steinberg (1963) predicted that in an aggregate of two cell types, in which the heterotypic energy is greater than the homotypic, the lowest energy configuration has the more cohesive cell type in a sphere completely surrounded by the less cohesive cell type. This classic cell sorting is seen in *hydra* endoderm and ectoderm, and in embryonic chicken neural and pigmented retinal cells. If the heterotypic energy is less than the homotypic energy, the lowest energy pattern is the checkerboard as seen in Japanese quail oviduct (Honda *et al.*, 1986). If the cell-medium energy is less than the heterotypic energy but greater than the homotypic energy then the cells divide into discrete homotypic clusters (as seen in mixtures of *Hydra viridiana* and *Hydra vulgaris* cells in which it results in self-nonself identification). Complications arise because the cell adhesion molecules may change both in quantity and identity, *e.g.* in

Dictyostelium, at least four different adhesion molecules are significant at different stages of aggregation and mound formation (Bozzaro and Ponte, 1995). We model all these complex changes as variations in cell specific adhesivity.

Membrane Fluctuations: How do cells move from their initial random positions to the lowest energy configuration? Do they have time to reach the lowest energy configuration? In mixtures of liquid droplets, the thermal fluctuations of the droplet surfaces cause diffusion (Brownian motion) leading to energy minimization. Cytoskeletally driven cell membrane ruffling of a few μm have no need to be thermal and the dynamics might depend sensitively on their fluctuation spectrum. In addition, we are neglecting cells like keratocytes as described by Anderson (this volume), which move in straight lines over long distances. Such highly correlated cell motion seems to depend on the presence of a substrate. In aggregates, cell motion is more random. In our ignorance of the actual spectrum and neglecting cytoskeletal structures, the simplest assumption is that the cell membrane fluctuates thermally with an **effective temperature**, T, of about a million degrees. In the model we use Monte Carlo Boltzmann dynamics: if a proposed change in configuration produces a change in effective energy, ΔE, we make it with probability:

$$P(\Delta E) = \begin{cases} 1 & : \Delta E \leq 0 \\ e^{-\Delta E/T} & : \ E > 0 \end{cases} . \tag{5}$$

One consequence of this dynamics is that if the energy fluctuates up and down greatly as cells rearrange, as it does in a random aggregate, then the cells will not be able to reach a globally optimal configuration if T is too small, *i.e.* the liquid freezes into a glass. Freezing, or partial cell sorting, occurs experimentally when cell membrane fluctuations are eliminated either with cytochalasin-B or by holding the cells at 4 °C. In our model, low values of T also lead to partial sorting.

Chemotaxis: Chemotaxis requires an additional field to describe the local concentration of the chemoattractant, $C(\vec{x})$, which diffuses in extracellular space (or, in the presence of gap junctions, through cells as well) with diffusion constant, d, decays at a rate, Γ, and is secreted or absorbed at the surface of cells in a complicated history dependent way, $s_C(\sigma, \vec{x}, t)$, which is difficult to measure and requires a detailed description of the internal processes of the cell. The equation for the field then is:

$$\frac{\partial C(\vec{x})}{\partial x} = d\nabla^2 C(\vec{x}) - \Gamma C(\vec{x}) + \sum_\sigma s_C(\sigma, \vec{x}, t). \tag{6}$$

Fortunately, a few simple approximations to $s_C(\sigma, \vec{x}, t)$ work well for *Dictyostelium* (Levine *et al.*, 1996; Hofer *et al.*, 1995). We then describe the cell's response to the chemical field by an effective chemical potential, $\mu(\sigma)$, which may be time dependent, *e.g.* in a refractory period, μ will be nearly 0. $\mu > 0$

yields repulsion, and $\mu < 0$ attraction. The effective chemical energy is:

$$E_{chemical} = \sum_{\sigma} \oint_{\substack{surface\\ of\ \sigma}} \mu(\sigma) C(\vec{x}) ds. \tag{7}$$

The cell executes a biased random walk, which averages to directed motion in the direction of the gradient:

$$\vec{v}(\sigma) = -\mu(\sigma)\vec{\nabla} C(\vec{x}). \tag{8}$$

which has the same form as equation (1), justifying our energy treatment. Unfortunately, the final configuration for an aggregate composed of two cell types is the same whether they exhibit differential adhesion or differential chemotaxis. Only the kinetics differs. Experiments on slug phase *Dictyostelium* suggest that the cell sorting that occurs between prestalk and prespore cells is primarily chemotactic (Takeuchi *et al.*, 1988; Traynor *et al.*, 1992), but the quantitative experiments to check the kinetics have not been done.

Differentiation: Classic reaction-diffusion equation models of differentiation look at pairs of continuous fields rather than individual cells. However, their formalism carries over to our cellular model. If A and B are diffusing morphogens, which evolve like the diffusants in chemotaxis (Turing, 1952):

$$\frac{\partial A(\vec{x})}{\partial t} = d_A \nabla^2 A(\vec{x}) - \Gamma_A A(\vec{x}) + f\big(A(\vec{x}), B(\vec{x})\big), \tag{9}$$

$$\frac{\partial B(\vec{x})}{\partial t} = d_B \nabla^2 B(\vec{x}) - \Gamma_B B(\vec{x}) + g\big(A(\vec{x}), B(\vec{x})\big). \tag{10}$$

A is excitatory if $\partial f(\sigma, \vec{x}, t)/\partial A > 0$ and inhibitory if $\partial f(\sigma, \vec{x}, t)/\partial A < 0$. Turing (1952) showed that if A is excitatory and B inhibitory and B diffuses faster than A, $d_A < d_B$, then an initial uniform distribution of A and B is unstable and the concentration field will evolve into domains. Cells, instead of moving, change their parameters, *e.g.* λ, in response to A and B, as a function, h, of the surface concentrations of the chemotractants:

$$\frac{\partial \lambda_\sigma}{\partial t} = h\left(\oint_{\substack{surface\\ of\ \sigma}} A d^2 s, \oint_{\substack{surface\\ of\ \sigma}} B d^2 s \right). \tag{11}$$

If the cells differentiate primarily as a result of their history and not due to positional signals (as appears to be the case with prestalk/prespore selection in *Dictyostelium* (Takeuchi *et al.*, 1988)), then the cell carries order parameters with it which evolve according to a set of internally defined ordinary differential equations.

Reaction Kinetics A gradient based on purely reaction diffusion mechanisms can be maintained only in small tissues ~ 1 mm, otherwise the time required for diffusive exchange of molecules is too long. In larger organisms and tissues, particular genes are activated in response to signals, setting off a gene cascade. This cascade requires autocatalytic activation of genes (Koch and Meinhardt, 1994). A simple example of a such a biochemical switch is:

$$\frac{\partial y}{\partial t} = \rho_y \frac{y^2}{1 + \kappa_y y^2} - \mu_y y + \sigma^{ext}. \tag{12}$$

In this equation ρ_y, μ_y and κ_y are constants; σ^{ext} describes the external signal. In the absence of such a signal, the system has two external steady states. If the external signal exceeds a threshold, the system switches from the low to the high state. More complex interactions allow the space-dependent activation of several genes under the influence of a single gradient (Koch and Meinhardt, 1994). To understand mechanisms controlling development, it is essential to recognize patterns of coordinated gene activity that ultimately alter the phenotype.

Gene regulation and cell signaling during development can be represented mathematically using reaction kinetics. Typically, a signaling network connecting cells at different stages of differentiation is identified. The signals controlling the different transitions are themselves regulated by a complex network of interactions. Once the transition rates are known, and assuming that transitions are first order, differential equations can be written for the evolution of the active cell species. Cell signaling has been studied widely in *Dictyostelium* (Loomis, this volume). Tang and Othmer (1995) have modeled cAMP responses of these cells - adaptation to stimuli, amplification of extracellular cAMP and the time scale of response to stimuli, as well as traveling wave chemotaxis. In a model, variation of a biochemical parameter can simulate changes in developmental state. Such parameters can be introduced very naturally in the Potts model. Concentrations of chemicals or gene products are assigned to each cell.

Gravity: Gravity is usually too weak to be important, but may play a role in *Dictyostelium* slug formation. It is simply a potential energy depending on the local cell mass, $\rho(\vec{x})$, and the vertical height, z:

$$E_{gravity} = \int_{space} \rho(\vec{x}) \; z \; d^3\vec{x}. \tag{13}$$

Extracellular Matrix: In many tissues, non-cellular materials provide much of the mechanical stability, *e.g.* bone in vertebrates, mesoglea in *Hydra*, and the slime sheath in *Dictyostelium*. In some cases cell sorting requires the secretion of ECM compounds (*e.g.* sorting between chick embryo cardiac mesenchyme and myocytes, Armstrong, 1985). ECM is similar to a secreted signaling molecule except that the secreted material does not diffuse. Instead of treating ECM as an auxiliary field, we treat it as a new cell with the property that its target area is always fixed to its current area. If a cell secretes a unit of ECM, the target area is increased by one unit. Similarly if a cell absorbs it, the target area decreases by one. If the ECM is solid, then local concentrations of ECM cannot fluctuate or move, if liquid, it behaves exactly like a normal cell. As before, we need to define the interaction energy between each cell type and the ECM.

Cell Polarity: Our assumption of a uniform distribution of adhesion molecules is an oversimplification. In many cases, *e.g.* aggregating *Dictyostelium* or *myxobacteria* (in which the apical/basal cell ends are much more cohesive than the sides, Bozzaro and Ponte, 1995; Kuspa *et al.*, 1992) or in vertebrates (where surface cells in embryos or epithelial cells in adults both are non-adhesive on the apical/luminal surface, Robertson *et al.*, 1980), the cell's membrane adhesivity varies over the surface of the cell and may have local maxima at junctions. Such cell polarity is difficult to treat in our framework. Two possible approaches are to create a physical cell membrane, by defining patches of membrane at the surface of each cell (computationally expensive), which allows locally variable properties, or assigning an orientation vector to each cell, which gives a cell axis and hence allows angular variation of parameters in compact form.

Putting it All Together–Simulation: If we sum the effective energies in equations (2, 4, 7, 13) we obtain:

$$E_{total} = E_{volume} + E_{contact} + E_{chemical} + E_{gravitational} + E_{other}. \qquad (14)$$

To simulate, we simply superimpose a lattice on the cells, where the value at a lattice site is σ if the site lies in cell σ (Srolovitz *et al.*, 1984; Grest *et al.*, 1984). We select a lattice site at random and propose to reassign it from cell σ to a cell σ' (usually a neighbor). We calculate the change in effective energy caused by the proposed reassignment and accept it according to equation (5) (Glazier and Graner, 1992). Similarly we discretize any auxiliary fields and evolve them according to equations (6) or (9, 10). The parameters of each cell evolve according to equations (3, 11), and any other equations we may have defined for the internal states of cells.

3. EXPERIMENTAL VERIFICATION

How well does it work? Extensive previous work has shown that this simulation method reproduces the evolution of fluids and foams subject to energy gradients (Glazier *et al.*, 1990; Jiang and Glazier, 1996). The simple mitosis model (Eqs. (2-5)) has been used by Mombach *et al.* (1993) to quantitatively reproduce the observed cell arrangements in a variety of plant tissues. Drasdo (in this volume) has presented his approach to tumor growth using a reduced version of this formalism, while Stott of University of Bath (private communication) is using the complete formalism to model tumor growth and angiogenesis as a function of nutrient supply.

The DAH component of the simulation (Eqs. (2, 4, 5)) has been verified in experiments using embryonic chicken cells (Mombach *et al.*, 1995, Fig. 2). In this case, if we begin with a random aggregate of neural and pigmented retinal

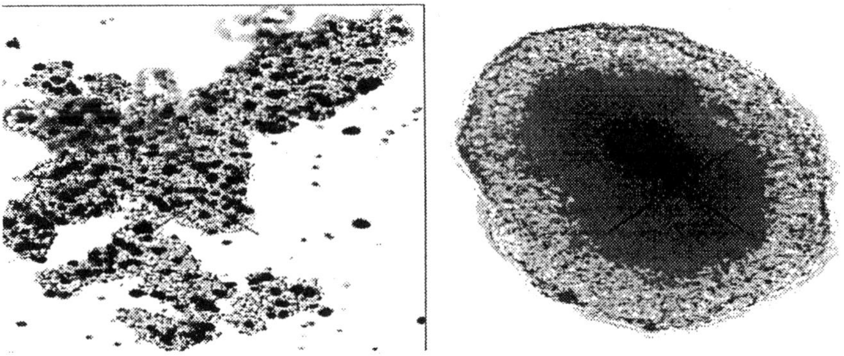

Fig. 2. — Sorting of an aggregate of neural retinal (light) and pigmented retinal (dark) cells from chick embryos. (a) Aggregate 5 h after mixing. (b) Aggregate after 72 h.

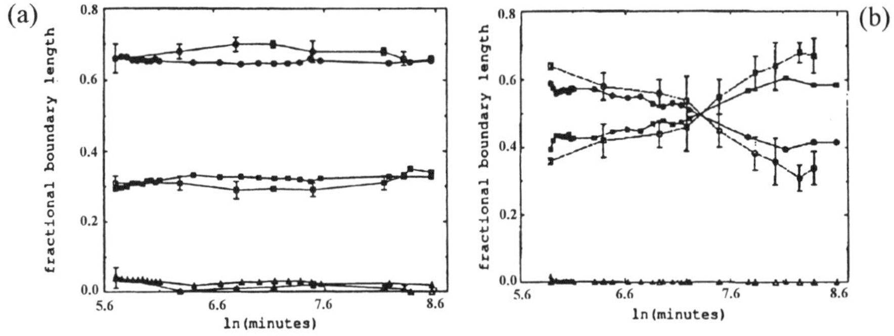

Fig. 3. — Time evolution of boundary lengths for partial (a) and complete (b) sorting for experiment (open symbols) and simulation (closed symbols). Circles, boundary between dark and light cells. Squares, boundary between light cells and medium. Triangles, boundary between dark cells and medium.

cells, and use the best available values for cell volumes, fluctuation amplitudes and surface energies, we are able to quantitatively reproduce the evolution of the dark-light contact length (Fig. 3).

In addition, as we hypothesized, if we reduce the fluctuation temperature in the simulation, or suppress membrane fluctuations in the experiment (with cytochalasin-B) the sorting halts (freezes) (Fig. 3a). Restoring the fluctuations leads to normal sorting as in Figure 3b. While we have not yet been able to verify that the cell membrane really does undergo thermal fluctuations, we have

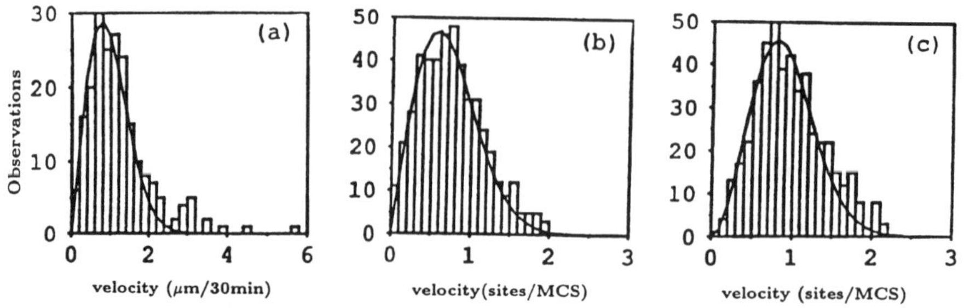

Fig. 4. — Velocity histograms for a single pigmented cell in a neural retinal aggregate: experiment (a), 2D simulation (b), 3D simulation (c). Results are fitted to a Maxwell-Boltzmann distribution.

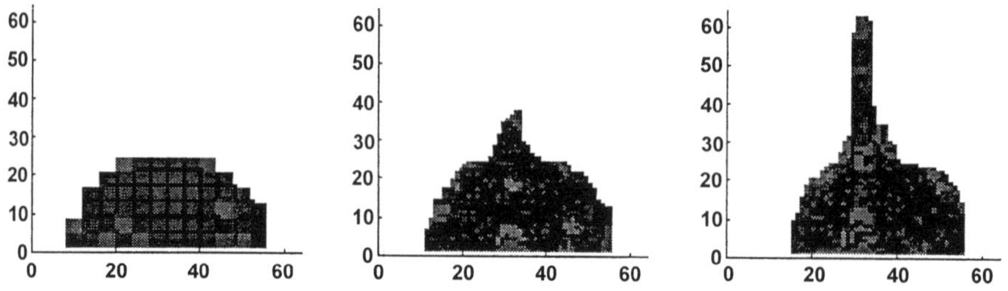

Fig. 5. — Tip formation in simulated *Dictyostelium* aggregate. Dark cells are prespore and light cells are prestalk. The strength of chemotaxis increases from left to right causing progressively larger tips to form.

shown that the spectrum of velocities of a single pigmented cell in a neural cell aggregate is identical to that of a simulated cell when the fluctuation amplitudes match (Mombach and Glazier, 1996), *i.e.* the experimental cells execute a thermodynamic random walk (Fig. 4).

More complex processes have been qualitatively verified in *Dictyostelium*. Levine has shown that the chemotactic mechanism presented here qualitatively reproduces the patterns of aggregating amoebae. Jiang (private communication) has shown that the peculiar tipped shape and cell sorting in the *Dictyostelium* mound is due to the competition between chemotaxis and differential adhesion (Fig. 5).

Finally, Savill and Hogeweg (1997) have used equation (12) to simulate qualitatively the entire aggregation process from single amoebae through migrating slug. The continuum reaction-diffusion equations (10, 11) have been successful in describing a large number of differentiation phenomena, *e.g.* head formation in *Hydra*, but have not yet been simulated in conjunction with equation (12).

4. REMAINING PROBLEMS

The main problem with modeling biological processes is the vast number of parameters that must be experimentally determined to have any hope of other than accidental agreement. The particular task is to determine which mechanisms combine in any developmental processes and their relative rates. The energy formalism provides insight into which parameters are important, *e.g.* cell-cell adhesivities, rates of cell motion in response to specified chemotactic gradients, cell membrane amplitude fluctuations, *etc.* In addition, our understanding of basic thermodynamic processes like phase separation and freezing gives us confidence that even if our parameters are somewhat inaccurate the qualitative behavior of the model will be correct. We should not observe phenomena wildly divergent from experiment. This robustness allows us the chance to begin with rough qualitative models and gradually refine them into quantitative models as our experimental measurements improve. In this sense, though we are still far from being able to reproduce even such basic phenomena as gastrulation, we have made a true step towards a comprehensive model of cellular patterns.

ACKNOWLEDGMENTS

This research has been supported by National Science Foundation grant DMR 92-57011, and the American Chemical Society/Petroleum Research Fund.

REFERENCES

Agarwal P., *J. Theor. Biol.* **176** (1995) 79-89.
Anderson M.P., Srolovitz D.J., Grest G.S. and Sahni P.S., *Acta. Metall.* **32** (1984) 783-791.
Armstrong P.B., *Exp. Biol. Med.* **10** (1985) 222-230.
Bozzaro S. and Ponte E., *Experientia* **51** (1995) 1175-1188.
Foty R.A., Pfleger C.M., Forgacs G. and Steinberg M.S., *Development* **122** (1996) 1611-1620.
Glazier J.A. and Graner F., *Phys. Rev. E* **47** (1992) 2128-2154.
Glazier J.A., Anderson M.P. and Grest G.S., *Phil. Mag. B* **62** (1990) 615-645.
Graner F., *J. Theor. Biol.* **164** (1993) 455-476.
Grest G.S., Srolovitz D.J. and Anderson M.P., *Phys. Rev. Lett.* **52** (1984) 1321-1324.
Hayles J. and Nurse P., *J. Cell. Sci.* **4** (1986) 155-170.
Hofer T., Sherratt J.A. and Maini P.K., *Proc. R. Soc. Lond. B* **259** (1995) 249-257.
Honda H., Yamanaka H. and Eguchi G., *J. Embryol. Exp. Morph.* **98** (1986) 1-19.

Hug C., Jay P.I., Reddy I., McNally J.G., Bridgman P.C., Elson E.L. and Cooper J.A., *Cell.* **81** (1995) 591-600.

Jiang Y. and Glazier J.A., *Philos. Mag. Lett.* **74** (1996) 119-128.

Kuspa A, Plamann L. and Kaiser D., *J. Bacteriol.* **174** (1992) 7360-7369.

Koch A.J. and Meinhardt H., *Rev. Mod. Phys.* **66** (1994) 1481-1507.

Levine H., Aranson I., Tsimring L. and Truong T.V., *Proc. Nat. Acad. Sci. USA* **93** (1996) 6382-6386.

Mombach J.C.M., de Almeida R.M.C. and Iglesias J.R., *Phys. Rev. E* **48** (1993) 598-602.

Mombach J.C.M., Glazier J.A., Raphael R.C. and Zajac M., *Phys. Rev. Lett.* **75** (1995) 2244-2247.

Mombach J.C.M. and Glazier J.A., *Phys. Rev. Lett.* **76** (1996) 3032-3035.

Robertson M., Armstrong J. and Armstrong P., *J. Cell. Sci.* **44** (1980) 19-31.

Sato-Maeda M., Uchida M., Graner F. and Tashiro H., *Dev. Biol.* **162** (1994) 77-84.

Savill N. and Hogeweg P., *J. Theor. Biol.* **184** (1997) 229-235.

Sawada, *J. Theor. Biol.* **164** (1993) 477-506.

Srolovitz D.J., Anderson M.P., Sahni P.S. and Grest G.S., *Acta Metall.* **32** (1984) 793-802.

Steinberg M.S., *Science* **141** (1963) 401-408. See also references in Graner and Sawada (1993).

Takeuchi I., Kakutani T. and Tasaka M., *Dev. Gen.* **9** (1988) 607-614.

Tang Y. and Othmer H.G., *Phil. Trans. R. Soc. Lond. B* **349** (1995) 179-195.

Traynor D, Kessin R.H. and Williams J.G., *Proc. Natl. Acad. Sci. USA* **89** (1992) 8303-8307.

Turing A.M., *Philos. Trans. Roy. Soc. Lond. B* **237** (1952) 37-72.

LECTURE 14

Networks of Droplets Induced by Coalescence: Application to Cell Sorting

D.A. Beysens, G. Forgacs[1] and J.A. Glazier[2]

Département de Recherche Fondamentale sur la Matière Condensée, CEA Grenoble, 17 rue des Martyrs, 38054 Grenoble, France
[1]*Department of Physics and Biology, Clarkson University, Potsdam, NY 13699-5820, USA*
[2]*Department of Physics, University of Notre-Dame, Notre-Dame, IN 46556, USA*

1. INTRODUCTION

A common goal in physical, chemical and life sciences is to understand the connection between the growth and the morphology of developing patterns. A general process is fusion or coalescence of interacting domains, especially in mixtures of immiscible liquids or tissues, where the coalescence of two drops or clusters is the basic phenomenon which governs the morphology and the kinetics of developing patterns (Steinberg, 1963).

In liquids, coalescence is a hydrodynamic process driven by the interfacial tension σ between phases A and B and damped by the viscosity η of the more viscous phase (Siggia, 1979; Onuki, 1994). By experimenting with fluids under reduced gravity, some very general results have been recently obtained for fluid droplets in interaction by coalescence (Nikolayev *et al.*, 1996). Because the volumes of the two phases remain constant in the process, the volume fraction ϕ of the minority phase is a measure of the average interdroplet distance, that is the range of droplet interactions. It was found that, when the volume fraction of the minority phase exceeds a "critical" value ϕ_H ($\approx 30\%$ for fluids with near equal viscosity), a cascade of collisions are induced by attractive hydrodynamic interactions (Nikolayev *et al.*, 1996). These interactions are caused by the

Fig. 1. — Typical pattern of interconnected droplets during the phase transition of a binary liquid of cyclohexane and methanol. This mixture exhibits a miscibility gap; at high temperature, it is completely miscible, at low temperature it shows up as two phases. Gravity effects are suppressed.

coalescence process itself and create a real dynamical network of droplets (Fig. 1). The evolution of such patterns can be described in terms of a characteristic length, the pseudo-period L_m between the phases. In this case, the growth follows the linear behavior

$$L_m = b(\sigma/\eta)t, \tag{1}$$

where b is a universal coefficient, expected of order of 0.24 (San Miguel *et al.*, 1985). Measured values are of order of 0.07 for binary liquids and 0.03 for simple fluids, with a large uncertainty difficult to estimate (Guenoun, 1987; Garrabos *et al.*, 1992).

In this paper, we show evidences that mixtures of tissues can behave exactly like liquid mixtures, qualitatively and quantitatively, with a dynamic network of coalescences driving the evolution. With the assumption that the energy per binding sites between cells plays the role of thermal energy in fluids, it becomes possible to accounts for the pattern morphology and its evolution and, similarly to liquids, to give several relations to quantitatively connect the binding energy to surface tension, viscosity and diffusion constant.

2. EXPERIMENTAL OBSERVATIONS

The experiments have already been described (Mombach and Glazier, 1996) and involve two embryonic chick tissues, pigmented retina (tissue #1) and neural retina (tissue #2). We simply recall the main steps. Dissected tissue

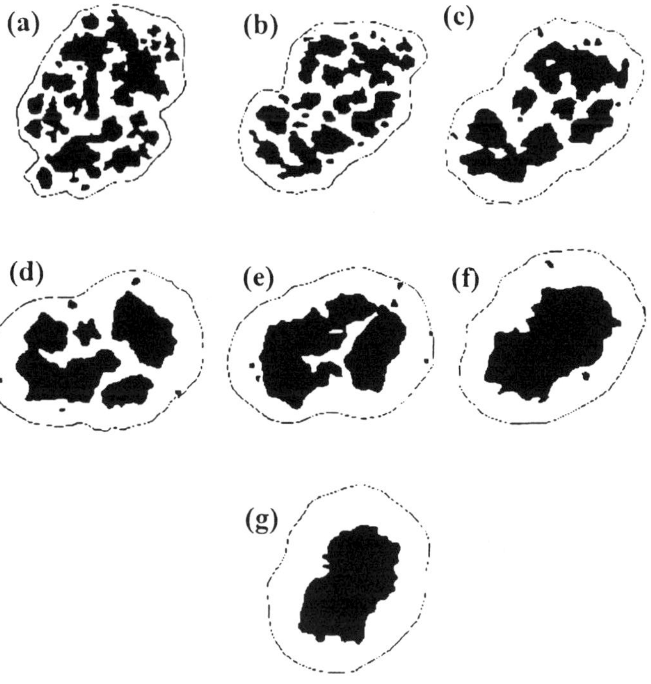

Fig. 2. — Binarized pictures showing the evolution of embryonic chick pigmented (black) and neural (white) retinal cells. (a): 17 h after the aggregate formation, (b): 22 h, (c): 30 h, (d): 42 h, (e): 51 h, (f): 64 h, (g): 73 h.

fragments are first dissociated into single cells. Then, cells from each tissue are diluted in two stock solutions. The solutions of tissues #1 and #2 are mixed in a shaker, which produces aggregates. These aggregates are then placed in optical cells for observation under a microscope. There is only one aggregate per cell. To prevent sedimentation and sticking on the cell walls, the cells are gently stirred. The volume fraction of tissues is determined optically, from the surface fraction, as discussed below. We show in Figures 2 a complete set of pictures showing the evolution of a particular aggregate.

3. IMAGE ANALYSIS

The digitized images (256 grey levels) are analyzed in order to determine the evolution of (i) the volume fraction ϕ and (ii) the typical pattern wavelength L_m.

From the images we first determine the surface fraction ε^2 of the minority phase. For this purpose, domains have to be defined. We adopt the following procedure. First, we define a grey level treshold about which the domains are white and below which they are black. We have to smooth out the images in order to remove the domains of the order of a single cell. In this treatment, the tissue border disappears. It can be redefined by adding it to the former image by the same operation with a different threshold. The value of the threshold which defines black and white domains is somewhat arbitrary; the precise value of the surface fraction ε^2 will thus depend on this threshold (see the discussion below).

The periodicity of the grey levels on the digitized images gives directly the pattern wavelength. The periodicity can also be deduced from the Fourier transform of the images. Because of the small number of domains in the aggregate, we adopt the following procedures. We determine directly the domain period L_m on a line and then average along lines in several directions. As a check of this procedure, a 1-D Fourier transform was performed on these lines, which gives a typical spatial frequency which, in turn, can be transformed into L_m. A 2D Fourier transform was also performed in a square inside the domains. All the procedures agree with each other. The data are displayed in Figure 3. On this figure, the evolution of L_m is clearly linear in time.

3.1. Volume fraction

The surface area ε^2 can be estimated from pattern analysis. However, ε^2 is dependent on the criteria used to define the domains, and especially on the grey level threshold. For instance, the surface area of domains in pattern (a) of Figure 2 can be found to be either of the following values $\varepsilon^2 = 0.492$, 0.367 or 0.346. This implies a variation of more than 30% depending on the choices of the grey threshold. We report in Figure 4 the evolution of ε^2 when using the same grey treshold for all patterns. As expected, ε^2 is constant with time around the value $\varepsilon^2 = 0.42$.

There is another difficulty, associated with the relation between the surface area ε^2 and the volume fraction ϕ. This connection is not simple for interconnected clusters as those found in the pictures (a) to (e) in Figure 2. It is not simpler even in the case of the two bulk phases in the last two pictures (f) & (g) in Figure 2, where assumptions have to be made about the symmetry of the phases. For two concentric spheres, we obtain

$$\phi = \varepsilon^{3/2}. \tag{2}$$

When one applies this relation to the average value $\varepsilon^2 = 0.42$, one finds $\phi \approx 0.27$.

3.2. Pattern evolution

Except for the last two images in Figure 2, where the domains have coarsened into a large, single droplet, the pattern of domains remains interconnected during its evolution. This is in agreement with the theory described in the

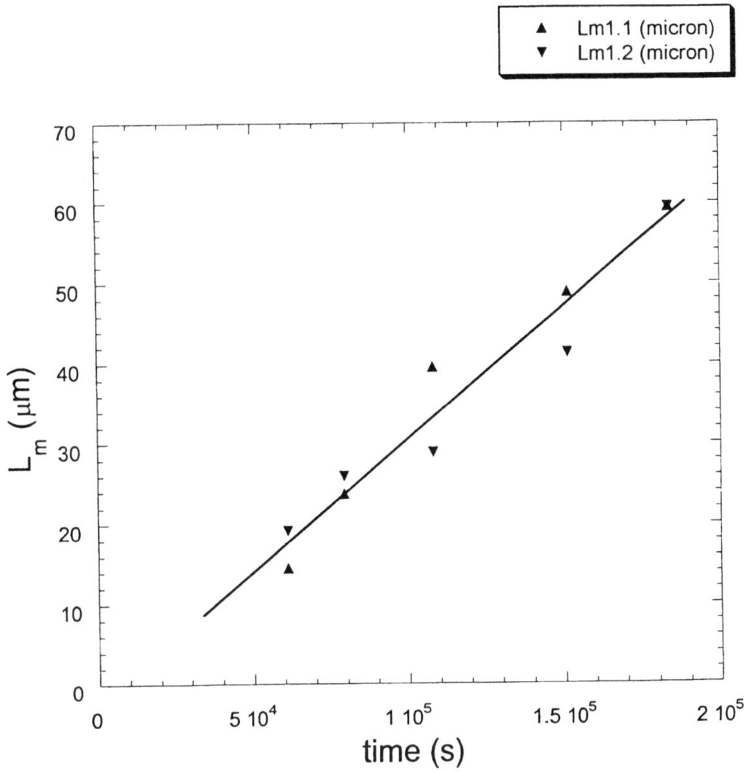

Fig. 3. — Evolution of the typical pattern wavelength $L_m(\mu m)$. t is time (s). (▲: directly from the pictures; ▼: from Fourier transform).

introduction concerning the evolution by coalescence of a drop pattern at large volume fraction. The linear growth of L_m with time (see Fig. 3) is in accordance with the behavior found in fluids. The data can be fitted to

$$L_m = A + Bt \tag{3}$$

with $A = -2.4 \pm 3$ μm and $B = (3.3 \pm 0.3)10^{-4}$ μm s^{-1}.

The process of cell sorting can qualitatively be understood as the consequence of the coalescence of cell clusters. It can quantitatively be described in terms of tissue viscosity and interfacial tension, by assuming that these quantities are the manifestations of cell adhesion (Foty *et al.*, 1994; Forgacs *et al.*, 1997). In this process, the flow of cells is accompanied by the continuous rupture and formation of adhesive bonds.

The realization that sorting out seems to exhibit the same universal behavior

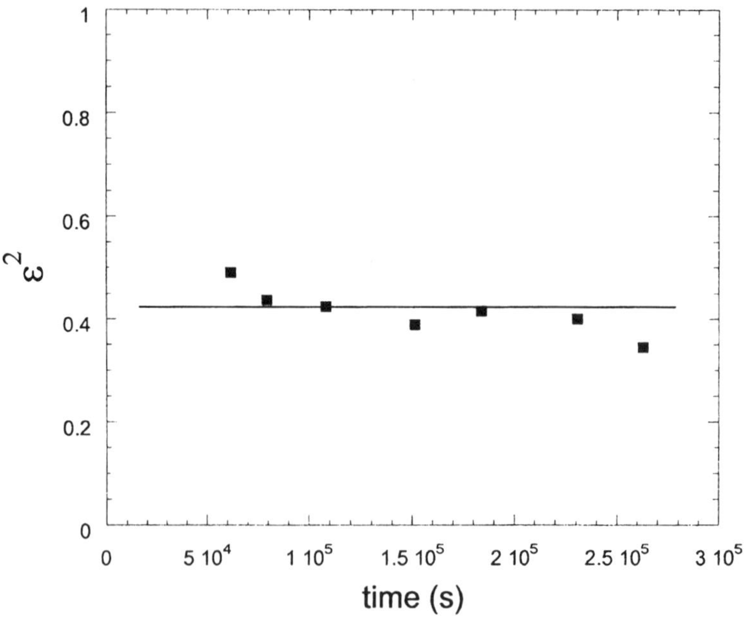

Fig. 4. — Evolution of the surface fraction ϵ^2 of the domains. t is time (s).

as the coalescence of liquid drops enables us to quantify this process in terms of effective surface tension and viscosity. The interfacial tension of chick pigmented tissue (#1) and neural retinal tissue (#2), σ, can be evaluated from the Foty *et al.* measurements (Foty *et al.*, 1994). These authors measured the interfacial tension between these tissues and the surrounding tissue culture medium. One obtains $\sigma \simeq 10$ dyne cm^{-1}. From their measurements of the viscoelastic properties of these tissues, Forgacs *et al.* (1997) deduced a value for the effective viscosities, $\eta \simeq 1\ 10^6$ Po (see also Gordon *et al.*, 1972, who found for heart tissues $\eta \simeq 1\ 10^7$). These values, together with the determination of the parameter B in equation (3), enables the universal parameter b in equation (1) to be estimated: $\mathrm{b} = \mathrm{B}\eta/\sigma \simeq 0.003$. Although the viscosity and surface tension of tissues are 10^9 times larger than that of fluids and liquid mixtures, the value for b remains of the same order of magnitude. Unfortunately, any further comparison with the value found with fluids is hampered either by the large uncertainties in the measurement of η and σ or the lack of a precise value for fluids.

3.3. Surface tension and binding energy

In liquids, thermal agitation is at the origin of (i) interface capillary fluctuations, damped by surface tension, and (ii) random (Brownian) motion of droplets. In the capillary wave theory (see *e.g.* Beysens and Robert, 1987), the interfacial thickness is given by the mean displacement of a microscopically sharp interface which undergoes thermal wave-like vibrations restored by gravity and surface tension. The density profile perpendicular to the interface can be understood as the time or ensemble average of these fluctuations. With z the distance perpendicular to the interface, the profile is described by an error function with a characteristic length L:

$$f(z) = \mathrm{erf}\left(\frac{\sqrt{\pi}z}{L}\right). \tag{4}$$

In the linearized regime of small interfacial distortions one finds:

$$L^2 = < z^2 > = \frac{1}{\pi}\int_{q_m}^{q_M} \frac{kT}{\sigma q^2 + g\Delta\rho} dq_x dq_y \tag{5}$$

where x and y are the cartesian coordinates in the interfacial plane, k is Boltzmann's constant, $\Delta\rho$ is the difference $\rho_1 - \rho_2$ between the bulk phase densities, g is the earth gravity and q is the wave vector of a capillary wave. Here q_M is the maximum wave vector corresponding to the smallest wavelength of allowable capillary waves, *i.e.* the interfacial thickness L itself, and q_m is the minimum wave number, determined by the linear dimension of the sample.

The integral in equation (5) is readily evaluated for experiments under zero gravity, with the results:

$$< z^2 > = \frac{kT}{\sigma} \ln\left(\frac{q_M}{q_m}\right). \tag{6}$$

This analysis can be applied to a cluster of tissue 1 in tissue 2, where the interfacial profile is the result of time averaging over the fluctuations in adhesion of the external cells. Such fluctuations, which exhibit a near-thermal spectrum (Mombach and Glazier, 1996) can be observed as a ruffling of the cell membrane, with amplitude of order 20% of the cell dimension. We thus write

$$< z^2 > = 4h^2a^2 \tag{7}$$

with 2a the length of a cell and $h \approx 0.2$ the proportionality constant. The fluctuations are driven by the rupture or creation of a cell adhesion site, corresponding to a total binding energy E for the cell in analogy to the thermal energy in fluids. With 2R the linear dimension of the cell aggregate, equation (7) can be written as:

$$< z^2 > = \frac{E}{\sigma} \ln\left(\frac{R}{a}\right), \tag{8}$$

where one has made $q_m = \pi/R$ and $q_M = \pi/a$, with 2a being the effective diameter of pigmented retinal cell. For the experiment analyzed above, 2a = 8 μm and the cluster diameter is 2R $\approx$ 200 μm. One readily obtains $E \approx 2$

10^{-8} erg. Since the number of binding sites is of order N = 10^4, the energy e per binding site is obtained readily, e = E/N $\approx$ 5 Kcal mol^{-1}. This value is consistent with recent estimates (5.6 Kcal mol^{-1}) of the strength of a cadherin bond (Koch *et al.*, 1997) of which there are many on the surface of both pigmented and neural retinal cells. It is also worth noting that by varying the number of cadherin bonds on the surface of genetically engineered cells, Steinberg and collaborators established that the surface tension of the investigated tissues varied linearly with N (Steinberg, private communication).

The adhesion energy per site seems therefore to play the same role as the thermal energy per degree of freedom in a fluid. In favor of this interpretation is the random motion of a cell of tissue 1 immersed in a tissue 2 as studied by Glazier *et al.* (Mombach and Glazier, 1996). The cell is seen to obey a random walk which allows the diffusion coefficient D ($\approx$ 5.5 10^{-12} cm^2 s^{-1}) to be determined. If we write D as a Stockes-Einstein diffusion coefficient for Brownian motion,

$$D = \frac{kT}{6\pi\eta a} \tag{9}$$

with E playing the role of kT, one arrives at

$$E = 6\pi\eta a D = 4\ 10^{-8}\ \text{erg}, \tag{10}$$

that is, a value close to that determined above (2 10^{-8} erg). This value is clearly very close to that derived above using an entirely different approach.

4. CONCLUSIONS

The above experiments and analysis show that the sorting out of certain embryonic cell populations is governed by a dynamic network of coalescence which exhibits the same morphology and obey the same general growth laws as phase ordering in fluids. This is a direct evidence for the effect of effective interfacial tension between tissues and effective tissue viscosity. The origin of surface tension has to be found in cell adhesion and the presented data (as well as the results of other investigators) show that embryonic tissues in many respect behave as liquids, with the adhesion energy playing the role of thermal energy.

There is an obvious need for more precise values of surface tension and viscosity. It remains also to investigate the particularity of the evolution process in such a viscoelastic medium. It would be interesting to determine whether the evolution is driven by the same hydrodynamic correlation process as in fluids or to a simple geometric constraint. The effect of (membrane) fluctuations in tissues is also a question which desserves further investigation.

ACKNOWLEDGMENTS

We would like to thank C. Chabot and Y. Garrabos from *Institut de la Matière Condensée de Bordeaux (France)* for their help in analysing the images. This study has been partly granted by CNES.

REFERENCES

Beysens D. and Robert M., *J. Chem. Phys.* **87** (1987) 3056; *ibid. J. Chem. Phys.* **93** (1990) 6911 (*erratum*).
Forgacs G., Foty R.A., Shafrir Y. and Steinberg M.S., *Biophys. J.* (1997) to appear.
Foty R.A., Forgacs G., Pfleger C.M. and Steinberg M.S., *Phys. Rev. Lett.* **14** (1994) 2298.
Garrabos Y., Le Neindre B., Guenoun P., Khalil B. and Beysens D., *Europhys. Lett.* **19** (1992) 49.
Gordon R., Goel N.S., Steinberg M.S. and Wiseman L.L., *J. Theor. Biol.* **37** (1972) 43.
Guenoun P., Gastaud R., Perrot F. and Beysens D., *Phys. Rev. A* **36** (1987) 4876.
Koch A.W., Pokutta S., Lustig A. and Engel J., *Biochemistry* **36** (1997) 7697.
Mombach J.C.M. and Glazier J.A., *Phys. Rev. Lett.* **76** (1996) 3032.
Nikolayev V.S., Beysens D. and Guenoun P., *Phys. Rev. Lett.* **76** (1996) 3144.
Onuki A., *Europhys. Lett.* **28** (1994) 175.
San Miguel M., Grant M. and Gunton J.D., *Phys. Rev. A* **31** (1985) 1001.
Siggia E.D., *Phys. Rev. A* **20** (1979) 595.
Steinberg M.S., *Science* **141** (1963) 401.

LECTURE 15

A Monte-Carlo Approach to Growing Solid Non-Vascular Tumors

D. Drasdo

Max-Planck-Inst. f. Kolloid-und Grenzflächenforschung, Kantstr. 55, 14513 Teltow, Berlin, Germany

1. INTRODUCTION

Cancer is still one of the most frequent causes of death in industrial countries despite the rapid progress in medicine. The formation of cancer is well known to present a multi-step process. It is believed that each step is accompanied by mutations. The cascade starts from an initial genetic mutation, probably in only a single cell followed by a hyperplasia where the cell division rate is increased and the structure of the tissue is still approximately conserved. In a third step the increased cell division rate is accompanied by a perturbation of the cell arrangement in the tissue and an abnormal shape of the cells (dysplasia). After a further mutation, increasing again the cell division rate, a solid in-situ tumor with pushing edge (preinvasive stage) may appear. If additional genetic changes allow for invasion into the surrounding tissue a malignant tumor is formed (*e.g.* Weinberg, 1996; Alison and Sarraf, 1997). To overcome a critical size of about 200 – 600 μm, dependent on the type of the tumor, the tumor cells secrete angionese factors and induce the growth of new vessels from the surrounding tissue which sprout towards and than gradually penetrate the tumor and provide it with a blood supply (*e.g.* Folkman, Klagsbrun, 1987). Throughout this paper, we focus on preinvasive stages and do not distinguish between dysplasia and in-situ tumors.

Cultured spherical aggregates of malignant cells, i.e. multicellular spheroids, have become an *in vitro* test system to study tumor microregions in avascular stages (Freyer and Sutherland, 1985, 1986; Müller-Klieser, 1987; Sutherland,

1988). Among many other properties their morphology and growth kinetics as, *e.g.* the increase of the cell number N and of the tumor diameter d_{cl}, have been intensively studied as functions of time t. As multicellular spheroids become sufficiently large, they form a necrotic core, surrounded by a layer of quiescent cells followed by a layer of proliferating cells. While the necrotic cells have suffered apoptosis, the quiescent cells can reenter the proliferation cycle if the outer ring of proliferating cells is removed. The number of cells $N(t)$ reveals initially exponential growth later crossing over into some sub-exponential growth. The data found for $N(t)$ in Freyer and Sutherland (1985) are in good agreement with an effective power law $N \sim t^{3.6 \pm 0.15}$, although a careful determination of a power law would require three decades of magnitude not only in N, but also in t. However, the power law fit seems not worse than the fit by the Gompertz distribution (Gompertz, 1825) that is usually and also here assumed to describe the tumor growth. The Gompertz distribution is based on a heuristic approach and assumes exponential growth of cells combined with exponential decay of the growth rate. The authors argue that the diameter $d_{cl}(t)$ of the tumor in the sub-exponential growth regime in (Freyer and Sutherland, 1985) obeys a linear law, but, assuming that its behaviour is related to that of $N(t)$, it can be also be approximated by a power law $d_{cl} \sim t^{1.2}$.

Power laws can also describe the increase of cell number in culture experiments (*e.g.* *cf.* Westermark, 1973, with Drasdo, 1996), at least in a intermediate regime. Recently, a power law growth-behavior has been found in breast cancer (Hart *et al.*, 1997). The authors of this article clearly point out that they find a growth law that can neither be fitted to a Gompertz distribution nor to an exponential but only to a power law.

The occurrence of a power law in physics is often a symptom of scaling, *i.e.*, certain features of the system behave in a characteristic way (*e.g.* Cardy, 1996). Even systems that may look very different at first sight can have the same scaling exponents independent of microscopic details, if they share certain characteristic features, a phenomenon called universality. From the above discussion one may hope that important growth features of tumors, at least in some stages of tumor growth, are governed by a common underlying mechanism that may be understood from a simple model in which many microscopic details are neglected.

The enormous increase of computing power during recent years facilitates a study of possible growth mechanisms relevant to tumors by mathematical modeling. Accordingly many attempts have been made to understand growth properties of tumors by mathematical models. Some representative examples will be briefly mentionend here.

Distinguished by the phenomenon they describe, important examples may be multi-stage and pharmacokinetical models of carcinogenesis (Murdoch *et al.*, 1987), models which focus on the growth laws of tumors (*e.g.* Gompertz, 1825; Rashevsky, 1945; Eden, 1961; Williams and Bjerkness, 1972; Richardson, 1973) and more recently models which take into account different stages and aspects of tumor growth. The latter range from tumor spheroids (see below) to models which account for angiogenesis (Byrne and Chaplain, 1995a-c; Chaplain, 1995, 1996), immune response (*e.g.* Kuznetsov, 1991; Owen and Sherrat, 1997) or to the behavior of tumors under radiation or chemotherapy (*e.g.* Coldman and Goldie, 1987; Thames, 1987; Düchting *et al.*, 1996).

Tumor spheroids have been modeled as an elastic solid continuum (Chaplain and Sleeman, 1992) and as incompressible fluid-like droplets with a surface tension balancing the internal pressure thus insuring the integrity of the tumor (Greenspan, 1976; Byrne and Chaplain, 1996a; Chaplain, 1996). Both predict a linear growth of the tumor diameter d_{cl} for small times saturating at large times. A further approach which takes into account a density limitation of growth is the logistic equation for $N(t)$ (Verhulst, 1838; Rashevsky, 1945). The solution of this equation reveals exponential growth for small times and saturation as $t \to \infty$. A simple extension of the logistic equation is to include spatial diffusion of cells (Fisher - Kolmogoroff - equation, *e.g.* Fisher, 1937; Kolmogoroff *et al.*, 1937; Lefever and Erneaux, 1984). In this case, $N(t) \sim \exp(\lambda t)$ (λ: growth rate) for small times and $d_{cl} \sim t$ for large times.

Among all these models one may roughly distinguish between continuum models, which average over mesoscopic length scales much larger than the typical diameter of the cells (or even over the whole space neglecting any influence of space) and microscopic models, which consider each cell as a single unit. Many phenomena on small length and time scales, such as extra-and intracellular transport of nutrients, oxygen, ions, cell metabolism *etc.*, may appear as noise and cause fluctuations in the quantities explicitly appearing in the model. Microscopic and continuum models could further be subdivided into stochastic models (that include noise) and deterministic models (without noise).

Most of the space-dependent models of tumor growth cited above are deterministic continuum models. Unfortunately, local properties of cells on small length scales, as the detachment of a single cell from a primary tumor that may precede metastasis formation or (de-)differentiation and apoptosis (if only a small number of cells at special positions in the population is considered), cannot be covered by a continuum model. Further, deterministic approaches often hide characteristic features of stochastic systems that correspond to higher statistical moments and can also be observed experimentally, *e.g.* fluctuations of the surface width of a growing surface (*e.g.* Halpin-Healy and Zhang, 1995), spatial fluctuations in the mitotic index, the subclone sizes, *etc.* In the vicinity of critical points, deterministic treatments completely fail to give a correct description of the dynamic or thermodynamic behavior of a system if the magnitude of the fluctuations overcomes that of the averages, as for example in cell populations, where the birth and death rates become identical (Mikhailov, 1989). Thus if death processes as apoptosis or necrosis are to be studied during normal immune response, medical treatment of cells with drugs or radiation, in most cases a stochastic model is more appropriate. Unfortunately, continuum models with noise are often very hard to deal with, and reliable understanding of the models often requires additional treatment by computer simulations.

A direct approach to cell population growth considering the cell as a single unit, allows for the measurement of generation numbers, mitotic indices, marking subclones etc. without changing the level of description and thus provides data which can at least in part be directly compared with experiments. Most of the microscopic models used to study cancer growth so far are lattice models (*e.g.* Eden, 1961; Williams and Bjerkness, 1972; Richardson, 1973; Landini and Rippin, 1993). A general disadvantage of lattice models is, that on large scales, the observed morphologies are often determined by the lattice symmetry (*e.g.* Batchelor and Henry, 1991).

In particular, the mentioned models do not cover the growth regimes observed in experiments. In the Eden model, which is the prototype of the first three above mentioned models, cell division only can take place, if at least one neighboring site on the lattice is not occupied by a cell (after a cell has divided, its daughter cells will occupy twice the orginal space in the next time step). This assumption leads to a nearly immediate crossover to a pure surface growth (at about $O(10)$ cells in $d = 2, 3$ dimensions) and no macroscopic exponential growth regime can be found.

In the lattice model described in Landini and Rippin (1993), which describes the other extremal case, also cells in the interior of a cluster are allowed to divide infinitely often by shifting a whole row of cells by one lattice constant into a randomly chosen direction. Because cell division is always allowed in this model the population size grows exponentially for all times. Since the activation energy necessary to break all cell-cell connections transversal to the direction of row movement scales as the number of cells of the row, the energy needed diverges as $N \to \infty$. Thus this approach can only be valid for small cell populations.

In contrast to these models the microscopic approach employed here is a lattice-free model and allows for quasi-continuous migration, rotation and growth. This is shown to have important consequences on the dynamic properties of the growing populations at least in an intermediate growth regime, that however can be quite long and probably dominates in experimental observations.

2. THE MODEL

One may regard a tumor as a disordered colloid-like solid with attractive interaction between its units (*i.e.* cells) which resists static shear. It is reasonable to assume that the solid structure is mainly a consequence of the cell-cell attraction due to cell adhesion molecules (CAMs). These are well known to play an essential role in cell-cell recognition processes as known *e.g.* from immune responses (Edelman, 1987), from fatemaps in early development of vertebrates (*e.g.* Edelman, 1988) and cell-cell ordering phenomena (*e.g.* Townes and Holfreter, 1955; Steinberg, 1970; Graner and Glazier, 1992). Often, a reduced number of cell-cell adhesion molecules have been found at the surface of tumor cells compared to those on normal cell surfaces resulting in a reduced cell-cell

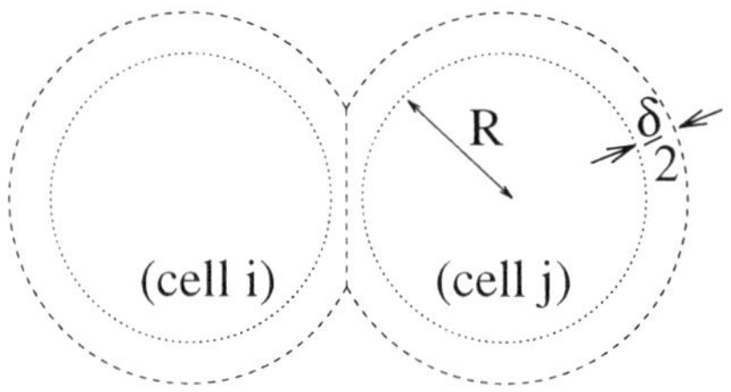

Fig. 1. — Two cells in interaction in two dimensions: the small circles with radius R describes the radius of maximum compressibility (*cf.* text), the large circles of radius $R+\delta/2$ the radius above which a cell will not feel any attraction from another cell. The position of the membrane is between these extrema and not specified in the model. Thus cell shape is described in a statistical sense: To each cell a region in space is attached where most of the cell volume is located with overwhelming probability.

adhesion. The closer two cells approach each other in space, the more their CAMs are able to interact and the stronger the attraction between these cells will be. Competing with this attracting mechanism, the limited cell compressibility and the loss of steric entropy due to a reduction of the possible membrane conformations (Helferich, 1978) cause repulsive effects as cells approach each other which become very large at sufficiently small distances. We denote the minimal diameter a cell can have under pressure, *e.g.* between two plates without destroying the cell by $2R$ (Foty and Steinberg, 1995) and assume that R is mainly determined by the cytoskeleton. It is clear, that in general, R may depend on the orientation of the cell between the two plates, *i.e.* $R = R(\phi, \theta)$ (ϕ: polar, θ: azimuthal angle). $R = R(\phi, \theta)$ then describes a surface of maximum compressibility (which encloses some "hard core" of the cell) that we refer to as the inner surface in the following.

The above described interactions are absorbed into an effective short-range-interaction potential V_{ij}: Let d_{ij} be the shortest distance between the inner surfaces of cell i and cell j. For distances $d_{ij} < 0$, $V_{ij} \to \infty$ because $d_{ij} = 0$ corresponds to a maximum compression of the cells. For distances d_{ij} larger than some interaction range δ the potential $V_{ij} = 0$ indicating no interaction. In an intermediate regime $d_0 < d_{ij} \leq \delta$, neighboring cells attract each other $V_{ij} \leq 0$ while for small distances $0 \leq d_{ij} \leq d_0$ (indicating a possible deformation of the cell), the repulsive part dominates the attractive part $V_{ij} \geq 0$ (Fig. 1). In the simulation outlined below, we choose V_{ij} as a square well potential, that is, $V_{ij} \to \infty$ for $d_{ij} < d_0 (\equiv 0)$, $V_{ij} = \epsilon \leq 0$ for $d_0 < d_{ij} \leq \delta$ and $V_{ij} = 0$ if $d_{ij} > \delta$ (for other choices see Drasdo, 1994; Drasdo, *et al.*, 1995). For the simulations we focus on cells with a spherical inner surface in a quiescent phase which changes as outlined in (Fig. 2) in the proliferation phase. It is clear that this excludes strongly concave cells and cells with long offshoots. However,

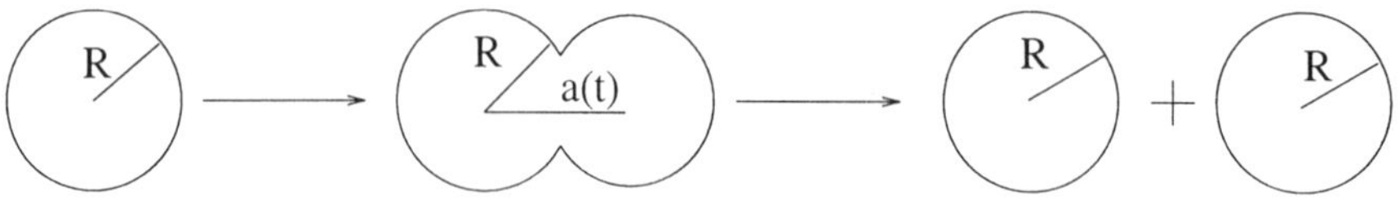

Fig. 2. — Development of the inner cell surface during a proliferation cycle: during proliferation, the cell deforms from spherical shape of radius R to a dumbbell by elongation of the axis from $a = 0 \to a = 2R$ in small steps of size ξ_g where $\xi_g \in [0, \delta a)$ is an uniformly distributed random number, and $\delta a = R/12$ is the maximal growth step size. The average number of growth steps to pass once through the cell cycle for a free cell is $\approx 2R/(\delta a/2)$ with corresponds to the biological cell cycle time τ.

about 80% of tumors stem from epithelial tissue (*e.g.* Alison and Sarraf, 1997), which consists of cells with a strong cytoskeleton and often appears under the microscope as approximately polygonal in shape and thus may well be approximated by a convex shape.

Cell division within a tumor changes the density locally and exerts a force on the neighboring cells. These move (migrate or rotate) to escape the perturbation (Principle of Le Chatelier, *e.g.* Landau and Lifschitz, 1987). It is reasonable to assume that for cell migration, inertia terms are small compared to dissipative terms and can be neglected. Thus the dynamics can be modeled by a Monte-Carlo-approach. Following the Metropolis algorithm (Metropolis *et al.*, 1953), a migration, rotation or division step of a cell i is accepted, if the step decreases the free energy $V_i = \sum_{j \neq i} V_{ij}$ or rejected with probability $P_i = [1 - \exp\{-(V_i^{after\ step} - V_i^{before\ step})\}]$ if the free energy is increased by the step. Thus cells move in the direction of the free energy gradient following a fuzzy decision process. Cell growth is assumed to be an active process which can only take place, if a growth step doesn't lead to overlaps with neighboring cells (for details, see Fig. 2).

A timescale can be introduced by relating the number of Monte-Carlo action trials, *i.e.* trials for migration, rotation, growth or division, to the timescale that is known from experiments. An interaction free cell has a diffusion length of $\xi = \sqrt{4\, D_0\, \tau}$ between two cell divisions, where D_0 is the diffusion constant of a cell in a surrounding medium, τ is the average cell cycle time. Both parameters can be adjusted to experimental data. In the model, D_0 is determined by the step sizes of migration. The diffusion law relates the timescale of cell division to a length scale ξ. In the following we will use $r_g \sim (D_0 \tau)^{-1/2} \sim 1/\xi$ as measure for the *intrinsic* growth rate r_g. In the simulations r_g has been varied over a wide range such, that $0.6R \leq \xi \leq 20R$. This choice has been found to cover the whole range of growth phenomena (Drasdo *et al.*, 1995). Throughout this paper we further assume for the interaction range $\delta \approx 0.1\, R$ (the qualitative behavior doesn't change as long as approximately $\delta \leq R$ (Drasdo *et al.*, 1995; Drasdo, 1996)) and the interation strength ϵ is either zero or $-\infty$ corresponding to the limiting cases of no attraction or very strong attraction (in the latter case, a cell will never perform a move that decreases its number of interaction partners).

We focus here on growth properties of a population of single cells where each cell is in principle capable of undergoing infinitely many cell divisions, that is, assuming no shortage of nutrients or oxygen. Although this seems to be a very crude assumption, we will see, that it leads nevertheless to a reasonable dynamic behavior which will be compared to that found in experimental systems described above. It is believed that some of the properties found in the model are universal, that is, they do not depend on microscopic details of the model but rather relate to a whole class of models.

After a study of the growth behavior of the model, we will introduce a simple death process inspired by a chemotherapeutic treatment with antimitotic drugs. Here again, the advantages of the quasi-continuum single cell approach will become obvious.

3. RESULTS

If not explicitly mentioned, the results reported below have been derived from simulations in $d = 2$ dimensions.

It is generally assumed, that a tumor is a monoclone from one precursor cell. Thus in the simulations, we start with a single cell and stop the simulation at population sizes between $N \sim O(10^3) - O(10^5)$ cells (depending on r_g).

In order to study the growth characteristics of the tumor cell populations, the number of cells $N(t)$ and the radius of gyration R_{gyr} are measured. R_{gyr} allows us to determine the (fractal) dimension d_N of a cluster. It is defined by $R_{gyr}(N) = \sqrt{1/N \sum_{i=1}^{N} (\underline{r}_i - \underline{r}_{cm})^2} \sim N^{1/d_N}$ where $\underline{r}_i$ denotes the position vector of cell i and $\underline{r}_{cm} = 1/N * \sum_{i=1}^{N} \underline{r}_i$ denotes the center of mass. $R_{gyr}(N)$ is proportional to the cluster diameter d_{cl} with the advantage of small statistical errors.

3.1. The macroscopic growth laws

In the **initial time regime**, $N \sim \exp(\lambda t)$ independent of the choice of parameters (Fig. 3) while the behaviour of the radius of gyration R_{gyr} depends on the strength of the cell-cell attraction: for small times and zero-attraction ($\epsilon = 0$), $R_{gyr} \sim t^{\nu}$ with $\nu \approx 0.75$. This may correspond to a regime in which the growth of the diameter is determined basically by diffusion of the (hard core) cells.

For strong attraction $\epsilon = -\infty$, we find $R_{gyr} \sim N^{1/d_N}$ with a non-fractal dimension $d_N \approx 2$ thus $R_{gyr} \sim \exp(\lambda t/2)$. Here due to the cell-cell attraction, no cell is able to escape the growing cluster. Diffusion can act only on very small scales $\ll \delta$ and relax the cluster.

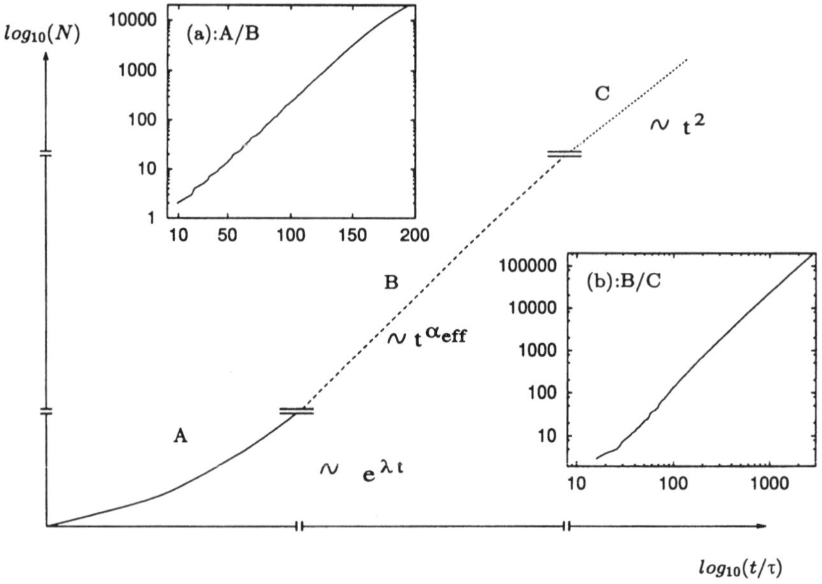

Fig. 3. — Schematic picture of the growth law $N(t)$: for small t, $N \sim \exp(\lambda t)$ (A) crossing over to an effective power law $N \sim t^{\alpha_{eff}}$ (B) ($\alpha_{eff} \geq 2$ at $t = t_{cr}$ (see text)). For $t_{cr} \ll t \to \infty$, $\alpha_{eff} \to 2$ (C). The insets show simulation data for N(t) for $\epsilon = 0$ and (a) small r_g (b) large r_g ((a) is a semi-, (b) a double-logarithmic plot, the time is measured in units of τ). (a) for small r_g, $t_{cr} \approx 140$ and the initial exponential behavior can be observed over a wide range for N. (b) to proceed far into the asymptotic regime, r_g has to be chosen sufficiently large. Here, $\alpha_{eff} \approx 2.6$ at $t = t_{cr} \approx 20$ and $\alpha_{eff} \approx 2.04$ at $t = 2000$, $N(t) = 200\,000$. If the local cell density is large as in (b), the real growth rate can be much smaller than the intrinsic growth rate r_g.

The initial regime is followed by a large **transient regime**. This regime can be characterized by three exponents defined by $N \sim t^{\alpha_{eff}}$, $N \sim (R_{gyr})^{d_N}$ and $R_{gyr} \sim t^{\zeta_{eff}}$, where due to the scaling relation $\zeta_{eff} = \alpha_{eff}/d_N$ only two exponents are independent. α_{eff} is determined from the simulation data using the relation $\alpha_{eff} \equiv \partial \log(N)/\partial \log(t)$. In the simulations, we find $\alpha_{eff} \geq 2$, $\partial\alpha_{eff}/\partial t < 0$, and $d_N = 2$. To characterize the crossover from one regime to another regime, one typically introduces the crossover time t_{cr}. Then, $N(t_{cr})$ determines the population size at the crossover. (Note, that t_{cr} and $N(t_{cr})$ can be observed in experiments, too.) The simulations as well as analytical arguments (Drasdo *et al.*, 1995) show, that α_{eff} and the crossover size from the initial to the transient regime $N(t_{cr})$ decrease for (i) increasing growth rate, (ii) increasing interaction strength and (iii) decreasing interaction range.

Asymptotically as $t \to \infty$ both, R_{gyr} and $N(t)$ become parameter-independent and for the exponents, $\alpha_{eff} = 2$ and $d_N = 2$ thus $\zeta_{eff} = 1$. The behavior of $N(t)$ and R_{gyr} are the consequence of the excluded volume effect: as long as the excluded volume is irrelevant, *i.e.* for small densities, all cells can divide, and when it does become important, asymptotically only cells at the surface can divide. Thus independent of whether necrosis is assumed to occur as an active (apoptotic) process, cells in the interior of the cluster will not contribute to its increase for large times.

Preliminary results from simulations (Drasdo, to be published) along the same line in three dimensions are compatible with existing experimental data (*e.g.* with those in Freyer and Sutherland, 1985)

3.2. A chemotherapy-inspired death process

In the following, we will consider a growth phase dependent death process. We will see, that it is the quasi-continuum steps (which imply moves of lengths $\ll \delta$) that are responsible for a regime, that could not be observed macroscopically in the microscopic models mentioned above and that may be observed in appropriate experiments. The modeled death process is inspired by growth-phase-dependent chemotherapeutic treatment of small micrometastasis with anti-mitotic drugs, as, *e.g.* vinblastine or vincristine, which bind to tubulin and disrupt the assembly of the mitotic spindle in the metaphase triggering cell apoptosis. The dead cells then are assumed to be phagocytosed by their neigbhors and macrophages. The latter present about 40% of the total tumor mass in some tumors.

The simulation again starts with a single cell. At a total population size of about $N_0 = 2000$ which corresponds to a two-dimensional patch of diameter $d_{cl} \approx 500$ μm (for larger N, necrosis and later, angiogenesis may become important), a death process is introduced as follows: a cell dies and is removed from the system with probability p_d directly before its division assuming, that a dead cell is phagocytosed sufficiently fast by its neighbors or macrophages.

In the following simulations, p_d and r_g have been varied, while $\epsilon \to -\infty$ and $\delta \approx 0.05\ R$ according to strong attraction between the cells. At some critical probability $p_d = 0.5 = p_c$ where death and birth rate are equal the population remains constant on the average. At this (critical) point a second order non-equilibrium phase transition occurs (Drasdo, unpublished). For $p_d < p_c$, the population size increases when observed over sufficiently long periods of time. The biologically relevant regime for chemotherapy is the one with $p_d > p_c$, where, on the average, the population size decreases. Typical morphologies for $p_d > p_c$ are shown in Figure 4.

For large growth rates, the crossover to pure surface growth takes place nearly instantaneously.

If $p_d = 1$, each dividing cell is killed with probability one, thus the population is melted from its boundaries and remains a single connected cluster until it is depleted (Figs. 4b and 5a). It can be shown, that the cluster has a dimension

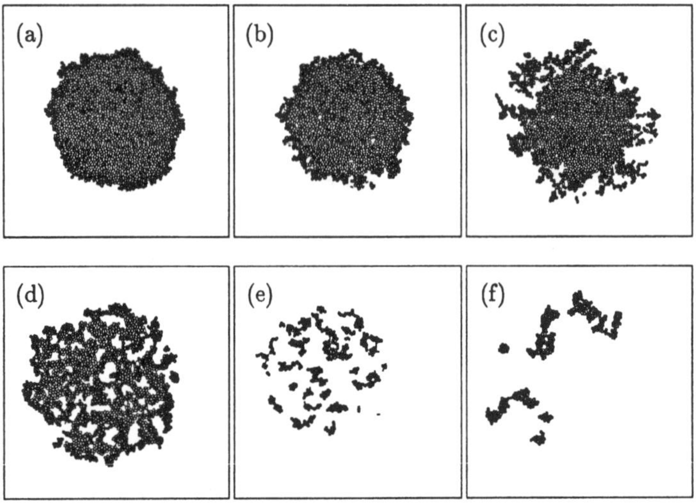

Fig. 4. — (a) Shows the starting configuration with $N_0 = 2000$, the parts (b-f) typical morphological configurations at different N. (b) Large r_g, $N \approx 1700$, $p_d = 1$: in the surface growth regime, every cell is killed and removed before it divides, thus the cluster is pealed like an onion. (c) Large r_g, $N \approx 1700$, $p_d = 0.51$: for death rates near the critical point, some cells at the surface can produce offsprings, others die. This induces large density fluctuations at the surface. At some threshold holes can percolate in the boundary layer of the cluster and destroy its connectivity. (d) Small r_g, typical configuration at $N \approx 1700$: all cells are in the proliferation phase. (e) Small r_g, $p_d = 1$, $N \approx 300$ far below the percolation threshold: the clusters fragmentate into smaller and smaller clusters. A killed and removed cell leaves a hole of size $l_0 \approx 2R$ in the cluster. The hole grows exponentially fast for $p_d > p_c$ with a characteristic growth rate $1/\tau_{rel} \sim r_g|(p_c - p_d)/p_c|$. This has the largest value at $p_d = 1$, where the configurations are nearly frozen and every hole appearing in the cluster grows quickly dividing the population into smaller and smaller clusters. (f) Small r_g, and $p_d = 0.51$: here, the relaxation time τ_{rel} is 50-times larger than in (e), thus the small clusters have sufficient time to merge into larger clusters and eroded clusters with many fingers can adopt a more compact shape. The process is driven by the minimization of free energy. For the applicability of the model it is not necessary that the empty space in the pictures are really empty, they will in fact be filled by extracellular material, macrophages *etc.*

of $d_N = 2$. The number of cells, N, obey a decay law which is characteristic for a process, in which cells only die at the surface.

If p_d is decreased towards p_c, a strong increase of fluctuations at the boundary of the cluster leading to an invasion of holes from the cluster surface into the cluster is observed with decreasing N: at sufficiently small $N \equiv N_{th}$, the holes

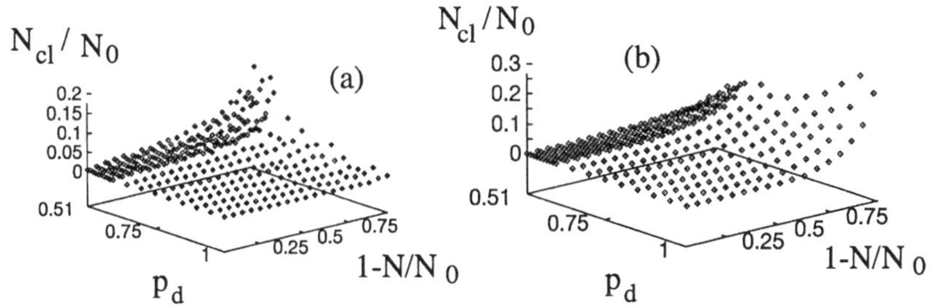

Fig. 5. — Fraction of clusters N_{cl}/N_0 *versus* p_d and $1-N/N_0$ for population (a) with very large growth rates r_g and (b) very small r_g. In (b), the original monocluster starts to fragmentate at a size of $N_{th} \approx N_0/2$ (for N_{th}, see text). Below N_{th}, N_{cl} decreases with p_d for large r_g and increases with p_d for small r_g. The fraction of clusters at fixed p_d increase with decreasing total population size N.

percolate and destroy the connectivity of the cluster (Fig. 5a). The number of disconnected clusters N_{cl} is the larger, the smaller is p_d (Fig. 5a). The size distribution of the clusters below the percolation threshold can be shown to be inhomogeneous and contain many small clusters (Drasdo, unpublished).

As we have seen in Figure 3, the model reveals a large exponential regime in $N(t)$ with large crossover time if r_g is small. In this regime all cells can divide and consequently die as the death process is switched on. This leads to an exponential decrease of the number of cells, $N = N_0 e^{\lambda_d t}$ where $\lambda_d \sim r_g(p_c - p_d)/p_c < 0$. Holes appear everywhere in the cluster and percolate below the percolation threshold $N_{th} \approx N_0/2$. In contrast to populations with large intrinsic growth rates r_g, we find here a stronger fragmentation with increasing death rate p_d (Figs. 5e,f). Further for $p_d = 1$, the cluster distribution is more inhomogeneous than for $p_d = 0.51$.

4. DISCUSSION

The model presented above may give alternative explanations for the growth behavior of cell cultures and spheroid, non-vascularized tumors in $d = 2$ (and also $d = 3$) dimensions (*cf.* Drasdo, 1996 and Drasdo, unpublished) and makes predictions on the expected qualitative behavior of a growth-phase dependent death process as it may occur during chemotherapeutic treatment by antimitotic drugs.

The model predicts that the growth of tumor spheroids should follow a universal behavior which starts with an exponential growth crossing over to an effective power law behavior $N \sim t^{\alpha_{eff}}$ with $\alpha_{eff} \geq d$, whereas for asymptotically large times α_{eff} tends to d (= 2 or 3). The diameter of the tumor should essentially behave as $R \sim N^{1/d} \sim t$. This is also compatible with the data found in Freyer and Sutherland, 1985. The asymptotic behavior does not depend on the range and the strength of interaction or on the growth rate and has also been shown to be independent of the details of the description of the cell shape or the form of the potential (as long as only short range potentials are considered, Drasdo, 1994; Drasdo *et al.*, 1995). Only the initial growth behavior for the tumor diameter has been found to depend on the strength of cell-cell interaction. It is clear that this simple approach can give an appropriate description of the growth process, only as long as nutrients and oxygen shortage are irrelevant. The predictions of the theory may be studied in experiments, *e.g.* for several cell cultured tumor spheroids (with cell types that have a short mitotic cycle and a low mobility) and eventually pulmonary tumors which are known to have a good blood supply where it may be reasonable to assume, that they can grow to a relatively large size before inducing angiogenesis (Steel, 1977). However, the influx of nutrients can be added easily to the model. (*E.g.*, preliminary simulations show a crossover to a fractal dimension if a homogeneous nutrient source is placed on a ring around a growing cluster of cells and if the activity of the cells depends on the local nutrient concentration (Meschkat *et al.*, 1996).)

If a cell-cycle-phase dependent death process is studied, which may be considered as a caricature of a chemotherapeutic treatment of micrometastasis with antimitotic drugs, the death behavior has been found to depend on the growth regime of the cluster before the death process was switched on. In the exponential regime, a small concentration of drugs corresponding to a small death rate slightly above the critical point allows for reorganizations in the cluster that reduce the probability of fragmentation into smaller clusters. In the asymptotic regime, the opposite seems to be appropriate: here it may be advantageous to have a large death rate corresponding to a treatment with high concentrations of chemotherapeutic drugs to suppress large fluctuations in the cluster surface that otherwise would increase the probability of cell detachments and fragmentation into smaller clusters. Note that in the latter case, the size of the smaller clusters that arise from one large precursor cluster, may fall below the crossover size $N(t_{cr})$ resulting again in an exponential growth regime.

The occurence of small clusters and single cells may provide a mechanisms of metastasis formation by chemotherapy (note that the death behavior of the clusters in the near of the critical point in $d = 3$ dimensions could be different from that in $d = 2$ dimensions).

The model for the apoptosis neglects many processes, among which the effects of the immune response and of genetic changes of the tumor cells under chemotherapeutic treatment may be the more important ones. Despite this simplifications it is likely, that the described scenaria are also present in real tumors although it is unclear if they are dominant.

In further steps, the model may serve as a reference for versions extended to incorporate other relevant processes. Some of them, as *e.g.* mitose-phase dependent cell-cell attraction, mitose-phase dependent cell diffusion (*e.g.* to model the influence of focal contacts), mutual growth phase dependent inhibition of growth by short range diffusive chemical substances have been considered elsewhere (Drasdo, 1994; Drasdo *et al.*, 1995). A slightly different version of the model which includes a more detailed description of the mitotic cycle allows for direct comparison with experiments where the mitotic index is measured (Drasdo, 1994; Drasdo *et al.*, 1995, in these references also further quantities as *e.g.* the subclone statistics and the generation number have been studied). Effects of the immune system can also easily be included in the model, however it seems reasonable, that the generic cases have been studied before overwhelming the model with too many parameters.

To proceed to $\sim O(10^9)$ cells in three dimensions (corresponding to a tumor of diameter $\sim$ 1 cm, a typical size of many tumors at the stage of clinical manifestation) on the present generation of computers, one would still have to rely on lattice models (*e.g.* Drasdo, 1997a) where, however, the model suggested above may again serve as reference model.

ACKNOWLEDGEMENTS

I thank G. Gompper, R. Kree and R. Lipowsky for stimulating discussions and O. Theissen for the critical reading of the orginal manuscript. I am indebted to G. Forgacs who is mainly responsible if the paper is also comprehensible to biologists and physicians.

REFERENCES

Alison M. and Sarraf C., Understanding cancer (Cambridge Univ. Press, 1997).
Batchelor M.R. and Henry B.I., *Phys. Lett.* **157** (1991) 229.
Byrne H.M. and Chaplain M.A.J., *Math. Bioscience* **130** (1995a) 151.
Byrne H.M. and Chaplain M.A.J., *Bull. Math. Biol.* **57** (1995b) 461.
Byrne H.M. and Chaplain M.A.J., *Appl. Math. Lett.* **8** (1995c) 71.
Chaplain M.A.J. and Sleeman B.D., *Math. Bioscience* **111**, (1992) 169.
Chaplain M.A.J., *Acta Biotheoretica* **43** (1995) 387.
Chaplain M.A.J., *Mathl. Comput. Modelling* **23** (1996) 47.
Cardy J., *Scaling and Renormalization* (Cambridge University Press, 1996).
Coldman A.J. and Goldie J.H., Cancer Modeling, edited by J.R. Thompson, B.W. Brown (Marcel Dekker Inc., NY, 1987).
Drasdo D., Monte-Carlo-Simulation in zwei Dimensionen zur Beschreibung von Wachstumskinetik und Strukturbildungsphänomenen in Zellpopulationen (Verlag Shaker, Aachen, 1994).
Drasdo D., Kree R. and McCaskill J.S., *Phys. Rev. E* **52** (1995) 6635.
Drasdo D., *Self-Organization of Complex Structures: From Individual to Collective Dynamics*, edited by F. Schweitzer (Gordon and Breach, London,

1996).
Drasdo D., 1997a, To be able to observe a macroscopic initial regime and a asymptotic regime with power-law growth, one (a) may introduce a critical number of cells N_c, that can be pushed away during cell division in a lattice model. Then, $N_c = 1$ corresponds to the Eden model while $N_c \to \infty$ correspond to the model used by Landini and Rippin, 1993. If N_c is assumed to be Gaussian distributed with $\langle N_c \rangle > 1$, the crossover between exponential and surface growth becomes variable. (b) An alternative approach would be, to assign an energy to each division on a lattice and perform a division step of cell i with probability $P_i \sim \exp(-\beta N_{i,shift})$ where $N_{i,shift}$ denotes the number of cells that would have to be moved by one lattice constant if cell i divides and β is some parameter.
Düchting W., Ulmer W. and Ginsberg T, *Europ. J. Cancer.*
Edelman G., *Immunological Rev.* **100** (1987) 11.
Edelman G., Topobiology (Basic Books, Inc., Publishers, New York, 1988).
Eden M., In *Proc. of the 4th. Berkeley Symposium on Mathematics and Probability*, Vol. IV., edited by J. Neyman (University of California Press, 1961).
Fisher R.A., *Ann. Eugenics* **7** (1937) 353-369.
Kolmogoroff A., Petrovsky I. and Piscounoff N., *Moscow Univ. Bull. Math.* **1** (1937) 1.
Lefever R. and Erneaux T., edited by W.R. Adey and A.F. Lawrence (Plenum, New York, 1984).
Landau L.D. and Lifschitz E.M., *Statist. Phys.* (1987).
Folkman J. and Klagsbrun M., *Science* **235** (1987) 442.
Foty R.A. and Steinberg M.S., Interplay of Genetic and Physical Processes in the Development of Biological Form, edited by D. Beysens, G. Forgacs, F. Gaill (World Scientific, 1995).
Freyer J.P. and Sutherland R.M., *J. Cell. Physiol.* **124** (1985) 516.
Freyer J.P. and Sutherland R.M., *Cancer Res.* **46** (1986) 3504.
Gompertz B., *Philos. Trans. Roy. Soc. (Lond.)* (1825) 115.
Graner F. and Glazier J.A., *Phys. Rev. Lett.* **69** (1992) 2013.
Greenspan H.P., *J. Theor. Biol.* **56** (1976) 229.
Halpin-Healy T. and Zhang Y.C., *Phys. Rep.* **254** (1995) 215.
Hart D., Shochat E. and Agur Z., The growth law of primary breast cancer as inferred from mammography screening trials, preprint (1997).
Helfrich W., *Z. Naturforsch.* **33A** (1978) 205.
Kuznetsov V.A., *Biomed. Sci.* **2** (1991) 465.
Landini G. and Rippin J.W., *Fractals* **1** (1993) 239.
Meschkat S., Drasdo D. and Schimansky-Geier L. (1996) unpublished.
Metropolis N., Rosenbluth A.W. und M.N. and Teller A.H., *J. Chem. Phys.* **21** (1953) 1087-1092.
Mikhailov A.S., *Phys. Rep.* **5 & 6** (1989) 307.
Müller-Klieser W., *J. Cancer Res. Clin. Oncol.* **113** (1987) 101.
Murdoch D.J., Krewski D.R. and Crump K.S., Cancer Modeling, edited by

J.R. Thompson and B.W. Brown (Marcel Dekker Inc., NY, 1987).
Owen M.R. and Sherratt J.A., *Pattern formation and Spatiotemporal Irregularity in a model for Macrophage-Tumor Interactions*, preprint (1997).
Rashevsky N., *Bull. Math. Biophys.* **7** (1945) 69.
Richardson D., *Proc. Camb. Phil. Soc.* **74** (1973) 515.
Steel G.G., Growth Kinetics of Tumours (Clarendon Press, Oxford, 1977).
Steinberg M.S., *J. Exp. Zool.* **173** (1970) 395-434.
Sutherland R.M., *Science* **140** (1988) 239.
Townes P.L. and Holfreter J., *J. Exp. Zool.* **128** (1955) 53.
Verhulst P.F., Notice sur la loi que la population suit dans son accroissement, Correspondance Mathématique et Physique, edited by A. Quetelet, Brussels 10, pp. 113-121.
Weinberg R.A., Scientific American on cancer medicine (1996).
Westermark B., *Int. J. Cancer* **12** (1973) 438.
Williams T. and Bjerknes R., *Nature* **236** (1972) 19-21.

Genetic, Immune, Molecular and Metabolic Networks

LECTURE 16

Intracellular Communication *via* Protein Kinase Networks

D.L. Charest and S.L. Pelech

Department of Medicine, Faculty of Medecine, University of British Columbia, 1779 W., 75th avenue Vancouver, B.C., V6P 6P2, Canada

1. INTRODUCTION

Cells respond to a diverse repertoire of hundreds of extracellular mediators that regulate homeostasis, growth, proliferation, induction of differentiation or promotion of programmed cell death. Incoming signals that first act at the cell surface ultimately trigger the deployment of a battery of cytoplamic and nuclear effectors that coordinate the appropriate cell responses (Robbins *et al.,* 1994). Signals are transmitted across the selectively permeable plasma membrane following engagement of ligands with their appropriate transmembrane receptors. Post-receptor signals are amplified and transmitted within the cytoplasm by recruitment of signal transducing molecules such as guanine nucleotide-binding (G) proteins, enzymes that generate second messengers, protein kinases and protein phosphatases (Fig. 1). Some of the proteins that are targeted by kinases include transcription factors, that migrate to the cell nucleus to regulate the expression of specific genes. Thousands of distinct proteins participate in intracellular communications, and they operate within highly sophisticated networks that receive, integrate, process and transmit information.

Covalent modification by reversible phosphorylation impinges on almost every cellular function (Campbell *et al.,* 1995). At least a third of the proteins inside a cell are subject to phosphorylation, typically at multiple sites. Regulation of a specific biological activity involves the opposing actions of protein kinases and protein phosphatases. Protein kinases control the activity of other proteins by catalyzing the transfer of a high energy phosphate from adenosine triphosphate to the hydroxyl group of amino acid residues on target proteins. Often the substrate of a protein kinase is another kinase that becomes activated upon its phosphorylation. In this way, long cascades of sequentially activating kinases can operate within signalling pathways. Over 1500 different protein kinases are predicted to be encoded by the human genome.

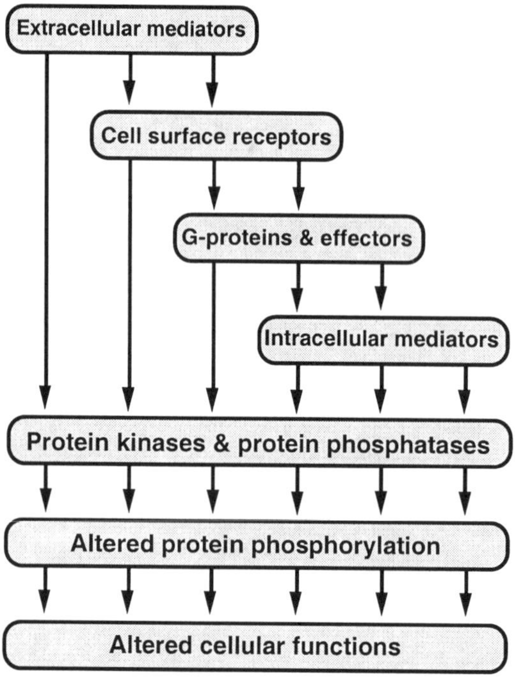

Fig. 1. Hierarchy of cell signalling leading to the regulation of cellular functions.

Discrete protein kinase pathways are sequestered through specific protein interactions. For example, signal transduction cascades in the budding yeast *Saccharomcyes cerevisiae* are organized on the principle of linearity such that little or no cross talk occurs between pathways. However in mammalian cells, kinase cascades are organized into dynamical intracellular communication networks where information flow is more fluid and can converge or branch off at different points along pathways. This intracellular circuitry is reminiscent of an intranet (Pelech, 1996). These integrated signalling systems act as innate sensors that can monitor and coordinate inputs from a variety of extracellular stimuli that are present in the milieu surrounding the cell. The biological readout is a consequence of the integration and processing of this information. The MAP kinase signalling pathways in eukaryotes have served as a paradigm for understanding how protein kinase cascades operate in parallel or within integrated circuits and will be the focus of this review.

2. PROTEIN KINASE SPECIFICITY AND REGULATION

2.1 Substrate specificity

Approximately 90 and 10 percent of the phosphorylation of proteins in eukaryotes occurs on seryl and threonyl residues, respectively. Tyrosyl, histidyl, arginyl and lysyl residues are more rarely phosphorylated. The substrate specificity of a protein kinase is also determined by the amino acid sequence surrounding the phosphorylated amino acid residue. For example, the mitogen-activated protein kinase (MAP kinases) family of kinases phosphorylate seryl/threonyl residues where the amino acid proline is often present at the –2 and +1 positions surrounding the phosphoacceptor site (*i.e.* $PX^{S}/_{T}P$ where X is any amino acid). The MAP kinase isoforms, Erk1 and Erk2 (extracellular signal regulated kinase), are broad specificity kinases that can phosphorylate scores of different substrates *in vitro*. In contrast, the MAP kinase activators Mek1 and Mek2 (MAP kinase/ Erk kinase) only phosphorylate threonyl and tyrosyl residues (TXY) in native forms of MAP kinase to elicit their stimulation. Therefore, Mek phosphorylation requires recognition of the three-dimensional structure of MAP kinases in addition to the phosphorylation motif.

2.2 Kinase topology

All protein kinases feature a 250 to 300 amino acid catalytic domain with a molecular mass in the range of 30 kDa (Hanks *et al.,* 1988). The kinase catalytic region is organized into eleven highly conserved subdomains that are important for catalytic function. These canonical subdomains are separated by stretches of protein sequence with lower conservation that are necessary for regulation and maintenance of the overall architecture of these enzymes. For example, the activating phosphorylation sites in MAP kinases and Mek's are located between conserved subdomains VII and VIII in these kinases. Sequences located at the extreme amino- and carboxy-terminal regions of these protein located outside the catalytic domain also play important roles in the regulation of enzyme activity, interactions with other signalling proteins and molecules, and their subcellular localization.

2.3 Kinase regulation

There are a number of regulatory mechanisms that stimulate kinase catalytic activity. Many of the extracellular growth factors (*e.g.* platelet-derived growth factor and insulin) bind to cell surface receptors that are protein-tyrosine kinases, which become catalytically active on engagement with their specific ligand. These receptor kinases typically phosphoryate themselves on tyrosine, and these sites are now able to recruit additional signalling proteins to the plasma membrane. Some surface receptors (*e.g.* for transforming growth factor-β (TGF β) and activins) are ligand-activated protein-serine/threonine kinases. Cytoplasmic protein kinases may be stimulated by receptor-induced increases in the intracellular concentration of secondary mediators termed second messengers (*e.g.* cAMP, cGMP, diacylglycerol and Ca^{2+}). Protein kinase C (PKC) is activated by binding lipid derivatives and Ca^{2+} in the two cysteine-rich regions located in its amino-terminal regulatory domain. Another kinase that is activated by Ca^{2+} is calmodulin-dependent protein kinase 2, which occurs in large multimeric complexes. In the case of cAMP-dependent protein kinase (PKA), cAMP induces the inactive heterodimeric complex to dissociate and liberate two active catalytic subunits. Finally, many cytoplasmic kinases, like the MAP kinases, are phosphorylated and activated sequentially by one or more upstream kinases within signalling pathways. In summary, although protein kinases share similar architecture, these cellular enzymes have also evolved unique regulatory domains that enables the cell to respond appropriately to widely diverse extracellular stimuli.

3. ERK1 AND ERK2 PATHWAYS IN MAMMALS

The MAP kinases, Erk1 and Erk2, are activated by over 50 extracellular signals (Pelech and Charest, 1995 and references therein). The mitogen, platelet-derived growth factor (PDGF) is an example of a ligand that is released by activated platelets at the site of a wound to stimulate the proliferation of fibroblastic cells. Interaction of dimeric PDGF ligand with specific binding sites located on the extracellular portion of the receptor triggers a myriad of bichemical events just beneath the surface of the plasma membrane (Fig. 2).

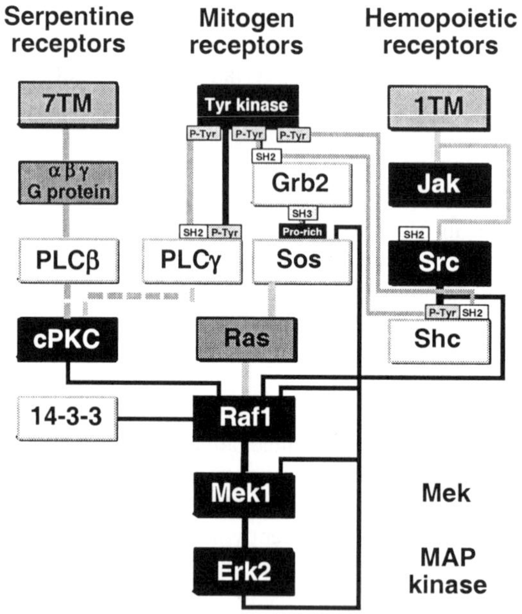

Fig. 2. Diverse receptor activation of the Mek-MAP kinase module.

Receptor juxtaposition allows the cytoplasmic kinase catalytic subdomains to phosphorylate on tyrosyl residues via an intermolecular autophosphorylation reaction. These phosphorylated residues are contiguous with specific recognition sequences that together serve as critical binding sites for phosphotyrosyl-directed peptide binding Src homology 2 (SH2) domains present in anchoring proteins such as the growth receptor binding-2 (Grb2). Grb2 serves as a molecular scaffold for binding other proteins through the two SH3 domains that flank the SH2 domain. Interaction of the proline-rich region of the guanine nucleotide exchange protein (GNEP) Son-of-sevenless (Sos) with the two SH3 domains in Grb2 promotes the translocation of this protein. Once recruited to the cell membrane, Sos stimulates the exchange of GDP for GTP in the monomeric G protein Ras. The intrinsic GTPase activity within Ras converts GTP into GDP and limits the duration of the Ras effector signal.

Activated Ras serves as a key intermediary between activities occurring at the inner plasma membrane and signal transduction events that are initiated with cytoplasmic effectors like the seryl/threonyl-kinase Raf1. The precise mechanism of Ras activation of Raf1 is the subject of intense debate and research. Cell stimulation promotes the complex formation between Ras and Raf1 proteins at the cell surface. Interaction of the amino-terminal domain of Raf1 (residues 52-132) occurs with the effector binding domain of Ras (residues 26-48) in a GTP-dependent manner (Pelech and Charest, 1995). Once Raf1 is translocated to the plasma membrane, the protein must undergo one or more modifications, including phosphorylation, complex oligomerization and lipid interaction, to become fully active (Hall, 1994; Pelech and Charest, 1996; Marshall, 1996). Among the kinases implicated in the phosphorylation of Raf1 are protein kinase C, Ksr1 (kinase suppressor of Ras) and the tyrosyl kinase Src. In the case of Ras activation of Raf1, Stokoe and McCormick (1997) demonstrated that the farnesyl lipid moiety which anchors Ras to the plasma membrane in conjunction with the Ras protein was sufficient to potently activate Raf1.

Raf1 phosphorylation of two seryl residues (Ser-218 and Ser-222) activates Mek1 and Mek2. (Pelech and Charest, 1996). In turn, activated Mek's stimulate Erk activity by phosphorylation of neighbouring residues in a threonyl-glutamyl-tyrosyl (TEY) motif. Mek's are the only known physiological activators described for Erk1 and Erk2. Mek1 and Mek2 are unable to phosphorylate denatured Erk proteins or synthetic peptides patterned after the TEY phosphorylation site. In fact, a 20 amino acid MAP kinase-binding site is located at the very amino-terminal of Mek that is essential for tight association between the two proteins and may explain, in part, why Mek only phosphorylates the native form of Erk (Fukuda *et al.,* 1997). Activation of Mek and Erk by extracellular stimuli cause both kinases to translocate from the cytoplasm to the nucleus. However, Mek becomes rapidly exported from the nucleus. The mechanism appears to involve a leucine-rich nuclear export signal located in the amino-terminal of Mek1. In contrast to Mek1, Erk1 and Erk2 display a broad substrate specificity and consequently phosphorylate a large number of substrates including other kinases such as Rsk, Mek1, Raf1 and Sos. Effector molecules such as the transcription factor Elk1 are also important nuclear targets that regulate the expression of genes necessary for full biological response.

4. RAS-RAF-DEPENDENT MAP KINASE PATHWAYS IN FROGS AND NON-VERTEBRATES

The utility of the Ras-Raf-dependent MAP kinase module is evident in the number of different organisms that have adapted the signalling pathway for diverse developmental purposes (Fig. 3). In fruit flies and nematodes, extracellular differentiation signals engage receptor protein-tyrosyl kinases to activate nearly identical signal transduction hardware to that which mammalian cells utilize for cell

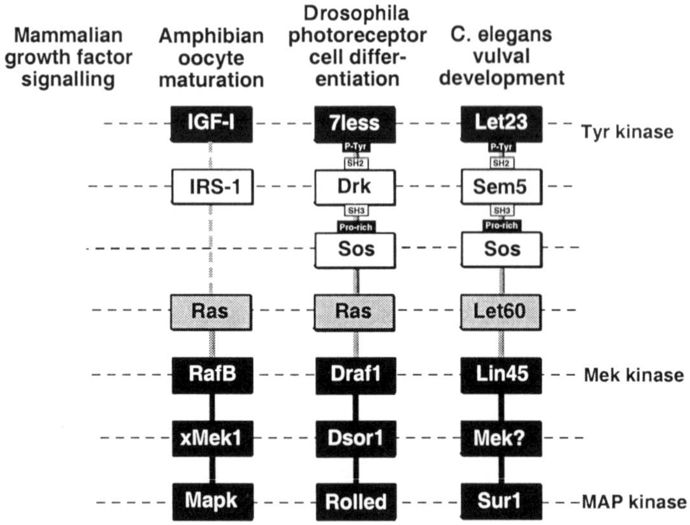

Fig. 3. Adaptation of Ras-Raf MAP kinase module for diverse signalling pathways.

proliferation and differentiation. The same molecular machinery is reiterated in the meiotic maturation of frog oocytes.

4.1 Regulation of oocyte maturation in frogs

Insulin treatment of *Xenopus laevis* oocytes arrested in prophase of the first meiotic cell division causes the resumption of maturation via activation of the insulin-like growth factor 1 (IGF-1) receptor (Pelech and Charest, 1995). The tyrosine autophosphorylation sites on the IGF-1 receptor like that of the PDGF receptor described earlier serve as binding sites for phosphotyrosyl-directed docking proteins. The insulin receptor substrate 1 (IRS-1) docking protein may be tyrosyl phosphorylated by the IGF-1 receptor. The intermediary link between the activated IRS-1 and stimulation of the Ras-RafB MAP kinase module remains to be defined. Whatever the molecular events, RafB, a Mek2-like MAP kinase kinase and an Erk2-like MAP kinase become sequentially activated near the onset of germinal vessel breakdown (Fig. 3).

The frog maturation hormone, progesterone, employs another strategy to induce oocyte maturation. This pathway requires the *de novo* synthsis of the seryl/threonyl-kinase Mos. Mos can directly phosphorylate and active Mek2 *in vitro*. Therefore, Mek serves as a convergence point for both Ras-Raf- and Mos-induced activation of a MAP kinase module.

4.2 Induction of eye development in flies

The compound eye of the fruit fly *Drosophila melanogaster* is composed of 800 20-cell units called ommatidia (Ferrel, 1996 and references therein). The cellular specialization of the eight photoreceptor cells relies on a sequential series of inductive messages transmitted by intercellular communication (Pelech and Charest, 1995; Ferrel, 1996). Neuronal differentiation of the R7 precursor cell is induced when the epidermal growth factor-like (EGF-like) protein Boss (Bride-of-sevenless) located on the R8 photoreceptor cell engages the Sev (Sevenless) EGF-like receptor protein-tyrosyl kinase. Sev activation triggers the association of Drk1 (Grb2-like) and Sos on the cytoplasmic tail of the receptor through SH2 and SH3 domains. Sos, the nucleotide exchange protein, colocalizes and activates the membrane anchored Enhancer of Sevenless (which is Ras-like). Activated GTP-Ras promotes the expression of R7 neuronal-specific genes through stimulation of the MAP kinase module consisting of DRaf, and the *Drosophila* Mek (Dsor1) and Erk (Rolled) homologues (Fig. 3). Two other EGF-like receptors, DER (*Drosophila* EGF receptor) and Torso use the identical Ras-Raf1 MAP kinase module to establish dorsoventral polarity and terminal structures in the developing embryo, respectively.

4.3 Induction of vulval development in worms

Developmental biologists studying sexual differentiation in the nematode *Caenorhabditis elegans* identified a Ras-Raf1 MAP kinase module that regulates vulval induction (Pelech and Charest, 1995; Ferrel, 1996). Three of the six vulval precursor cells receive an intercellular message, Lin3 (cell *lin*eage abnormal), from the adjacent gonadal anchor cell. The inductive signal stimulates the three pluripotent cells to proliferate and differentiate into 22 specialized cells that form the vulva, which is critical for egg laying in the adult hermaphrodite. The diffusible Lin3 protein activates the EGF-like tyrosyl kinase receptor Let-23 (lethal) expressed on the surface of the three vulval precursor cells. Sem5 (sex muscle defective) is a Grb2-like protein that links cell surface events with cytoplamic transducers through activation of the Ras homologue Let-60. How Grb2 activates Ras in *C. elegans* has not been defined. Whatever the mechanism, Ras activation of the cellular transducers Lin45 (lineage, alias Raf1), Mek2 and Mpk1 (MAP kinase) produces the appropriate development outcome (Fig. 3). Although null mutants have been uncovered for Lin45 and Mek2, the absence of dominant-negative mutants for Mpk1 implies the existence a second redundant MAP kinase isoform.

5. PARALLEL MAP KINASE MODULES IN BUDDING YEAST

The unicellular yeast *Saccharomyces cerevisiae* like individual cells in multicellular organisms differentially activate distinct but structurally related MAP kinase modules in response to disparate signals (Herskovitz, 1995; Pelech and Charest, 1996). Various MAP kinase family members that recognize a similar phosphorylation consensus sequence still have somewhat different substrate specificities. Additionally, variances in the size of the proteins (40- to 100-kDa) and the presence of additional protein sequence allow the kinases to possess unique regions that promote discrete complexes with upstream regulators and downstream effectors. This may impart a certain degree of fidelity on the pathway so as to prevent cross-talk between the various MAP kinase pathways (Fig. 4).

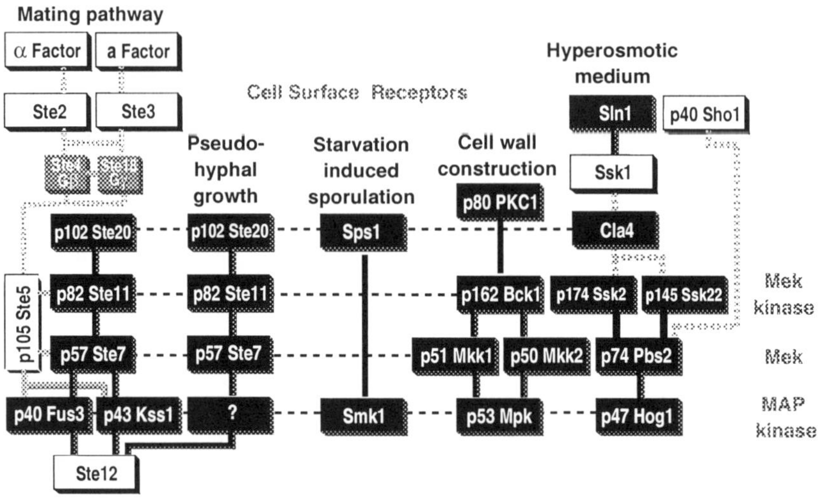

Fig. 4. MAP kinase-dependent signalling pathways in budding yeast.

5.1 Mating-response MAP kinase pathway

The first MAP kinase signal transduction pathway was elucidated from genetic studies into the regulation of mating in haploid budding yeast (Herskovitz, 1995). For a haploid cell to exit the cell cycle prior to DNA replication and initiate mating through conjugation, the mating-pheromone-regulated MAP kinase pathway is stimulated by engagement of specific heterotrimeric G protein-linked, seven transmembrane receptors (Ste2 or Ste3) with their cognate mating pheromone ligands (MATα-factor for Ste2 and MATa-factor for Ste3) that are released by the opposite mating type. Release of the G protein dimeric β- (Ste4) and γ- (Ste8) subunit complex allows the Gβ-subunit portion to interact with and perhaps recruit Ste20 to the inner cell surface of the plasma membrane where this kinase is stimulated by an unknown mechanism (Peter *et al.,* 1996; Leberer *et al.,* 1997). Activated Ste20 subsequently promotes the sequential phosphorylation and stimulation of Ste11, the Mek Ste7 and the MAP kinases Fus3 and Kss1 in a linear fashion (Fig. 4). The non-kinase protein Ste5 may also be essential to maintain signalling integrity in haploid cells induced to undergo meiotic arrest and cell conjugation (Herskovitz, 1995). Ste5 may acts as a molecular scaffold to sequester Ste4, Ste11, Ste7, Fus3 and Kss1 into an oligomeric signal transducing complex. This may prevent cross-talk between similar MAP kinase modules that control disparate biological responses such as cell wall biosynthesis during bud formation or homeostasis during exposure to a hypertonic environment. Additionally, Ste5 may serve to localize the MAP kinase complex, and in particular Ste11, in the vicinity of the activated Ste20 so as to increase the velocity at which the message is transduced to the nucleus (Pelech and Charest, 1996). However, discreet protein associations also occur between kinase-pairs within the signalling pathway in the absence of Ste5. In fact, Ste7 forms stable, high affinity (Kd = 89 5 nM) protein complexes with the mating pheromone pathway MAP kinases Fus3 and Kss1 while displaying no interactions with *S. cerevisiae* MAP kinase isoforms Mpk1 (Bardwell *et al.,* 1996). This may explain why the MAP kinases from yeast are not functionally redundant. The specificity of the Mek/MAP kinase complexes are such that the Mek's may act as molecular nodes through which the flow of information must traverse (Mordret, 1993).

5.2 Stress-activated MAP kinase pathway

There exists at least four other distinct MAP kinase modules in *S. cerevisiae* that contribute to the regulation of different cellular responses (Herskovitz, 1995 ; Pelech and Charest, 1995) One of these is the stress-activated Hog (high osmolarity glycerol response) MAP kinase pathway. Exposure of budding yeast to elevated environmental osmolarity induces a protective and compensatory pathway to increase the expression of genes required for glycerol synthesis (Herskovitz, 1995). *S. cerevisiae* has evolved a two-component osmosensor coupled to a MAP kinase module that is distinct from the mating-response pathway. The yeast Hog MAP kinase pathway is regulated by a multistep phosphorelay mechanism involving the Sln-Ypd-Ssk1 osmosensor similar to the prokaryotic two-component system (Maeda *et al.,* 1995; Posas *et al.,* 1996) The extracellular cell surface histidyl-kinase sensor, Sln1, is activated under condition of normal osmolarity and through a relay mechanism inactivates Ssk1 (suppressor of sensor kinase). The Sln1 autophosphorylates on a histidyl residues and then immediately transfers the phosphate to an aspartyl residue located in the receiver domain of Sln1. The phosphate on aspartyl residue in Sln1 is then transferred to a histidine on Ypd1, a second receiver protein and finally to aspartate in Ssk1. When the extracellular osmolarity increases, Sln1 is inactivated, thereby allowing the accumulation of dephospho-Ssk1. The ensuing activation of the Hog pathway is initiated by activation of a pair of redundant Mek kinases, Ssk2 and Ssk22, followed by the MAP kinase kinase, Pbs2 (polymyxin B sensitive) and the MAP kinase, Hog1 (Fig. 4).

6. STRESS SIGNALLING MAP KINASE PATHWAYS IN MAMMALS

The stress-activated protein (SAP) kinases (SAP kinases α, β & γ alias Jun N-terminal kinases or Jnk) are activators of the transcription factor c-Jun, a major component of the AP-1 transcription factor complex that controls immediate early gene expression. SAP kinases are stimulated by dual phosphorylation in a TPY motif found in the kinase activation loop between subdomains VII and VIII; similar to the phosphorylation sites (TEY) present in the Erk1/Erk2 family of MAP kinases. The substitution of a prolyl residue for a glutamyl residue, which is expected to create a kink in the peptide structure of the activation loop, may contribute to the regulation of SAP kinases by distinct signalling pathways. Indeed, Sek1 (SAPK/Erk kinase alias Mkk4) and Mek7 are the only known activators of SAP kinases *in vivo.*

Like Mek1 and Mek2, Sek1 and Mek7 are phosphorylated and activated on two equivalent residues located in the kinase activation loop. Similarly, Sek1 and Mek7 act as convergent points for many Sek kinases including the budding yeast Ste11-like Mek kinase family (Mekk1, 2, 3 & 4), TGFβ-activated protein kinase 1 (TAK1), tumour progression locus 2 (Tpl2 alias Cot) (Fanger *et al.*, 1997). Alternatively, Mekk1 through Mekk3 as well as the oncoprotein Tpl2 have been demonstrated to activate the Erk1/Erk2 family. Several Mek kinase kinases are also known to lead to indirect activation of SAP kinases (Fanger *et al.*, 1997). Comparisons with the yeast mating pheromone MAP kinase pathway reveal that the Mek kinases are mammalian homologues of Ste20 and thus by analogy may operate at the same level. Two TAK1-binding proteins (TAP), the germinal centre kinase (GCK), and three budding yeast Ste20-like p21-activated kinases (PAK) appear to mediate SAP kinase activation without direct phosphorylation of Sek1. In contrast to TAP, GCK and PAK which require at least one intermediary Mek kinase for activation of Sek1 and SAP kinase, the mixed-lineage kinase 3 protein (MLK3 alias SPRK) and the thousand and one amino acid kinase (TAO) activation of the SAP kinase-dependent pathway may be mediated by direct phosphorylation of Sek1 (Fanger *et al.*, 1997). Several other Ste20-related kinases may act in the same pathway as SAP kinases or operate within homologous but distinct MAP kinase pathways; they include Ste20/oxidant stress response kinase (SOK), kinases responsive to stress (Krs1 and Krs2) and MAP kinase upstream kinase (MUK) (Robinson and Cobb, 1997). It is evident from recent studies that the SAP kinases are regulated by a diverse group of upstream kinases. The roles of these protein kinases as physiological transducers connected to the Sek1-SAP kinase module still remain to be established, since many of the connections that have been proposed are based on their over-expression in cell culture systems. For instance, it is not entirely clear whether PAK and the monomeric G proteins Cdc42 and Rac1, which bind and activate PAK, truly regulate the SAP kinase-dependent pathway. Additionally, it remains to be established how disparate stimuli (*e.g.* TGF-α, DNA damaging agents and protein synthesis inhibitors) can activate SAP kinases. Putative relationships amongst these kinases are depicted in Figure 5.

The second stress-activated MAP kinase pathway was identified by several research teams and as a consequence is known as RK (reactivating kinase), Mpk2 (MAP kinase 2) and Csbp (cytokine-suppressive anti-inflammatory drug binding protein kinase), but since it is also activated in response to changes in extracellular osmolarity, Hog1 is the generally accepted appellation. Hog1 is activated by distinct signalling pathways from those that act on Erk1/Erk2 and SAP kinases in the same cell types (Pelech and Charest, 1995; Robinson and Cobb, 1997). This is supported by the fact that the Hog MAP kinase family members are smaller in size and regulated by the nearly identical dual phosphorylation site TGY, in which the glutamyl residue in Erk1 is replaced by glycine. The difference in the phosphorylation site motif in Hog1 indicates that the regulation of its enzyme activity is likely to be unique from other MAP kinases. While Sek1 is able to activate Hog1 *in vitro*, transient transfection studies have shown that Mekk specifically activates SAP kinase. Mek1 had no effect on Hog1 activity in similar experiments (Pelech and Charest, 1995). Two separate Mek's (Mkk3 and Mkk6), when rendered constitutively active by substitution of the two regulatory phosphorylation sites in the activation loop with glutamyl residues are able to stimulate the Hog1 pathway and induce transformation when overexpressed in mammalian cells. The kinases responsible for the activations of Mkk3 and Mkk6 remain obscure. However, TAK1 and TAO are reported to stimulate their activity (Robinson and Cobb, 1997). Mxi2, Erk6 and p38b are several recently characterized Hog1-like kinases that may define new MAP kinase modules, since each isoform may be regulated by a distinct pathway (Robinson and Cobb, 1997). Finally, there are additional MAP kinase-Mek modules such as Erk5 (alias BMK for big MAP kinase) and Mek5, the function of which has been elusive.

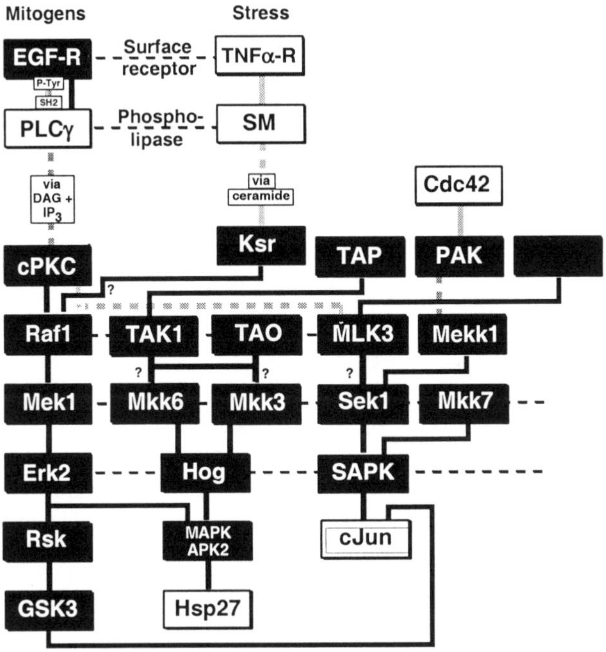

Fig. 5. Mek-MAP kinase modules in mitogen and stress signalling.

With the complete sequencing of the human genome, soon all of the components in the cell's signalling network will be known. The lessons learned so far from studies of MAP kinase-dependent pathways will permit subclassifications of many kinases at different tiers of protein kinase modules. However, a major challenge will be the elucidation of the physiologically relevant connections among these kinases.

ACKNOWLEDGMENTS

DLC was the recipient of a Medical Research Council of Canada (MRCC) Studentship Award and SLP was an MRCC Industrial Scientist. Operating support to SLP from Kinetek Pharmaceuticals, Inc., the MRCC, the National Cancer Institute of Canada, and the B.C. and Yukon Heart and Stroke Foundation is gratefully acknowledged.

REFERENCES

Bardwell L., Cook J.G., Chang E.C., Cairns B.R. and Thorner J., *Mol. Cell. Biol.* **16** (1996) 3637-3650.

Campbell J.S., Seger R., Graves J.D., Graves L.M., Jensen A.M. and Krebs E.G., "The MAP kinase cascade", in Recent Progress in Hormone Research, Vol. 50, edited by C.W. Bardin (Academic Press Inc., San Diego, 1995).

Fanger G.R., Gerwins P., Widmann C., Jarpe M.B. and Johnson G.L., *Curr. Opin. Genet. Dev.* **7** (1997) 74-76.

Fukuda M., Gotoh Y. and Nishida E., *EMBO J.* **16** (1997) 1901-1908.

Ferrell J.E. Jr., MAP kinases in mitogenesis and development, in Current Topics in Developmental Biology, Vol. 33, edited by R.A. Pederson and G.P. Schatten (Academic Press London, 1996) pp. 1-60.

Hall A., *Science* **264** (1994) 1413-1414.

Hanks S.H., Quinn A.M. and Hunter T., *Science* **241** (1988) 42-52.

Herskowitz I., *Cell.* **80** (1995) 187-197.

Maeda T., Takekawa M. and Saito H., *Science* **269** (1995) 554-558.

Marshall C.J., *Nature* **383** (1996) 127-128.

Pelech S.L., *Current Biology* **6** (1996) 551-554.

Pelech S.L. and Charest D.L., "MAP kinase-dependent pathway in cell cycle control", in Progress in Cell Cycle Research, Vol. 1, edited by L. Meijer, S. Guidet and H.Y. Lim Tung (Plenum Press, New York, 1995).

Posas F., Wurgler-Murphy S.M., Maeda T., Witten E.A., Thai T.C. and Saito H., *Cell.* **86** (1996) 865-875.

Mordret G., *Biol. Cell.* **79** (1993) 193-207.

Robbins D.J., Zhen E., Cheng M., Xu S., Ebert D. and Cobb M.H., "MAP kinases Erk1 and Erk2: pleiotropic enzymes in a ubiquitous signaling network", in Advance in Cancer Research, Vol. 63, edited by G.F. van Woude and G. Klein (Academic Press Inc., San Diego, 1994) pp. 93-116.

Stokoe D. and McCormick F., *EMBO J.* **16** (1997) 2384-2396.

LECTURE 17

Molecular Networks that Regulate Development

W.F. Loomis, G. Shaulsky, N. Wang and A. Kuspa[1]

Center for Molecular Biology, UCSD, LaJolla, CA 92093, USA
[1]Department of Biochemistry, Baylor College of Medicine, Houston, TX 77030, USA

1. INTRODUCTION

Changing networks of genes and proteins underlie the development of all multicellular organisms. Specific cell types arise as the result of their developmental history and adapt their physiology to their function. Understanding the structure of the networks that gate and regulate the temporal sequence of events leading to the terminally differentiated state requires knowledge of the critical molecular components and their interconnections. Once the general architecture of a network is known with some confidence, detailed quantitative analyses of the individual reactions can lead to testable predictions (Loomis and Sternberg, 1995). However, embryogenesis of most metazoans is so complex that it will be some time before it is possible to have a complete understanding of the sequence of events leading from an egg to a newborn. At present we can only glimpse the workings of certain sub-systems. Luckily, the evolution of development has followed the path of least resistance and used pre-existing networks that were present in simpler ancestors. With some tinkering, networks selected for one function can be adapted to another.

Therefore, understanding the properties of a network in one system often lays the groundwork for understanding its derivatives in less tractable organisms. Studies with the soil amoeba, Dictyostelium discoideum, have shed some light on the molecular networks that are able to account for the behavior of the cells as they aggregate into mounds of 10^5 cells, differentiate into prespore and prestalk cells, and proceed through the stages of morphogenesis to generate fruiting bodies where the spores are held up on gently tapering stalks (Loomis, 1996).

2. AN AGGREGATION NETWORK

One of the central nodes of many different networks is the cAMP dependent protein kinase, PKA. This enzymes catalyzes the phosphorylation of either serine or threonine in a diverse set of proteins and can either activate or inhibit the function of its targets (Lee, 1991). It is a complex enzyme with a catalytic subunit, PKA-C, and a regulatory subunit, PKA-R, that inhibits the activity when bound to the catalytic subunit (Taylor *et al.*, 1990). PKA activity is stimulated when the level of cAMP in the cell rises and binds to two slightly different sites in the regulatory subunit leading to the dissociation of the complex. Studies with a variety of cell types have shown that PKA activity is inhibited when the level of cAMP falls. However, purified PKA-R binds cAMP so avidly (Kd = 4 nM) that it would never be expected to release the ligand *in vivo* (De Gunzberg and Veron, 1982). This paradox has been recently resolved by the discovery of a cAMP specific phosphodiesterase that binds PKA-R where it can reactivate the regulatory subunit by hydrolyzing the ligand (Shaulsky *et al.*, 1998). The activity of this phosphodiesterase, the product of the Dictyostelium *regA* gene, is stimulated 20 fold when bound to PKA-R from either Dictyostelium or mammals. Thus, PKA should be considered as a complex with catalytic, regulatory and phosphodiesterase subunits. Mutant strains lacking *regA* behave almost identically to those that lack *pkaR*; they both develop precociously and have defects in stalk formation (Shaulsky *et al.*, 1998).

Adenylyl cyclase is the enzyme that synthesizes cAMP needed for activation of PKA. The gene encoding adenylyl cyclase in Dictyostelium, *acaA*, is expressed for a few hours shortly after the initiation of development (Parent and Devreotes, 1996; Wang and Kuspa, 1997). Enzyme activity and cAMP level rise but much of the newly synthesized cAMP is secreted. Adjacent cells respond to extracellular cAMP when it binds to a seven transmembrane receptor, cAR1. The gene encoding cAR1, *carA*, is also transiently expressed shortly after the initiation of development. When cAMP is bound to cAR1, the signal is transduced via trimeric G proteins and an adapter protein, CRAC, to activation of internal adenylyl cyclase (Fig. 1). This feedback loop results in relay of the signal that cAMP has been secreted. Within 2 minutes of adding a pulse of 10 nM cAMP, the activity of adenylyl cyclase rises 10 fold and then returns to the basal level in the next several minutes. There is now direct evidence that PKA activity is responsible to turning off adenylyl cyclase activity.

The gene encoding the catalytic subunit of PKA, *pkaC*, is expressed following the initiation of development and continues throughout the remainder of development (Mann and Firtel, 1991). The expression pattern of the gene encoding the regulatory subunit, *pkaR*, is similar but shows subtle differences (Part *et al.*, 1985). During the first 8 hours of development, as the cells prepare to aggregate and then stream into mounds, the catalytic and regulatory subunits of PKA both accumulate. Therefore, PKA would be expected to be active only following the activation of adenylyl cyclase. After a time delay in the order of a minute or so after cAMP binds to the regulatory subunit, PKA brings the adenylyl cyclase activity back to the basal level. Direct support for this connection can be found by studying mutants lacking PKA activity as the result of disruption of *pkaC*. Stimulation of cells of these strains with external cAMP results in a normal rise in adenylyl cyclase activity but the activity does not return to the basal level for more than 15 minutes (Mann *et al.*, 1996). These *pkaC*$^-$ mutant cells are unable to aggregate or develop further (Mann *et al.*, 1992).

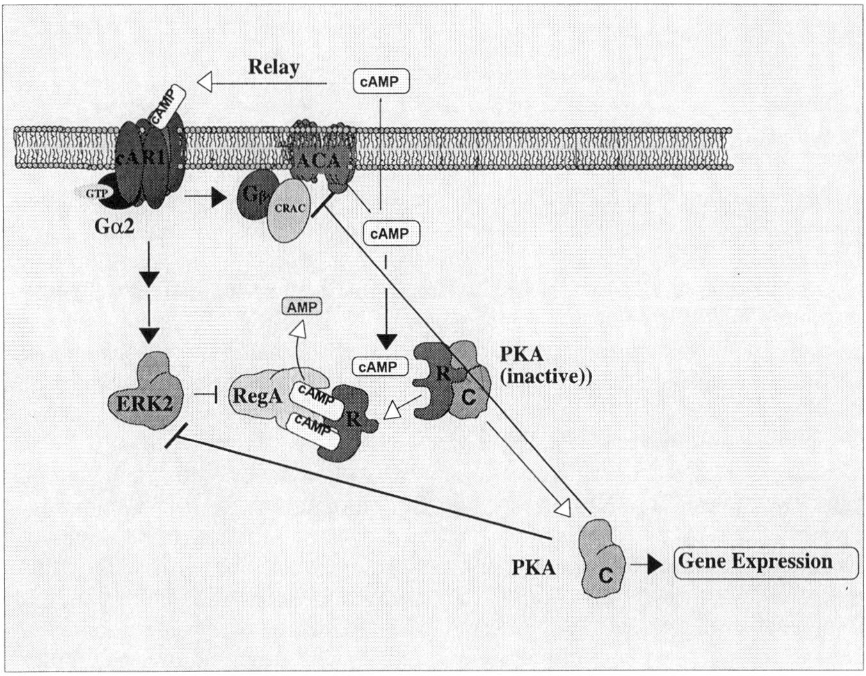

Fig. 1. A proposed aggregation stage network. External cAMP binds to the seven-transmembrane receptor, cAR1, leading to dissociation of the trimeric G protein into its Gα2 and Gβγ subunits. The cytosolic regulator of adenylyl cyclase (CRAC) then associates with the membrane and stimulates adenylyl cyclase (ACA) to synthesize cAMP. Lingand binding to cAR1 also activates the MAP kinase ERK2 that inhibits RegA, the cytosolic phosphodiesterase. cAMP can then accumulate and relay the signal. However, as the internal concentration of cAMP increases, it activates the protein kinase, PKA, that blocks further activation of ACA and also inhibits ERK2. RegA activity then reduces the cAMP levels and the cells enter a refractory period. This network results in stable oscillations in PKA and other activities with a periodicity of about 7 minutes.

The gene encoding the cAMP phosphodiesterase, *regA*, that resets PKA in its inactive complex is expressed 4 hours after the initiation of development and continues throughout the remainder of development (Shaulsky *et al.*, 1996). Genetic studies have suggested that this phosphodiesterase is inhibited immediately after cAMP binds to the cAR1 receptor as the result of activation of a MAP kinase encoded by *erkB* (Lu *et al.*, 1998). The activity of this protein kinase has been directly observed to increase rapidly following stimulation of cells with a pulse of cAMP reaching a peak at 90 seconds (Knetsch *et al.*, 1996; Aubry *et al.*, 1997). No stimulation of ERK2 activity is seen in mutants lacking the cAR1 receptor. Nor is cAMP produced after stimulation of *erkB*$^-$ mutant cells. However, *erkB*$^-$ *regA*$^-$ double mutants produce normal amounts of cAMP following stimulation (Lu *et al.*, 1998).While there is yet no direct biochemical evidence that ERK2 inhibits RegA, the fact that *regA*$^-$ null mutations can suppress the defect in cAMP production and the block to aggregation resulting from disruption of *erkB* indicates that RegA functions downstream of ERK2.

After reaching a peak of activity 90 seconds after stimulation of cells with cAMP, ERK2 activity decays with a half life of 2 minutes (Knetsch *et al.*, 1996; Aubry *et al.*, 1997). However, in mutants that have reduced PKA activity, either as the result of null mutations in *pkaC* or expression of a modified version of *pkaR* that makes a regulatory subunit unable to bind cAMP and therefore always able to inhibit PKA-C, the activity of ERK2 stays high for at least 15 minutes (Knetsch *et al.*, 1996; Aubry *et al.*, 1997). These results indicate that PKA activity is responsible for inhibition of ERK2 a few minutes after the cells are stimulated. Assuming that the inhibition of RegA activity by ERK2 is rapidly reversible, possibly by the action of a non-specific protein phosphatase, the phosphodiesterase would be expected to reduce internal cAMP such that the regulatory subunit inhibits the catalytic subunit of PKA once again. Neither ERK2 nor adenylyl cyclase would be inhibited any longer by PKA and spontaneous or catalyzed reactivation of these enzymes would reset the entire network to the state found prior to stimulation of the cells by external cAMP. During the time it takes to complete a cycle, the extracellular cAMP is degraded by a secreted phosphodiesterase, the product of the *pdsA* gene (Parent and Devreotes, 1996). When a new pulse is released, the cells can respond once again. The network that emerges from these observations (Fig. 1) has been confirmed by several other molecular genetic tests. For instance, mutants lacking CRAC have prolonged ERK2 activity following following a pulse of external cAMP (Aubry *et al.*, 1997). Adenylyl cyclase is not activated in these cells and cAMP levels do not rise (Insall *et al.*, 1994). PKA would not be expected to be activated or to inhibit ERK2. The network is broken by the lack of CRAC and these cells cannot develop.

One of the consequences of the network centered on PKA is that cAMP is produced in pulses. Adenylyl cyclase is active for a minute or so and then becomes refractory to further stimulation. It takes 2 or 3 minutes for the system to be reset such that the cells can respond to external cAMP by relaying the signal. Mathematical modeling of the relay and refractory states has been able to account for the generation and propagation of waves of excited cells and the entrainment of cells in a field by properly timed signals. These studies have helped to explain the observed patterns of concentric rings of cAMP that spread from a central pacemaker. However, until recently they were not able to account for the spiral patterns that are seen under certain conditions. Only *ad hoc* addition of cellular heterogeneity in the field generated spiral patterns. Now that we know more of the molecular biological details of Dictyostelium development it is possible to recognize where the heterogeneity comes from (Levine *et al.*, 1996).

Activation of PKA not only results in inhibition of adenylyl cyclase and ERK2 but also stimulates expression of genes in the network. Both *acaA* and *carA* mRNA accumulate to higher levels in cells stimulated by pulses of cAMP. Induction of these genes is dependent on the surface receptor, cAR1, the coupling components, Gα2, Gβ and CRAC, as well as PKA (Parent and Devreotes, 1996). In a wild type cell, the basal levels of cAR1 and adenylyl cyclase are sufficient for some cells to respond to cAMP in the environment resulting in further accumulation of these components. Some cells may be delayed in reaching the basal threshold and so will not enter the positive feedback loop for several hours. A field in which some of the cells are highly responsive and others show reduced relay properties is predicted to form spiral patterns able to bring more cells into an aggregate (Levine *et al.*, 1996).

3. CELL TYPE DIVERGENCE

As soon as the relay process becomes well developed, individual cells start to move chemotactically in the direction of the cAMP wave. Mutual adhesion and continued chemotactic aggregation breaks the field up into streams that collect in central mounds. At about this stage in development, individual cells can be seen to follow one or the other of two mutually exclusive patterns of gene expression. PKA plays an essential role in both of these pathways as shown by the phenotype of cells transformed with a vector carrying a modified version of the regulatory subunit gene, Rm, in which the cAMP sites have been destroyed. When expression of Rm is under the control of a prestalk specific promoter, the prestalk pathway is blocked; when expression of Rm is under the control of a prespore promoter, the prespore pathway is blocked (Zhukovskaya *et al.*, 1996; Hopper *et al.*, 1993b; 1995). PKA does not seem to determine the choice of pathway but merely facilitates the execution of other signals.

One of the prestalk specific differentiations is the accumulation of a pair of closely related membrane proteins, TagB and TagC, that appear to work together as an ATP driven exporter of a peptide used later in development to signal terminal differentiation (Shaulsky *et al.*, 1995). Mutants lacking either of these proteins due to disruption of *tagB* or *tagC* form both prestalk and prespore cells but neither cell type can undergo terminal differentiation. However, if they are mixed with wild type cells and allowed to develop as chimeras, the mutant prespore cells encapsulate normally when they receive the peptide signal from the wild type prestalk cells. The network that transduces the peptide signal in prespore cells appears to be built from many of the same components that are used in the aggregation network although the outcome is quite different.

4. A CULMINATION NETWORK

The phosphodiesterase gene, *regA*, was initially discovered from a screen of suppressor mutations that would allow *tagB*$^-$ null mutants to sporulate (Shaulsky *et al.*, 1996). Only later was RegA found to control PKA activity and function as well during aggregation. A full characterization of the phenotype resulting from inactivation of *regA* showed that spores are formed several hours earlier than in wild type strains and can do so in the absence of intercellular signaling (Shaulsky *et al.*, 1998). If *regA*$^-$ cells are treated with DIF, the chlorinated hexaphenone that induces stalk differentiation, they will also form stalk cells without direct cell-cell interaction (Kay, personal communication). As was seen in the stages of aggregation and cell type divergence, constitutive PKA activity resulting from disruption of either the *regA* gene or the *pkaR* gene affects the timing but not the nature of cell type specific differentiations. However, during culmination the signal that is responsible for inhibiting RegA emanates from prestalk cells and passes to the prespore cells where it leads to encapsulation (Richardson *et al.*, 1994).

A heat stable peptide has been purified from *regA*$^-$ cells that induces encapsulation of cells incubated at such a low density that they cannot aggregate (Anjard *et al.*, 1998). However, the peptide is not able to induce encapsulation in mutant strains lacking a receptor kinase encoded by the *dhkA* gene. Binding of the peptide to an extracellular loop in DhkA may activate it such that it modifies RegA and inactivates the phosphodiesterase (Fig. 2). The primary structure of *dhkA* indicated that its product is a member of the two component family of signal transduction systems first found in bacteria and more recently in eukaryotes (Wang *et al.*, 1996; Loomis *et al.*, 1997). In these sytems a histidine kinase phosphorylates the response regulator to affect a variety of processes. In Dictyostelium *dhkA* encodes a histidine kinase and *regA* encodes a response regulator. RegA has a domain near its N-terminus that has all the earmarks of a response regulator and is phosphorylatable. Although inactivation of *regA* suppresses the block to sporulation resulting from *dhkA*$^-$ mutations and so must function downstream of DhkA, the connection is probably indirect. Since a dominant gain of function mutation in *dhkA* can partially suppress the block to sporulation in *tagB*$^-$ cells, the pathway from prestalk cells where *tagB* is active, to *dhkA*, *regA* and *pkaC* in prespore cells is genetically supported (Tab. I). Since either overexpression of *pkaC* or inactivation of *pkaR* results in constitutive PKA activity, it appears that all that is need for encapsulation is PKA activity. DhkA adds a new connection to the outside such that a signal from prestalk cells can be received and result in inhibition of of RegA. It is not yet clear whether ERK2 plays a significant role at this stage but it certainly has not been ruled out.

Table I. Genetic suppressors*.

Genotype	**Viable Spores**
wild type	100%
tagB$^-$	$<10^{-10}$
tagB$^-$ *dhkA*$^{25-900}$	10^{-5}
regA$^-$	50%
regA$^-$ *pspA::R*$_m$	1.0%
tagB$^-$ *regA*$^-$	30%
tagB$^-$ *pkaR*$^-$	14%
dhkA$^-$	4%
dhkA$^-$ *regA*$^-$	35%
dhkA$^-$ *pkaR*$^-$	14%
dhkA$^-$ *pkaC::pkaC*	60%

*The number of spores is presented as a fraction of the original number of cells developed.

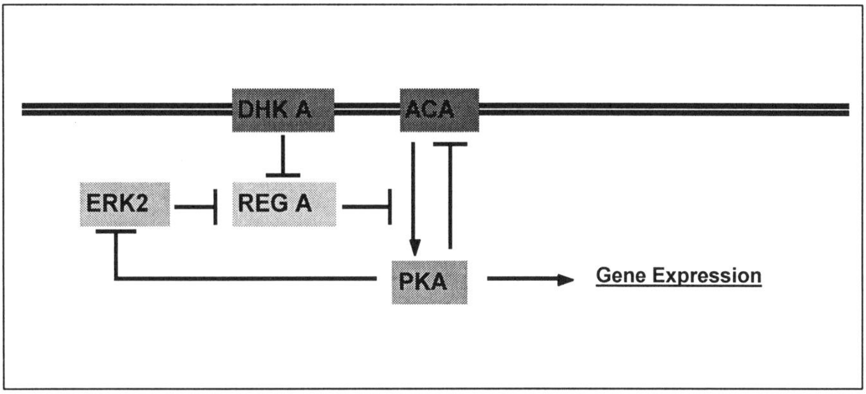

Fig. 2. A proposed culmination stage network. The membrane associated receptor kinase, DhkA, is activated by a peptide, SDF-2, released by prestalk cells. DhkA acts indirectly to result in the inhibition of the cytosolic phosphodiesterase, RegA. This signal transduction pathway might be mediated by activation of the MAP kinase ERK2 that inhibits RegA. When RegA is inhibited, cAMP generated by an adenylyl cyclase can accumulate and activate the protein kinase PKA that regulates gene expression. PKA activity also leads, either directly or indirectly, to inhibition of ERK2 and adenylyl cyclase.

Normally, changes in cAMP levels as well as the relative concentrations of the regulatory and catalytic subunits of PKA participate along with the phosphodiesterase in controlling the activity of PKA. However, it has recently been shown that mutants which never accumulate measurable levels of cAMP due to disruption of the *acaA* gene that encodes adenylyl cyclase will develop normally if *pkaC* is moderately overproduced from an artifical promoter (Wang and Kuspa, 1997). As long as the cell density is initially fairly high, aggregation can result from collisions of these mutant cells. Thereafter, the cell types diverge, sort out normally to form migrating slugs that will culminate to form properly proportioned fruiting bodies - all without any cAMP! Only PKA activity is required which in this case is ensured by having a slight excess of the catalytic subunit over the regulatory subunit.

Expression of *pkaC* in wild type strains occurs preferentially in prestalk cells following divergence of the cell types (Mann *et al.*, 1994). During the slug stage, PKA-C may accumulate in prestalk cells until constitutive PKA activity results from an excess of the catalytic subunit. PKA may then phosphorylate diverse proteins, one of which may facilitate accumulation of the peptide that triggers encapsulation of prespore cells. Thus, we can consider a feedback loop in which the peptide signal also activates DhkA in prestalk cells leading to inhibition of RegA phosphodiesterase activity and further stimulation of PKA such that more of the peptide is released. This would lead to a burst of peptide signal from the prestalk cells when they start to vacuolize into definitive stalk cells. The peptide would then rapidly diffuse to prespore cells and signal terminal differentiation. This loop would not function directly in prespore cells since they do not express the genes, *tagB* and *tagC*, that encode the ATP driven exporter.

5. EVIDENCE OF SIMILAR NETWORKS IN METAZOANS

cAMP dependent protein kinase (PKA) is a central component of several signal transduction pathways in Drosophila and vertebrates (Lane and Kalderon, 1993; Hammerschmidt *et al.*, 1996). In Drosophila these include inhibition of expression of *decapentaplegic* (dpp), *patched* (ptc) and *wingless* (wg) in pathways responsive to the secreted product of *hedgehog* (Lepage *et al.*, 1995; Pan and Rubin, 1995; Li *et al.*, 1995; Jiang and Struhl, 1995). In zebrafish PKA acts in a *hedgehog* responsive pathway during embryogenesis (Hammerschmidt *et al.*, 1996).

A role for PKA in higher cognitive processes such as long-term potentiation and memory in transgenic mice has been directly demonstrated by expressing a modified gene, R(AB), that encodes a PKA-R protein which is unable to bind cAMP at either of its two sites as the result of site directed mutations (Abel *et al.*, 1997). Expression of this dominant-negative gene under the control of the Ca^{++}/ calmodulin protein kinase IIα regulatory region in the hippocampus results in mice with defects in spatial and long-term memory. Likewise, mutations in the Drosophila genes that encode adenylyl cyclase (*rutabaga*), cAMP phosphodiesterase (*dunce*) or PKA-C (*DCO*) result in flies that are learning impaired and have additional developmental defects (Livingstone *et al.*, 1984; Levin *et al.*, 1992; Dudai *et al.*, 1976; Davis and Kiger, 1981; Kalderon and Rubin, 1988; Drain *et al.*, 1991; Skoulakis *et al.*, 1993). It seems that the memory defects in these mutant flies result from loss of cAMP control of PKA activity.

REFERENCES

Abel T., Nguyen P.V., Barad M., Deuel T.A., Kandel E.R. and Bourtchouladze R., Genetic demonstration of a role for PKA in the late phase of LTP and in hippocampus-based long-term memory, *Cell.* **88** (1997) 615-626.

Anjard C., Pinaud S., Kay R.R. and Reymond C.D., Overexpression of DdPK2 protein kinase causes rapid development and affects the intracellular cAMP pathway of Dictyostelium discoideum, *Development* **115** (1992) 785-790.

Anjard C., Zeng C. Loomis W.F. and Nellen W., Signal transduction pathways leading to spore differentiation in Dictyostelium discoideum, *Devel. Biol.* **193** (1998) 146-155.

Aubry L., Maeda M., Insall R., Devreotes P.N. and Firtel R.A., *J. Biol. Chem.* **272** (1997) 3883-3886.

Davis R.L. and Kiger J., *dunce* mutants of Drosophila melanogaster: mutants defective in the cyclic AMP phosphodiesterase enzyme system, *J. Cell. Biol.* **90** (1981) 101-107.

De Gunzburg J. and Veron M., A cAMP-dependent protein kinase is present in differentiating Dictyostelium discoideum cells, *EMBO J.* **1** (1982) 1063-1068.

Drain P., Folkers E. and Quinn W.G., cAMP-dependent protein kinase and the disruption of learning in transgenic flies, *Neuron* **6** (1991) 71-82.

Dudai Y., Jan Y.-N., Byers D., Quinn W. and Benzer S., *dunce*, a mutant of Drosophila deficient in learning, *Proc. Natl. Acad. Sci. USA* **73** (1976) 1684-1688.

Hammerschmidt M., Bitgood M.J. and McMahon A.P., Protein kinase A is a common negative regulator of Hedgehog signaling in the vertebrate embryo, *Genes Devel.* **10** (1996) 647-68.

Harwood A.J., Hopper N.A., Simon M.N., Bouzid S., Veron M. and Williams J.G., Multiple roles for cAMP-dependent protein kinase during Dictyostelium development, *Dev. Biol.* **149** (1992) 90-99.

Hopper N.A., Anjard C., Reymond C.D. and Williams J.G., Induction of terminal differentiation of Dictyostelium by cAMP-dependent protein kinase and opposing effects of intracellular and extracellular cAMP on stalk cell differentiation, *Development* **119** (1993a) 147-154.

Hopper N.A., Harwood A.J., Bouzid S., Veron M. and Williams J.G., Activation of the prespore and spore cell pathway of Dictyostelium differentiation by cAMP-dependent protein kinase and evidence for its upstream regulation by ammonia, *EMBO J.* **12** (1993b) 2459-2466.

Hopper N.A., Sanders G.M., Fosnaugh K.L., Williams J.G. and Loomis W.F., Protein kinase A is a positive regulator of spore coat gene transcription in Dictyostelium, *Differentiation* **58** (1995) 183-188.

Insall R., Kuspa A., Lilly P.J., Shaulsky G., Levin L.R., Loomis W.F. and Devreotes P., CRAC, a cytosolic protein containing a pleckstrin homology domain, is required for receptor and G protein-mediated activation of adenylyl cyclase in Dictyostelium, *J. Cell. Biol.* **126** (1994) 1537-1545.

Jiang J. and Struhl G., Protein kinase A and hedgehog signaling in Drosophila limb development, *Cell.* **80** (1995) 563-572.

Kalderon D. and Rubin G.M., Isolation and characterization of Drosophila cAMP-dependent protein kinase genes, *Genes Dev.* **2** (1988) 1539-1556.

Knetsch M.L.W., Epskamp S.J.P., Schenk P.W., Wang Y.W., Segall J.E. and Snaar-Jagalska B.E., Dual role of cAMP and involvement of both G-proteins and ras in regulation of ERK2 in Dictyostelium discoideum, *EMBO J.* **15** (1996) 3361-3368.

Lane M.E. and Kalderon D., Genetic investigation of cAMP-dependent protein kinase function in Drosophila development, *Genes Dev.* **7** (1993) 1229-1243.

Lee K.A.W., Transcriptional regulation by cAMP, *Curr. Opin. Cell. Biol.* **3** (1991) 953-959.

Lepage T., Cohen S.M., Diaz-Benjumea F. and Parkhurst S., Signal transduction by cAMP dependent protein kinase A in Drosophila limb patterning, *Nature* **373** (1995) 711-715.

Levin L.R., Han P.L., Hwang P.M., Feinstein P.G., Davis R.L. and Reed R.R., The Drosophila learning and memory gene rutabaga encodes a Ca^{2+}/calmodulin-responsive adenylyl cyclase, *Cell.* **68** (1992) 479-489.

Levine H., Aranson I., Tsimring L. and Truong T.V., Positive genetic feedback governs cAMP spiral wave formation in Dictyostelium, *Proc. Natl. Acad. Sci. USA* **93** (1996) 6382-6386.

Li W., Ohlmeyer J.T., Lane M.E. and Kalderon D., Function of protein kinase A in hedgehog signal transduction and Drosophila imaginal disc development, *Cell.* **80** (1995) 553-562.

Livingstone M.S., Sziber P.P. and Quinn W.G., Loss of the calcium/calmodulin responsiveness in adenylate cyclase of rutabaga, a drosophila learning mutant, *Cell.* **37** (1984) 205-215.

Loomis W.F. and Sternberg P.W., Genetic Networks, *Science* **269** (1995) 649.

Loomis W.F., Genetic networks that regulate development in Dictyostelium, *Microbiol. Rev.* **60** (1996) 135-150.

Lu S. Wang B., Sucgang R. and Kuspa A. (1998) submitted.

Mann S.K.O. and Firtel R.A., A developmentally regulated, putative serine/threonine protein kinase is essential for development in Dictyostelium, *Mech. Devel.* **35** (1991) 89-101.

Mann S.K.O., Yonemoto W.M., Taylor S.S. and Firtel R.A., DdPK3, which plays essential roles during Dictyostelium development, encodes the catalytic subunit of cAMP-dependent protein kinase, *Proc. Natl. Acad. Sci. USA* **89** (1992) 10701-10705.

Mann S.K.O., Richardson D.L., Lee S., Kimmel A.R. and Firtel R.A., Expression of cAMP-dependent protein kinase in prespore cells is sufficient to induce spore cell differentiation in Dictyostelium, *Proc. Natl. Acad. Sci. USA* **91** (1994) 10561-10565.

Mann S., Brown J., Briscoe C., Parent C., Pitt G., Devreotes P. and Firtel R., *Devel. Biol.* **183** (1997) 208-221.

Pan D. and Rubin G.M., Protein kinase and *hedgehog* act antagonistically in regulating *decapentaplegic* transcription in Drosophila imaginal discs, *Cell.* **80** (1995) 543-552.

Parent C.A. and Devreotes P.N., Molecular genetics of signal transduction in Dictyostelium, *Ann. Rev. Biochem.* **65** (1996) 411-440.

Part D., De Gunzburg J. and Veron M., The regulatory subunit of cAMP-dependent protein kinase from Dictyostelium discoideum: cellular localization and developmental regulation analyzed by immunoblotting, *Cell. Differ.* **17** (1985) 221-227.

Posas F., Wurgler-Murphy S.M., Maeda T., Witten E.A., Thai T.C. and Saito H., Yeast HOG1 MAP kinase cascade is regulated by a multistep phosphorelay mechanism in the SLN1-YPD1-SSK1 "two-component" osmosensor, *Cell.* **86** (1996) 865-875.

Richardson D.L., Loomis W.F. and Kimmel A.R., Progression of an inductive signal activates sporulation in Dictyostelium discoideum, *Development* **120** (1994) 2891-2900.

Shaulsky G., Escalante R. and Loomis W.F., Developmental signal transduction pathways uncovered by genetic suppressors, *Proc. Natl. Acad. Sci. USA* **93** (1996) 15260-15265.

Shaulsky G., Kuspa A. and Loomis W.F., A multidrug resistance transporter serine protease gene is required for prestalk specialization in Dictyostelium, *Genes Devel.* **9** (1995) 1111-1122.

Shaulsky G. Fuller D. and Loomis W.F., A cAMP-phosphodiesterase controls PKA-dependent differentiation, *Development* **125** (1998) 691-699.

Simon M.N., Pelegrini O., Veron M. and Kay R.R., Mutation of protein kinase-A causes heterochronic development of Dictyostelium, *Nature* **356** (1992) 171-172.

Skoulakis E.M., Kalderon D. and Davis R.L., Preferential expression in mushroom bodies of the catalytic subunit of protein kinase A and its role in learning, *Neuron* **11** (1993) 197-208.

Taylor S.S., Buechler J.A. and Yonemoto W., cAMP-dependent protein kinase: framework for a diverse family of regulatory enzymes, *Ann. Rev. Biochem.* **9** (1990) 971-1005.

Wang B. and Kuspa A., *Dictyostelium* development in the absence of cAMP, *Science* **277** (1997) 251-254.

Wang N., Shaulsky G., Escalante R. and Loomis W.F., A two-component histidine kinase gene that functions in Dictyostelium development, *EMBO J.* **15** (1996) 3890-3898.

Zhukovskaya N., Early A., Kawata T., Abe T. and Williams J., cAMP-dependent protein kinase is required for the expression of agene specifically expressed in Dictyostelium prestalk cells, *Devel. Biol.* **179** (1996) 27-40.

LECTURE 18

H-Bond Networks in Stability and Function of Biological Macromolecules

M.-C. Bellissent-Funel

Laboratoire Léon-Brillouin (CEA-CNRS), CEA Saclay, 91191 Gif-sur-Yvette, France

1. INTRODUCTION

Liquid water is a particularly complex molecular liquid. This is due essentially to the intermolecular hydrogen bond formation. The hydrogen bonding is established between two oxygen atoms through a proton and the electronic distribution called "lone pair" which is due to the hybridation of the electronic orbitals. The H-bonds have interaction energies comparable to ionic bonds and are characterised by a high directionality which gives rise to a locally tetrahedral coordination of water molecules (Fig. 1) (Walrafen, 1972). An other characteristic of liquid water is the short hydrogen bond lifetime (between 10^{-13} and 10^{-12} s).

The structural and dynamic properties of bulk water are now mostly well understood in some ranges of temperatures and pressures. In particular, many investigations using different techniques such as X-ray diffraction, neutron scattering, nuclear magnetic resonance (NMR), differential scanning calorimetry (DSC), molecular dynamics (MD), Monte Carlo (MC), simulations have been performed in the deeply supercooled regime (Angell, 1981; Lang and Ludemann, 1982; Chen and Teixeira, 1985; Dore, 1985; Bellissent-Funel, 1991; Chen, 1991; Poole *et al.*, 1992) and in a situation where the effects due to the hydrogen bonding are dominant.

However, in many common situations, water is not in its bulk form but instead attached to some substrates or filling some cavities. In particular, in living systems, essential water-related phenomena occur in restricted geometries in cells, and at active sites of proteins and membranes or at their surface. We know that this interfacial or confined water plays a crucial role in the stability and the catalytic function of biological macromolecules.

This paper recalls some properties of bulk water before focussing on the newly discovered properties of confined water. Some selected results relative to water-protein interactions as compared to water-model systems interactions are presented.

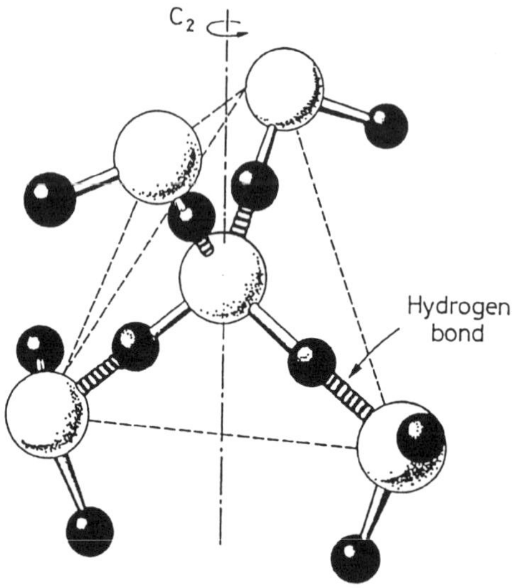

Fig. 1. Hydrogen bonding and local tetrahedral geometry of a water molecule. Small spheres, H atoms; large spheres, O atoms. Disks refer to hydrogen bonds (adapted from Walrafen, 1971).

2. LIQUID WATER AS A TRANSIENT GEL

At the microscopic level, liquid water can be seen as a transient gel or a network of hydrogen bonds with a local tetrahedral symmetry. This picture has been developed by the percolation model of Stanley and Teixeira (1980) which gives a good explanation of the enhanced anomalies of the thermodynamic and transport properties of liquid water observed at low temperatures. The connectivity properties of the network have been studied by computer molecular dynamics (CMD) simulations (Geiger and Stanley, 1982). It appears that the hydrogen-bond network includes tiny spatially-correlated "patches" of four-bonded molecules (Fig. 2), and that the local density near a "patch" is lower than the global density. The size of the patches has been estimated by evaluating their radius of gyration. The value is consistent with that of the correlation length of the density fluctuations as deduced by X-ray scattering (Bosio *et al.*, 1981) and neutron scattering (Bosio *et al.*, 1989). Within the Ornstein-Zernike theory, the correlation length has been estimated as equal to 8 Å at –20 °C, and it increases when the temperature decreases.

The microscopic structure of bulk and confined water is currently studied by using X-ray or/and neutron diffraction techniques which are complementary techniques. These diffraction techniques allow to access to the intermolecular pair correlation function g(r) (Lovesey, 1987) of a system that is the probability density of finding another atom lying in another molecule at a distance r from any atom. In X-ray measurements g(r) is the pair correlation function of the molecular centres, to a good approximation equal to the oxygen-oxygen correlation function. In neutron measurements g(r) is the weigthed sum of the three partial functions relative respectively to the oxygen-oxygen pairs, oxygen-deuterium pairs and deuterium - deuterium pairs. In particular, it is heavily weighted towards deuterium - deuterium and oxygen-deuterium partial correlation functions.

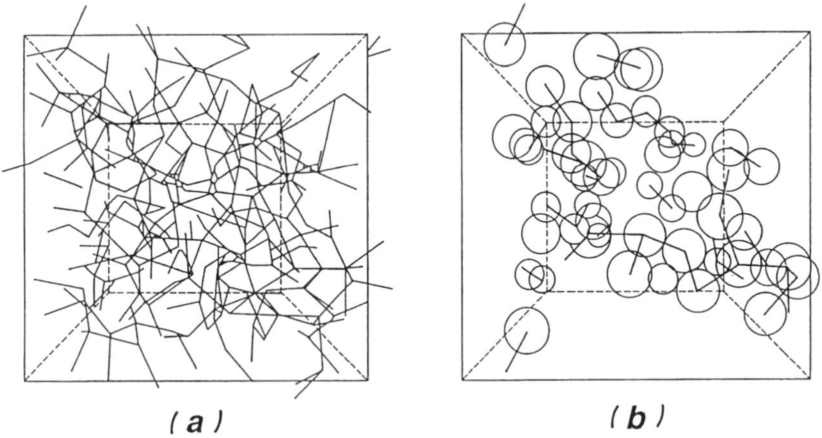

Fig. 2. The results of an instantaneous "snapshot" of a Rahman-Stillinger molecular dynamics simulation of 216 ST2 water-like particles. (a) The hydrogen bond network; (b) the tiny ramified clusters or patches in the network (adapted from Geiger and Stanley, 1982).

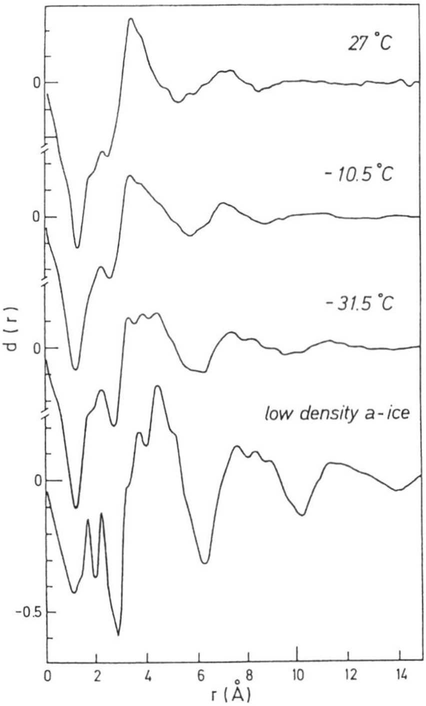

Fig. 3. Neutron pair correlation function d(r) of D_2O as a function of temperature. (Adapted from Bellissent-Funel, 1991).

Figure 3 gives the intermolecular d(r) function as defined by $d(r) = 4\pi\rho r[g(r) - 1]$, as a function of temperature; ρ is the molecular number density of water. Some comparison is made with the low density amorphous ice. We observe simultaneously an out of phase of the oscillations of the d(r) function and an increase of the large-r oscillations amplitude as the temperature is decreased. This demonstrates that the structure of deeply supercooled water is characterized by some progressive building of the H-bond network and as a consequence tends towards that of the low density amorphous ice.

3. INTERFACIAL WATER

Water in confined space has attracted a considerable interest in the recent years. It is commonly believed that the structure and dynamics of water are modified by the presence of solid surfaces, both by a change of hydrogen bonding and by a modification of the molecular motion; this depends on the distance of water molecules from the surface.

In the field of biology, the effects of hydration on equilibrium protein structure and dynamics are fundamental to the relationship between structure and biological function (Rupley and Careri, 1991; Smith, 1991; Colombo *et al.*, 1992; Steinbach and Brooks, 1993; Lounnas and Pettitt, 1994). In particular, the assessment of perturbation of liquid water structure and dynamics by hydrophilic and hydrophobic molecular surfaces is fundamental to the quantitative understanding of the stability and enzymatic activity of globular proteins and functions of membranes. Examples of structures that impose spatial restriction on water molecules include polymer gels, micelles, vesicles and microemulsions. In the last three cases since the hydrophobic effect is the primary cause for the self organization of these structures, obviously the configuration of water molecules near the hydrophilic-hydrophobic interfaces is of considerable relevance.

The structural and dynamic properties of water may be affected by both purely geometrical confinement and/or interaction forces at the interface. Therefore, a detailed description of these properties must take into account, the nature of the substrate and its affinity to form bonds with water molecules, and the hydration level or number of water layers. In order to discriminate betwween these effects, reliable model systems exhibiting hydrohilic or hydrophobic interactions with water are required. This appears the appropriate strategy to be developed to access to some understanding of the behaviour of water close to a biological macromolecule, as presented in the following.

In the last few years, computer simulations and theoretical treatments of the structure and dynamics of water in different kinds of environments have been undertaken (Levitt and Sharon, 1988; Rossky, 1994) and some important results are now available: for instance, molecular dynamics simulations indicate that the water density increases up to 1.5 g cm^{-3} in the first few angstroem of the shell around a protein and give information concerning the pair correlation functions and orientations of water molecules (Alary *et al.*, 1993). Figure 4 gives the hydration network of myoglobin as deduced from the simulation of Lounnas and Pettitt (1994).

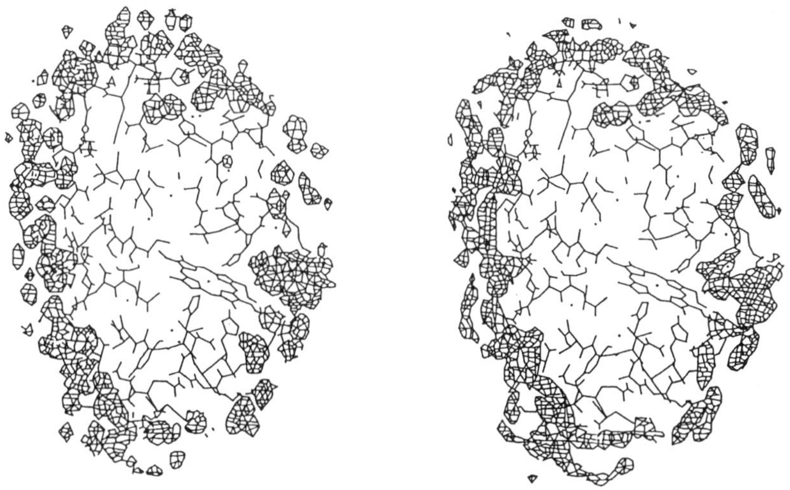

(a) (b)

Fig. 4. The hydration network of myoglobin reconstructed with the anisotropic harmonic model and the solvent density map obtained directly from the simulation. (Adapted from Lounnas and Pettitt, 1994).

4. WATER-PROTEIN INTERACTIONS

The amount of information about protein-water correlations is small. For neutron diffraction, deuterated samples are required and difficult to obtain. However, the first results have been obtained in the case of a photosynthetic C-phycocyanin protein. The X-ray crystallographic structure is now known to a good resolution of 1.66 Å (Duerring *et al.*, 1987).

C-phycocyanin is abundant in blue-green algae. Nearly 99% deuterated samples of this phycobiliprotein were isolated from the cyanobacteria which were grown in perdeuterated cultures (Crespi, 1977) (99% pure D_2O) at Argonne National Laboratory. This process yielded deuterated C-phycocyanin proteins (d-CPC) that had virtually all the ^{1}H-C bonds replaced by ^{2}H-C bonds. One can obtain a lyophilised sample which is similar to amorphous solids as determined by neutron diffraction (Bellissent-Funel *et al.*, 1993). As it has been defined in previous papers (Bellissent-Funel *et al.*, 1992, 1996), the level of hydration h = 0.5 corresponds to 100% hydration of C-phycocyanin which leads to a coverage of about 1.5 monolayers of water molecules on the surface of the protein (Middendorf, 1995). In the neutron scattering experiments, we prepared several samples of deuterated C-phycocyanin protein (d-CPC) with different levels of hydration from h = 0.5 to 0.1. The water content was measured by the increase in weight of the protein sample after exposing the protein to the vapor of water. The hydrated powders were wrapped in thin aluminium foil and placed inside the appropriate container. These samples contain these amount of water in addition to the 4% of water molecules (D_2O) which have to be considered like in many other proteins, as an integral part of the molecule.

The water (D_2O)-protein correlations at the surface of a fully deuterated amorphous protein C-Phycocyanin (d-CPC) have been studied by neutron diffraction as functions of temperature and hydration level (Bellissent-Funel *et al.*, 1993); the resulting pair correlation functions d(r) are given in Figure 5. In the case of the lowest hydrated sample (h = 0.175), the perturbation to the structure of protein due to water of hydration is not detectable. For the highest hydrated sample (h = 0.365), a definite peak appears at 3.5 Å in the pair correlation function. This distance is the average distance between the center of mass of a water molecule in the first hydration layer and amino-acid residues on the surface of the protein. This correlation distance of 3.5 Å measured in these diffraction experiments compared well with computer simulations work on polypeptides and proteins (Levitt and Sharon, 1988; Rossky and Karplus, 1979) and has been interpreted as resulting from some increase in the clustering of water molecules.

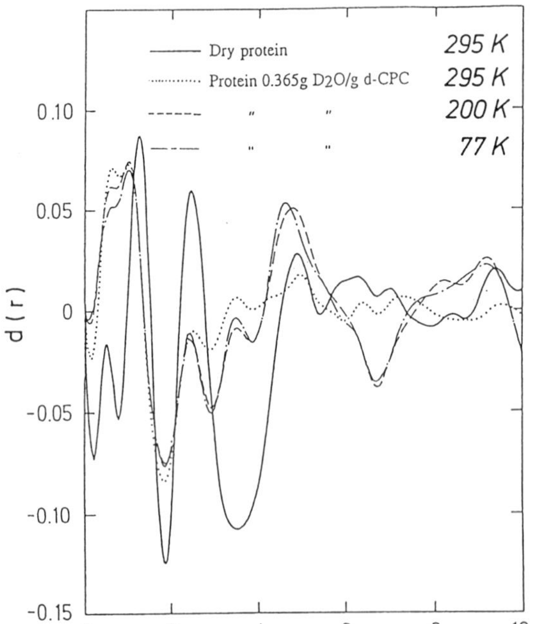

Fig. 5. Pair correlation function d(r) for a dry d-CPC protein at 295 K and for a D_2O-hydrated (h = 0.365) d-CPC protein at different temperatures. (Adapted from Bellissent-Funel *et al.*, 1993).

Some similarity between the behaviour of water close to C-phycocyanin protein and close to hydrophilic model systems such as a porous Vycor glass can be stressed. In fact, for low hydrated protein samples, no crystallisation of water is detectable which is also the case for low hydrated Vycor glass (Zanotti, 1997) while for more than one monolayer coverage there is appearance of the hexagonal crystalline ice. However, it should be noted that at the highest hydration level, water nucleates into hexagonal ice at low temperature; this is in contrast with hydrated Vycor where water nucleates into cubic ice. It is important to mention the recent findings about the structure of water from low hydrated hydrophilic sample; in fact, at room temperature, the structure of interfacial water looks like the structure of deeply supercooled water and exhibits a small temperature dependence (Zanotti, 1997).

Some other interesting findings come from the study of glass-liquid transition and crystallisation behaviour of water trapped in loops of methemoglobin chains (Sartor *et al.*, 1993).

5. DYNAMICS OF CONFINED WATER

Traditionally, the dynamics of interfacial water has been studied by nuclear magnetic relaxation techniques. Piculell and Halle (1986) have shown that ^{17}O magnetic relaxation in water is dominated by a quadrupolar coupling to the electric field gradient of intramolecular origin. Thus, it is a particularly suitable method for investigating the single-particle dynamics of interfacial water and thus the protein-water interaction. The main general findings from these techniques are a reduced lateral mobility of water molecules (10 – 100 times) and a long residence time (10 – 100 ps). With regard to the solvent diffusion constant near protein and silica surfaces there are reports from other groups that it is reduced by a factor of about 5 compared with that of bulk water (Polnaszek *et al.*, 1987).

An ideally microscopically detailed method for exploring the change in hydrogen-bonding patterns as well as the translational and rotational diffusion constants and residence times of water molecules, when they are near surfaces is computer molecular dynamics (CMD). Results of CMD simulations (Lee and Rossky, 1994) generally indicate that the dynamics of water molecules on protein and silica surfaces where hydrophilic interactions are dominant suffer only a mild slowing down compared to bulk water. More specifically, it has been reported that the retardation is by a factor of *ca.* 2 in the protein case and about a factor five in the silica case. Residence times of water in the first hydration layer are typically of the order of 100 ps.

These CMD results are still qualitative and somewhat conflicting with the available experimental data (Piculell and Halle, 1986), largely because of the simplified models used for the surfaces and more certainly due to difficulties in choosing suitable potential functions for the simulations. However, recently, molecular dynamics simulations of the hen egg-white lysozyme-Fab D1.3 complex have been reported; both the crystal state and the complex in solution were studied (Alary *et al.*, 1993). The findings are consistent with the observation by various experimentalists of a reduced water mobility in a region extending several angstroems beyond the first hydration layer (Piculell and Halle, 1986) as reported also from CMD simulations (Wong and McCammon, 1986).

From the above comparison it seems clear that there are considerable discrepancies in the degrees of slowing down between NMR experiments and CMD. This is especially true for the translational diffusion constant.

A way to resolve these discrepancies has been recently attempted by quasi-elastic and inelastic neutron scattering which is a powerful and unique tool for studying the self dynamics of interfacial water; actually the large incoherent scattering cross section of the protons yields unambiguous results about the individual motions of water molecules (Lovesey, 1987). In fact, this technique is a method for studying the diffusive motion of atoms in solids and liquids (Springer, 1972). It gives access to the correlation function for the atomic motions which are explored over a space domain of the order of a few Å and for times of the order of 10^{-12} s. This space and time domain is similar to that of CMD which makes the comparison between neutron scattering and CMD justified. This method has been used with success for studying the self dynamics of bulk water as a function of the temperature as previously reported (Teixeira *et al.*, 1985).

The studies of single particle dynamics of hydration water in proteins have been hampered by the fact that about 40% of the constituent atoms in a typical protein molecule are hydrogen atoms, present in the backbone and in the side chains. The elastic contribution is thus too large for an accurate determination of the dynamical parameters which are characteristic of hydration process. However, by working with a deuterated protein/H_2O system, it has been possible recently to focus on the water dynamics at and near the protein surface (Bellissent-Funel *et al.*, 1996).

Common features arise from the quasi-elastic neutron scattering studies for water close to some hydrophilic Vycor surface or for water close to a more complex surface such that of a protein. In the case of water close to different residues of a biological material several contributions have to be considered such as the hydrogen atoms of the protein itself, the possibility of their exchange with water molecules, the presence of hydrophilic and hydrophobic regions.

Results of a C-phycocyanin protein at a hydration level h = 0.4 are particularly important since this monolayer coverage of water molecules allows the protein to initiate its function (Rupley and Careri, 1991). Figure 6 gives the behaviour of the linewidth Γ of the Lorentzian line as deduced from the quasi-elastic spectra. While the linewidth of the bulk water is characteristic of a random jump diffusion model, that of water from hydrated protein exhibits a different behaviour. Γ is analysed at low Q in terms of a model of a diffusion inside a sphere of radius a; the diffusion coefficient inside the sphere is D_{local} (Volino and Dianoux, 1980). At high-Q, Γ is accounted by the random jump diffusion model characterized by the diffusion coefficient D_t and the residence times τ_0. Table I gives for hydrated protein (h = 0.4) the values of D_{local} and D_t, and those of the residence times τ_0 for confined water as compared with the corresponding values of bulk water.

For hydrated protein (Tab. I), the values obtained for D_{local} are lower than those of bulk water; they are close to those obtained at the same temperature for 25% H_2O-hydrated Vycor (Bellissent-Funel *et al.*, 1995) which demonstrates the influence of the hydrophilic groups on the water molecules when one reaches a

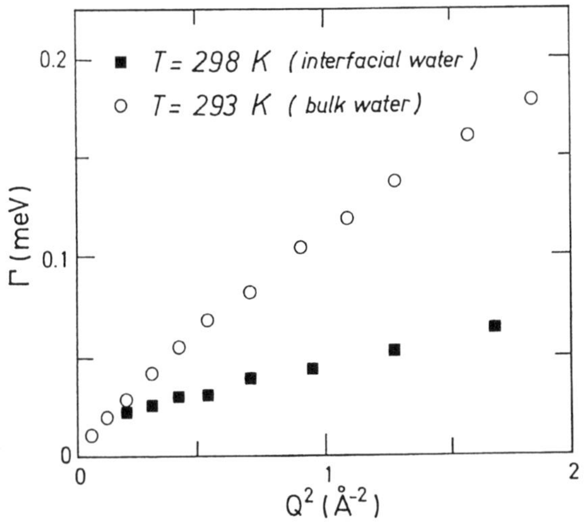

Fig. 6. Behaviour of the linewidth of the Lorentzian line as deduced from quasi-elastic neutron scattering for H_2O-hydrated d-CPC (h = 0.4), at 27 °C. The comparison is made with bulk water at the same temperature. (Adapted from Bellissent-Funel *et al.*, 1996).

Table I. Parameters for water near surface of a C-phycocyanin protein (hydrated lyophilised sample, h = 0.40).

T (°C)	a (Å)	D_{local} Confined water 10^{-5} cm² s⁻¹	D_t Confined water 10^{-5} cm² s⁻¹	D_t Bulk water 10^{-5} cm² s⁻¹	τ_0 Confined water (ps)	τ_0 Bulk water (ps)
40	4.5	1.28	1.52	3.20	5.9	0.90
25	4.3	0.97	1.20	2.30	6.6	1.10
0	4.0	0.84	0.76	1.10	8.2	3.00

monolayer coverage. This shows that the diffusive motion of water molecules is strongly retarded by interactions with a protein surface. The residence times τ_0 of confined water from 25% hydrated Vycor and from hydrated protein are always longer than the residence time of bulk water at the same temperature. They increase rapidly as either the temperature or the level of hydration decreases.

The values of the hydrogen-bond life time τ_1 for confined water are slightly higher than that of bulk water (Bellissent-Funel *et al.*, 1995); they have an Arrhenius

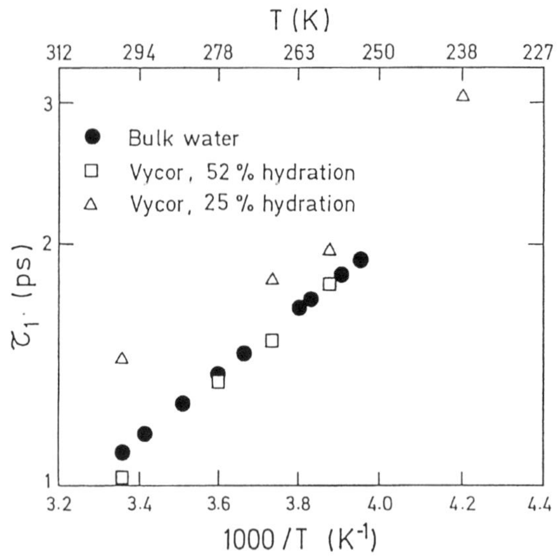

Fig. 7. Arrhenius plot of the hindered rotations characteristic time, τ_1. This time can be associated with the hydrogen bond life time (Adapted from Bellissent-Funel *et al.*, 1995).

temperature dependence while the residence time τ_0 does not exhibit such a behaviour (Fig. 7).

6. CONCLUSION

In this paper, we have presented the more recent up to date account of the structure and dynamics of water molecules on surface of a deuterated protein as compared with that of bulk water.

From the more recent findings combining neutron techniques and molecular dynamics simulations, it is now possible to have a more precise picture of confined water. Water in the vicinity of a hydrophilic surface is in a state equivalent to supercooled bulk water at a lower temperature which has been previously demonstrated. The nucleation properties depend on the degree of hydration of the sample. In particular, room temperature interfacial water exhibits some peculiar structural properties and shows a dynamic behaviour similar to that of bulk water at a temperature 25 K lower.

It appears that the short time dynamics of water molecules at or near hydrophilic model surface and at a soluble protein surface is much slower as compared to that of the bulk water. It is important to notice that the more significant slow dynamics of interfacial water is reflected in the long residence time for jump diffusion. This is the consequence of the confined diffusion theory which has been used to analyse the quasi-elastic neutron scattering data. This simple theory gives information on the confinement volume and the slow dynamics of the single particle motions. This suggests that there may be a common underlying mechanism for the slowing-down of the single-particle dynamics of interfacial water. In order to understand the microscopic origin of the confinement and slowing down of motions of water molecules and the exact role played in this context, the theory of kinetic glass transition in dense supercooled liquids (Gotze and Sjogren, 1992) has been recently used. This theory leads to some description of the dynamics of confined water in terms of correlated jump diffusion (Chen *et al.*, 1996) instead of jump diffusion (Teixeira *et al.*, 1985). This description looks consistent with recent molecular dynamics simulations of supercooled water (Gallo *et al.*, 1996).

This more sophisticated way shows a large distribution of residence times for water molecules in the cage formed by the neighbouring molecules, which is a more realistic view than the sharp separation of water molecules into two classes, according to their mobility (Bellissent-Funel *et al.*, 1996). Short time dynamics results about hydrated myoglobin have been recently interpreted by using this same theory of kinetic glass transition in dense supercooled liquids (Settles and Doster, 1996).

Finally, new progresses have recently been done to understand the origin of the temperature shift, of the order of 25 K, proper to water close to some hydrophilic surface. That can be explained within the reliable percolation model of water (Stanley and Teixeira, 1980), assuming that only three hydrogen bonds can be formed between interfacial water molecules instead of four in bulk water.

ACKNOWLEDGEMENTS

The author is grateful to H.L. Crespi for his continuing collaboration in the preparation of the perdeuterated protein C-phycocyanin and to L. Bosio, S.H. Chen and J. Teixeira for the collaboration and the fruitful discussions along the last many years.

REFERENCES

Alary F., Durup J. and Sanejouand Y.-H., *J. Phys. Chem.* **97** (1993) 13864.

Angell C.A., Water: *A comprehensive treatise*, edited by F. Franks, Vol. 7 (Plenum Press, New-York, London, 1981) Chap. 1.

Bellissent-Funel M.-C., in "*Hydrogen Bonded Liquids*", edited by J.C. Dore and J. Teixeira, NATO ASI Series C (Kluwer Academic Publishers, 1991).

Bellissent-Funel M.-C., Chen S.H. and Teixeira J., *J. Phys. I France.* **2** (1992) 995-1001.

Bellissent-Funel M.-C., Lal J., Bradley K.F. and Chen S.H., *Biophys. J.* **64** (1993) 1542.

Bellissent-Funel M.-C., Chen S.H. and Zanotti J.-M., *Phys. Rev. E.* **51** (1995) 4558.

Bellissent-Funel M.-C., Zanotti J.-M. and Chen S.H., *Far. Disc.* **103** (1996) 281-294.

Bosio L., Teixeira J. and Stanley H.E., *Phys. Rev. Lett.* **46** (1981) 597.

Bosio L., Teixeira J. and Bellissent-Funel M.-C., *Phys. Rev. A* **39** (1989) 6612.

Chen S.H. and Teixeira J., *Adv. Chem. Phys.* **64** (1985) 1.

Chen S.H., in "*Hydrogen-Bonded Liquids*", edited by J.C. Dore and J. Teixeira, NATO ASI Series C329 (Kluwer Academic Publishers, 1991) p. 289.

Chen S.H., Gallo P. and Bellissent-Funel M.-C., *Non Equilibrium Phenomena in Supercooled Fluids, Glasses and Materials* (World Scientific Publication).

Dore J.C. (1985), in "*Water Science Reviews*", edited by F. Franks F. (Cambridge University Press, 1996).

Colombo M.F., Rau D.C. and Parsegian V.A., *Science* **256** (1992) 655.

Crespi H.L., in *Stable Isotopes in the Life Science,* IAEA, Vienna **111** (1977).

Duerring M., Schmidt G.B. and Huber R., *J. Mol. Biol.* **217** (1987) 577.

Gallo P., Sciortino F., Tartaglia P. and Chen S.H., *Phys. Rev. Lett.* **76** (1996) 2730.

Geiger A. and Stanley H.E., *Phys. Rev. Lett.* **49** (1982) 1749.

Gotze W. and Sjogren L., *Rep. Prog. Phys.* **55** (1992) 241.

Lang E.W. and Ludemann H.D., *Angew. Chem. Int.* **21** (1982) 315.

Lee S. H. and Rossky P.J., *J. Chem. Phys.* **100** (1994) 3334.

Levitt M. and Sharon R., *Proc. Natl. Acad. Sci. USA: Biophysics* **85** (1988) 7557.

Lounnas V. and Pettitt B.M., *Proteins: Structure, Function and Genetics* **18** (1994) 148.

Lovesey S.W., *Theory of neutron scattering from condensed matter*, 3rd edition (Clarendon Press, Oxford, 1987).

Middendorf H.D., in "*Nonlinear Excitations in Biomolecules*", edited by M. Peyrard (Les Éditions de Physique, Les Ulis, 1995) p. 369.

Piculell L. and Halle B., in "*Water and Aqueous Solutions*", edited by G.W. Neilson and J.E. Enderby (Adam Hilger, Bristol, 1986) p. 219 and references therein.

Polnaszek C.F., Hanggi D.A., Carr P.W. and Bryant R.G., *Analyt. Chim. Acta* **194** (1987) 311.

Poole P.H., Sciortino F., Essman U. and Stanley H.E., *Nature* **360** (1992) 324 and references therein.

Rossky P.J. and Karplus M., *J. Am. Chem. Soc.* **101** (1979) 1913.

Rossky P.J. in "*Hydrogen Bond Networks*", edited by M.C. Bellissent-Funel and J.C. Dore, *NATO ASI C* **435** (Kluwer Academic Publ., 1994) p. 337.

Rupley J.A. and Careri G., *Adv. Prot. Chem.* **41** (1991) 37.

Sartor G., Hallbrucker A., Hofer H. and Mayer E., *Conf. Proc. Water Biomolecule Interactions* (SIF, Bologna, 1993) pp. 143-146

Settles M. and Doster W., *Far. Disc.* **103** (1996).

Smith J., *Quart. Rev. Biophys.* **24** (1991) 227.

Springer T., Quasi-elastic neutron scattering for the investigation of diffusive motions in solids and liquids, *Springer Ser. Modern Phys.* **64** (1972).
Stanley H.E. and Teixeira J., *J. Chem. Phys.* **73** (1980) 3404.
Steinbach P.J. and Brooks B.R., *Proc. Natl. Acad. Sci. USA* **90** (1993) 9135.
Teixeira J., Bellissent-Funel M.-C., Chen S.H. and Dianoux A.J., *Phys. Rev. A* **31** (1985) 1913.
Volino F. and Dianoux A.J., *Mol. Phys.* **41** (1980) 271.
Walrafen, Water. *A comprehensive treatise*, edited by F. Franks, Vol. 1 (Plenum Press, New-York, London, 1981) Chap. 5.
Wong C.F. and McCammon J.A., *Isr. J. Chem. Phys.* **27** (1986) 211.
Zanotti J.-M. (1997) Ph. D Thesis.

LECTURE 19

A Network of Cell Interactions Mediated by Frizzled is Essential for Gastrulation and Anteroposterior Axis Determination in *Xenopus* Embryos

S.Y. Sokol and K. Itoh

Department of Microbiology and Molecular Genetics, Harvard Medical School and Molecular Medicine Unit, Beth Israel Deaconess Medical Center, 330 Brookline Ave., Boston, MA, USA

1. INTRODUCTION

Generation of diversity among embryonic cells is essential for development of all multicellular organisms. In a nematode *Caenorhabditis elegans* many cells divide asymmetrically to give rise to pairs of different daughter cells (Horvitz and Herskovitz, 1992). Unequal cell division may depend on cell polarity, *i.e.* polarization of the mitotic spindle and cytoskeletal elements inside the cell. In the embryo, cell polarity is further translated into polarity of organs and tissues, leading to generation of biological patterns.

The process of asymmetric cell division is affected in several *C. elegans* mutants, and studies of the mutated genes may provide insights into how cell polarity is established at the molecular level. One of these mutations is in a gene *lin-17* which encodes a member of the Frizzled family of putative transmembrane receptors (Sawa *et al.*, 1996). *Frizzled* mutation has been initially identified in *Drosophila* and leads to disruption of the polarized pattern of hairs on the fly wing (reviewed by Adler, 1992). Similar to unequal cell division, the direction of hair growth may be based on polarization of cytoskeletal elements. Thus, the molecular and cellular pathways leading to generation of cell and tissue polarity appear to be conserved both in flies and in worms.

Polarization of cytoskeletal elements within each embryonic cell may be regulated by the neighbouring cells. Cell-to-cell signaling or induction is often responsible for establishment of embryonic polarity and tissue patterns. In fact, vertebrate body plan is thought to be specified through a cascade of inductive interactions. One class of signaling molecules that are active throughout animal development are Wnts (Parr and McMahon, 1994; Nusse and Varmus, 1992). Wnt genes are related to the Wnt1-proto-oncogene and are expressed in a variety of tissues and organs. More than a dozen of different Wnt genes have been described

(Parr and McMahon, 1994) and shown to be required for many developmental processes in which cell and tissue polarity is established, including segment polarity in *Drosophila* (reviewed by Perrimon, 1994), vertebrate limb polarity (Yang and Niswander, 1995; Riddle *et al.*, 1995; Parr and McMahon, 1994) and organogenesis (reviewed by Parr and McMahon, 1994).

Accumulating evidence indicates that determination of embryonic polarity by Wnt and Frizzled signaling may proceed through the same molecular pathways. Wnt products are secreted polypeptides, whereas Frizzled proteins consist of an extracellular cysteine-rich domain (CRD), seven transmembrane domains, and a short cytoplasmic tail (reviewed by Perrimon, 1996). Transfection of a cDNA encoding for *Dfz-2*, a novel *Drosophila* Frizzled homologue, into a non-responsive S2 cell line confers responsiveness to Wingless (Wg), a *Drosophila* Wnt1 homologue, and promotes Wg binding to the cell surface (Bhanot *et al.*, 1996). These experiments suggest that Frizzled may function as receptors for Wnt ligands.

Recent studies in vertebrates have identified two large Frizzled-related protein families: Frizzled transmembrane receptors which may be involved in the transduction of Wnt signals, and secreted polypeptides which are related to the Frizzled extracellular domain and antagonize Wnt signaling (Wang *et al.*, 1996; Leyns *et al.*, 1997; Mayr *et al.*, 1997; Rattner *et al.*, 1997; Wang *et al.*, 1997; Xu *et al.*, 1997). This review focuses on the recent experimental data suggesting that Frizzled and Frizzled-related proteins may coordinate morphogenetic movements during gastrulation and play a role in anteroposterior axis specification. A hypothesis is put forward that convergent extension movements are controlled by a pathway which is similar to the pathway involved in tissue polarity determination in *Drosophila*.

2. FRZA, A NATURALLY OCCURRING, SECRETED ANTAGONIST OF WNT SIGNALING

A screen for genes expressed by the bovine aortic endothelium led to identification of a novel protein, which shares a considerable similarity with Frizzled CRD, but lacks its transmembrane and intracellular domains (Duplaa and D'Amore, personal communication). This protein was named FrzA (Frizzled in Aorta) and shown to be actively secreted by transfected Cos cells (Xu *et al.*, 1998). Immunostaining and *in situ* hybridization revealed that FrzA is abundant in the aortic endothelium, brain and retinal neurons, lung and kidney epithelium, and cardiac myocytes. Since Wnts are known to be expressed in the majority of these tissues (Parr and McMahon, 1994), a hypothesis was tested that FrzA functions to modulate Wnt signaling.

To determine whether FrzA functions as a soluble Wnt receptor, we assessed whether FrzA is capable of *in vitro* binding to Wingless (Wg), a *Drosophila* Wnt-1 homologue. Immunoprecipitation of FrzA from cell supernatants containing soluble Wingless protein led to co-precipitation of Wg (Xu *et al.*, 1998). These data indicate that soluble Wg and FrzA proteins are bound to each other in cell culture supernatants.

To evaluate a possible role of FrzA in modulating Wnt pathways *in vivo*, the ability of FrzA to affect induction of secondary axes by Wnts (Sokol *et al.*, 1991) was also studied. FrzA inhibited formation of secondary axes triggered by Xwnt-8 and by human Wnt-2, but not by *Xenopus* Dishevelled, an intracellular component of the Wnt signal transduction pathway (Xu *et al.*, 1998). Furthermore, FrzA suppressed Xwnt-8-dependent activation of the early response genes in ectodermal explants (Xu *et al.*, 1998). These effects are consistent with direct interaction of FrzA and Wnt products *in vitro* and *in vivo*.

Together, these observations indicate that FrzA can bind to Wnts and inhibit their ability to signal. Naturally-occuring short forms of Frizzled similar to FrzA were independently discovered by several other groups (Leyns *et al.*, 1997; Mayr *et al.*, 1997; Rattner *et al.*, 1997; Wang *et al.*, 1997) and proposed to antagonize Wnt signaling. These inhibitors may be used as dominant negative forms of Frizzled to assess developmental role of Frizzled receptors in early pattern formation.

3. ROLE OF WNT/FRIZZLED SIGNALING IN ANTEROPOSTERIOR PATTERNING

3.1 Classical studies of anteroposterior axis formation

Determination of the anteroposterior body axis in vertebrate embryos has been one of the best known examples for the establishment of developmental polarity. Anteroposterior axis is most evident in the developing nervous system with the formation of the fore-, mid-, hindbrain and spinal cord. Studies of Spemann's group have shown that regional differences in the newly formed neural tissue are specified by signals emitted from the organizer (reviewed by Hamburger, 1988). Organizer is a special region of the early gastrula which elicits a secondary body axis formation upon transplantation to a recipient embryo. Inducing activities of the organizer vary considerably, depending on the precise source of the graft and on developmental stage, at which transplantations are carried out (Hamburger, 1988). These findings suggested that neural induction is a dynamic process which proceeds continuously throughout gastrulation. This was acknowledged in the early 1950s by Nieuwkoop and Eyal-Giladi, who proposed that neural tissue is induced and patterned in at least two steps (Fig. 1; reviewed by Slack and Tannahill, 1992). During the first step, dorsal ectoderm is “activated” (or “neuralized”) to specify anterior neural tissue. During the second step, the same cells which are initially specified to become anterior neural, are re-specified (“posteriorized”) to achieve more posterior fates (Slack and Tannahill, 1992; Sive *et al.*, 1989). Although the evidence for this mechanism remains circumstantial, the Nieuwkoop model illustrates that anteroposterior axis is specified progressively under the continuous action of inductive stimuli.

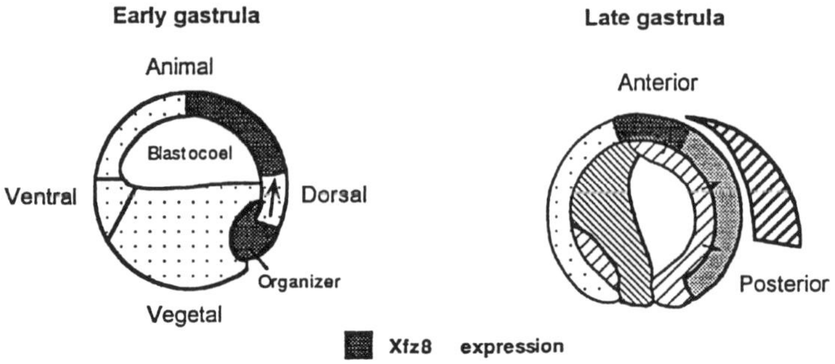

Fig. 1. Neural induction, anteroposterior patterning and Xfz8 expression.

3.2. Factors involved in anteroposterior axis determination

The model of progressive specification of anteroposterior axis has recently been supported by identification of factors which may mediate neuralizing activities of the organizer. Molecules implicated in anteroposterior patterning fall into three distinct classes based on the spectrum of region-specific markers that they induce in ectodermal cells (Fig. 2).

First class includes hedgehog (a secreted factor, Lai *et al.*, 1995), glycogen synthase kinase 3 (GSK3, an inhibitor of Wnt signaling, Itoh *et al.*, 1995) and Otx2 (a brain-specific transcription factor, Blitz and Cho, 1995). These factors activate anterior ectodermal cell fates (such as cement gland or hatching gland) in isolated ectoderm in the apparent absence of neural tissue. Since these proteins induce the identical spectrum of markers, they could function in the same signal transduction pathway. This pathway may operate during the initial stage of signaling from the organizer, before neural development is specified (Sive *et al.*, 1989). Consistent with this idea, only anterior ectoderm markers (XAG1, Otx2) in the absence of neural tissue were activated by GSK3 in ectodermal *explants*, whereas ectopic neural tissue was induced by GSK3 *in whole embryos* where additional signals may operate (Itoh *et al.*, 1995).

The second class of factors consists of noggin, chordin, follistatin, Xnr3 and cerberus which induce anterior neural tissue markers, thereby resembling the "activation" step, proposed by Nieuwkoop (Hamburger, 1988; Slack and Tannahill, 1992). They are thought to function by inhibiting anti-neuralizing effects of bone morphogenetic proteins (for reviews, see Sasai and DeRobertis, 1997; Wilson and Hemmati-Brivanlou, 1997). The connection between the first and the second class factors is not so clear, they may represent two independent ways to specify anteroposterior cell fates.

The third class of factors has "transforming" or posteriorizing activity (in Nieuwkoop's terms) because these molecules induce posterior neural cell fates. This class includes fibroblast growth factors (FGF), retinoids and Wnts. FGF has been proposed as a candidate posteriorizing factor based on its ability to induce posterior neural markers in ectodermal explants and to posteriorize anterior neural tissue generated by noggin (see Doniach, 1995, for review). Retinoids are also involved in posteriorization of neural tissue (Durston *et al.*, 1989; Blumberg *et al.*, 1997), but their effects appear to be more spatially restricted (Blumberg *et al.*, 1997).

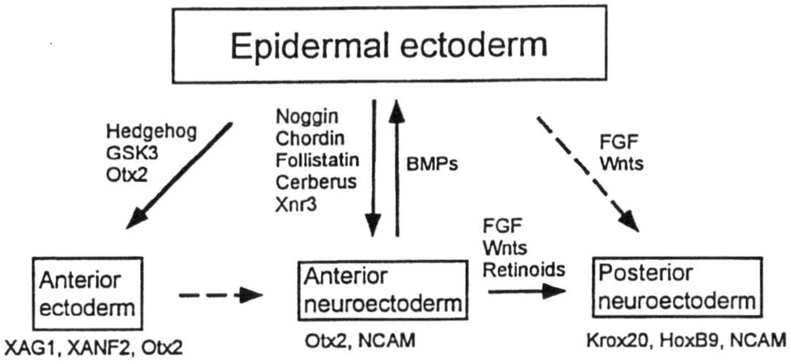

Fig. 2. Factors involved in anteroposterior patterning.

More studies are necessary to assess whether these activities are required for neural induction *in vivo.*

3.3 Wnt/Frizzled signaling and anteroposterior axis determination

Another class of posteriorizing factors are Wnts. Wnt3a was demonstrated to posteriorize anterior tissue induced by noggin, follistatin and hedgehog (McGrew *et al.*, 1995). Dishevelled, a component of the Wnt signal transduction pathway, was shown to mimic all three neuroectodermal cell states (Fig. 2) by inducing and posteriorizing ectodermal explants in a dose-dependent manner (Itoh and Sokol, 1997). Twofold increments in levels of Dishevelled mRNA caused a gradual shift in cell fates along the anteroposterior axis. Whereas lower doses of Xdsh mRNA activated anterior ectodermal and neural markers, higher doses induced more posterior neural tissue and posterior mesoderm markers (Itoh and Sokol, 1997). Together, these observations suggest that a Wnt-like activity may act early in embryogenesis to specify anteroposterior cell fates. Future experiments are required to evaluate the relative contribution of the FGF and Wnt signals to anteroposterior patterning and to determine whether FGF signals operate upstream, or downstream of Wnts, or in a parallel pathway.

Although several Wnts including Xwnt8, Xwnt5a and Xwnt11 are expressed at the gastrula stages (Moon *et al.*, 1993), at present it is not clear which Wnt protein may be responsible for the posteriorizing activity. It turns out that Wnts are able to bind to more than one Frizzled receptor, and that one Frizzled-like protein may inhibit the activity of several Wnts (Bhanot *et al.*, 1996; Xu *et al.*, 1998). Thus, Wnt-Frizzled interactions reveal considerable redundance. Since more than 15 different

Frizzled and a similar number of Wnts may be expressed during vertebrate development (Wang *et al.*, 1996; Parr and McMahon, 1994), the specificity of Wnt-Frizzled interactions remains an important issue to be addressed in the future studies.

3.4 Xfz8 is an anteriorly expressed homologue of Frizzled which may participate in axis formation

In a search for candidate molecules which may mediate neural induction and patterning by the organizer, we have isolated a cDNA encoding a novel *Xenopus* homologue of Frizzled (Itoh *et al.*, 1998). The new gene product was named Xfz8 since it is mostly similar to the mouse Mfz8 cDNA. *In situ* hybridization analysis demonstrated a biphasic expression pattern of Xfz8 mRNA (Itoh *et al.*, 1998). At the beginning of gastrulation, Xfz8 is expressed in dorsal marginal zone cells and in future neuroectoderm. Dorsal ectoderm-specific expression indicates that Xfz8 may be induced by neuralizing' signals from the organizer (Slack and Tannahill, 1992; Fig. 1). Later on, expression of Xfz8 becomes gradually restricted to the most anterior neural plate. The second phase of expression in telencephalon and anterior ectodermal tissue suggests that Xfz8 transcription is repressed by "posteriorizing" signals everywhere in ectoderm except the anterior margin (Fig. 1). Thus, Xfz8 expression pattern may reflect the response of ectoderm to organizer-generated signals that are involved in anteroposterior axis specification.

Possible functional activities of Xfz8 were revealed in ectopic expression experiments where a synthetic Xfz8 mRNA was microinjected into different embryonic blastomeres (Itoh *et al.*, 1998). Overexpression of Xfz8 mRNA in the ventral blastomeres triggers formation of a complete secondary axis in the injected embryos, indicating that Wnt/Frizzled signaling may be involved in determination of the body axis. This effect was specific to Xfz8, since several other Frizzled homologues did not generate this phenotype (Itoh and Sokol, unpublished data). Despite the overall similarity of secondary axes induced by different members of the Wnt signal transduction pathway, induction of secondary axes by Xfz8 appears to be generated by a different mechanism, because secondary axes were efficiently induced by the extracellular domain of Xfz8 (Xfz8-N, unpublished). Since Frizzled CRD and CRD-related proteins such as FrzA may participate in binding to Wnts (Bhanot *et al.*, 1996; Xu *et al.*, 1998; Moon *et al.*, 1997), we hypothesize that ectopic expression of Xfz8-N on the ventral side may inhibit action of a Wnt, which is a negative regulator of axis formation.

4. CONVERGENT EXTENSION IN VERTEBRATES AND TISSUE POLARITY DETERMINATION IN *DROSOPHILA* MAY BE CONTROLLED BY A SIMILAR MOLECULAR PATHWAY

Convergent extension during gastrulation and neurulation is another important example of coordinated regulation of tissue polarity. Convergent extension movements consist of dynamic reorganization of cytoskeleton, production of bipolar cell protrusions and subsequent mediolateral cell intercalations at the dorsal side of the embryo (Keller, 1991). Gastrulation is first characterized by tissue invagination and involution at the blastopore lip, followed by blastopore closure. Convergent extension movements commence at about midgastrula stage and are thought to be essential for elongation of embryonic body axis (Keller, 1991).

Involution of tissues through the blastopore requires signaling by FGF, because the blastopore remains open in embryos overexpressing a dominant negative FGF receptor (Amaya *et al.*, 1993). Pronounced dorsal convergence and extension start only later, at midgastrula stages. This later pocess may depend on Wnt signaling, because microinjections of mRNA encoding a dominant negative form of *Xenopus* Dishevelled (Sokol, 1996), or dorsal injections of Xwnt5a, Xwnt11, Xwnt4 mRNAs (Du *et al.*, 1995) lead to development of much shorter axis. Although the effect of FGF on tissue involution and subsequent morphogenesis is likely to depend on induction of posterior mesoderm (Amaya *et al.*, 1993), interference with Dishevelled function does not seem to interfere with mesoderm induction (Sokol, 1996), and this phenotype was interpreted as a block in convergent extension movements. Thus, axis extension could be separated from specification of mesodermal fate.

In agreement with these observations, microinjection of FrzA mRNA into the dorsal blastomeres of the early embryo resulted in a shortened body axis with severe posterior deficiencies (Xu *et al.*, 1998). Dorsal injections of mRNA encoding Xfz8 or Xfz CRD caused similar inhibition of axis extension (Itoh *et al.*, 1998). Furthermore, FrzA and Xfz8 blocked elongation of ectodermal explants in response to activin, a potent mesoderm-inducing factor. Since both FrzA and dominant negative Dishevelled were shown to interfere with Wnt signal transduction pathways, it is tempting to propose that convergent extension is controlled by a Wnt activity. So far it is not known which Wnt ligand may be responsible for this activity. One candidate is Wnt5a, because in mice lacking Wnt5a gene, a similar inhibition of axis extension is observed (T. Yamaguchi, personal communication).

These observations indicate that both Frizzled and Dishevelled may be involved in the control of convergent extension movements in *Xenopus*. Interestingly, GSK3, a downstream component of the Wnt signal transduction, does not seem to affect cell movements during gastrulation and neurulation, suggesting that it is not involved in this process (Itoh *et al.*, 1995).

Lack of requirement for GSK3 in convergent extension is reminiscent of regulation of tissue polarity in the fly. Several genes including *frizzled, dishevelled, prickle/spiny legs*, *inturned* and *fuzzy* were identified in *Drosophila* and hypothesized to function in the genetic pathway leading to determination of tissue polarity (Adler, 1992). Both the establishment of tissue polarity in *Drosophila* and convergent extension in vertebrates require cross-talk between neighboring cells which results in coordinate regulation of cytoskeletal elements. The same molecules including Frizzled and Dishevelled, but not shaggy/GSK3, are involved in both processes. Thus, it is possible that the molecular pathway operating in these two distinct events are conserved. If this hypothesis is correct, further analysis of tissue polarity mutations in *Drosophila* and identification of their vertebrate homologues will be helpful to understand convergent extension movements in the frog embryo.

5. CONCLUSIONS

Based on the activities of purified factors, three steps for anteroposterior ectodermal patterning were proposed: generation of anterior ectodermal, anterior neural and posterior neural/mesodermal cell fates. Wnts, a family of secreted polypeptides, and Frizzled, their putative receptors, are expressed in the prospective neuroectoderm and are able to affect axis determination in *Xenopus*. Accumulating experimental evidence indicates that Wnt/Frizzled signaling may be involved in anteroposterior axis specification, posteriorization of neural tissue and regulation of axis extension. A hypothesis is proposed that the convergent extension pathway in vertebrates is equivalent to the pathway which is required for determination of tissue polarity in *Drosophila*, but differs from the known Wg/Wnt pathways in the requirement for glycogen synthase kinase 3.

ACKNOWLEDGEMENTS

We thank V. Krupnik for comments on this manuscript. This work was supported by the grants from the March of Dimes Birth Defects Foundation and from National Institutes of Health.

REFERENCES

Adler P.N., *BioEssays* **14** (1992) 735-741.

Amaya E., Stein P.A., Musci T.J. and Kirschner M.W., *Development* **118** (1993) 477-487.

Bhanot P., Brink M., Harryman Samos C., Hsieh J.C., Wang Y.S., Macke J.P., Andrew D., Nathans J. and Nusse R., *Nature* **382** (1996) 225-230.

Blitz I.L. and Cho K.W.Y., *Development* **121** (1995) 993-1004.

Blumberg B., Bolado J. Jr., Moreno T.A., Kintner C., Evans R.M. and Papalopolu N., *Development* **124** (1997) 373-379.

Doniach T., *Cell.* **83** (1995) 1067-1070.

Du S.J., Purcell S.M., Christian J.L., McGrew L.L. and Moon R.T., *Mol. Cell. Biol.* **15** (1995) 2625-2634.

Durston A.J., Timmermans J.P.M., Hage W.J., Hendriks H.F.J., de Vries N.J., Heideveld M. and Nieuwkoop P.D., *Nature* **340** (1989) 140-144.

HamburgerV., The Heritage of Experimental Embryology, New York (Oxford University Press, 1988).

Horvitz H.R. and Herskovitz I., *Cell.* **68** (1992) 237-255.

Itoh K., Tang T.L., Neel B.G. and Sokol S.Y., *Development* **121** (1995) 3979-3988.

Itoh K. and Sokol S.Y., *Mech. Dev.* **61** (1997) 113-125.

Itoh K., Jacob J. and Sokol S.Y., *Mech. Dev.* **74** (1998) 145-157.

Keller R., In *Xenopus laevis*: Practical uses in cell and molecular biology. *Methods Cell. Biol.* **36**, edited by B.K. Kay and H.B. Peng (Academic Press, San Diego, 1991) pp. 61-113.

Lai C.J., Ekker S.C., Beachy P.A. and Moon R.T., *Development* **121** (1995) 2349-2360.
Leyns L., Bouwmeester T., Kim S.-H., Piccolo S. and De Robertis E.M., *Cell.* **88** (1997) 757-766.
Mayr T., Deutsch U., Kuhl M., Drexler H.C.A., Lottspeich F., Deutzmann R., Wedlich D. and Risau W., *Mech. Dev.* **63** (1997) 109-125.
McGrew L.L., Lai C.-J. and Moon R.T., *Dev. Biol.* **172** (1995) 337-342.
Moon R.T., Christian J.L., Campbell R.M., McGrew L.L., DeMarais A.A., Torres M., Lai C.J., Olson D.J. and Kelly G.M., *Development Suppl.* (1993) 85-94.
Moon R.T., Brown J.D., Yang-Snyder J.A. and Miller J.R., *Cell.* **88** (1997) 725-728.
Nusse R. and Varmus H.E., *Cell.* **69** (1992) 1073-1087.
Parr B.A. and McMahon A.P., *Curr. Opin. Genet. Dev.* **4** (1994) 523-528.
Perrimon N., *Cell.* **76** (1994) 781-784.
Perrimon N., *Cell.* **86** (1996) 513-516.
Rattner A., Hsieh J.C., Smallwood P.M., Gilbert D.J., Copeland N.G., Jenkins N.A.and Nathans J., *Proc Natl Acad Sci USA* **97** (1997) 2859-2863.
Riddle R.D., Ensini M., Nelson C., Tsuchida T., Jessell T.M. and Tabin C., *Cell.* **83** (1995) 631-640.
Sawa H., Lobel L. and Horvitz H.R., *Genes Dev.* **10** (1996) 2189-2197.
Sasai Y. and De Robertis E.M., *Dev. Biol.* **182** (1997) 5-20.
Sive H.L., Hattori K. and Weintraub H., *Cell.* **58** (1989) 171-180.
Slack J.M.W. and Tannahill D., *Development* **114** (1992) 285-302.
Sokol S., Christian J.L., Moon R.T. and Melton D.A., *Cell.* **67** (1991) 741-752.
Sokol S., *Curr. Biol.* **6** (1996) 1456-1467.
Wang Y., Macke J.P., Abella B.S., Andreasson K., Worley P., Gilbert D.J., Copeland N.G., Jenkins N.A. and Nathans J., *J. Biol. Chem.* **271** (1996) 4468-4476.
Wang S., Krinks M., Lin K., Luyten F.P. and Moos M. Jr., *Cell.* **88** (1997) 757-766.
Wilson P.A. and Hemmati-Brivanlou A., *Neuron.* **18** (1997) 699-710.
Xu Q., D'Amore P. and Sokol S.Y. (1998) submitted.
Yang Y. and Niswander L., *Cell.* **80** (1995) 939-947.

LECTURE 20

Fluid Lipid-Bilayer Membranes: Some Basic Physical Mechanisms for Lateral Self-Organization

L. Miao, P.L. Hansen and J.H. Ipsen

Department of Chemistry, The Technical University of Denmark, Building 206, 2800 Lyngby, Denmark

1. INTRODUCTION

All biomembranes share one universal construction principle: their basic structural element is a bilayer composed of amphiphilic lipid molecules, which primarily serves as a two-dimensional solvent for various proteins. In the plasma membrane of eucaryotic cells, the extracellular side of the lipid-bilayer is decorated by a macromolecular film called "glycocalix" extending into, or even attaching to, the extracellular matrix; the intracellular side couples to the membrane-associated cytoskeleton, a quasi two-dimensional macromolecular network which in turn couples to a three-dimensional network of actin filaments in most of cells. A large number of the talks in this workshop were devoted to the subjects both of networks of extracellular matrix and of cytoskeletal and actin networks. Our study of model lipid bilayers draws its inspiration from the lipid-bilayer component of biomembranes [1]. Lipid bilayers may form spontaneously when the types of amphiphilic lipid molecules constituting biomembranes are dispersed in an aqueous environment. They are the simplest model systems of biomembranes, which retain some of the essential physical properties, and can mimic at different levels the molecular complexity, of biological lipid bilayers. We will mainly discuss one aspect of *equilibrium* thermodynamic behavior of model lipid-bilayer membranes: their self-organization into lateral domains of different chemical or physical nature.

Two observations of the lipid-bilayer component of biomembranes underlie the motivation of our study. First, a biological lipid bilayer is composed of two opposing monolayers, each containing an astonishingly large number of types of lipids that are different in either the chemistry of their (hydrophilic) headgroups or that of their fatty-acid (hydrophobic) chains. This unique molecular architecture dictates that any physical characterization of such systems requires different types of "in-plane" degrees of freedom: *translational* degrees of freedom to describe the positions of lipid molecules within each monolayer; *conformational* degrees of freedom to describe the large number of conformations each lipid chain can assume; finally, *compositional* degrees of freedom to describe the lipid compositions of each monolayer. It may be expected that under different thermodynamic (and chemical) conditions these different types of degrees of freedom show characteristically different types of collective behavior, or *lateral ordering*. Indeed, biophysical studies of model lipid bilayers have been providing mounting evidence for this [1,2]. Lateral (in-plane) structural heterogeneity may arise naturally, associated with the ordering processes in these systems. For example, immiscibility between the components, or lateral phase separation, is expected to occur in an appropriate temperature and composition range, as in conventional systems of mixtures. Furthermore, even a single-component lipid bilayer may exhibit distinct microscopic or mesoscopic (length scales larger than molecular dimensions, *e.g.*, the bilayer thickness) domains where lipid chains assume conformations of different characters. The biologically relevant question is thus: does the self-organization of lateral structures within the lipid-bilayer moiety that is of purely physical nature play any active role in membrane-bound biological processes such as the formation of functional complexes or the activity of membrane-bound enzymes?

Secondly, a biological lipid bilayer is fluid in nature and can, therefore, relax a shear stress by lateral hydrodynamic flow of the material. Moreover, it is extremely flexible with respect to its bending, or "out-of-plane", deformations. Thermal energies under physiological conditions can easily excite such deformations. Hence, any physical description of a fluid lipid bilayer should also (at least, in principle) deal with the physics associated with "out-of-plane" deformations. The possibility of extending into the third dimension of the physical space makes a bilayer a truely unconventional two-dimensional system.

The purpose of this article is to present our study of a phenomenological, therefore, generic, model of some simplest lipid bilayers (pure and two-component systems) that describes some of the basic physics governing both the "in-plane" degrees of freedom and "out-of-plane" deformations. Through our presentation, we hope to illustrate that, due to their extraordinary molecular architecture, even the simplest lipid bilayers possess some basic physical mechanisms for self-organization in terms of lateral structures and are far from being "inert" matrices of right fluidity.

2. A PHENOMENOLOGICAL MODEL OF FLUID LIPID BILAYERS

The model we will describe in this Section is a phenomenological one, reflecting an approach that is often taken by physicists to study systems of large number of degrees of freedom, to which lipid bilayers certainly belong. Generically speaking, in such phenomenological descriptions microscopic (or molecular) details of a system are "coarsed grained" into a few "fields" and physical parameters. Explicitly in our model, the translational degrees of freedom are represented by a local density field; the chain conformational degrees of freedom may be described by a local chain-order-parameter field; and the chemical compositions are modeled by local concentrations of specific lipid species. Similarly, in describing "out-of-plane" deformations of a lipid bilayer, the thickness of the bilayer is neglected, and a particular geometical configuration of the bilayer is modeled as a geometrical surface, characterized (up to the lowest non-trivial order of approximation) by the area and two local principal curvatures. Consequently, our model is only valid in describing phenomena occuring on length scales larger than some coherence length of the "in-plane" microscopic variables and larger than the thickness of the bilayer.

2.1. "Out-of-plane" deformations and bending rigidity

In order to describe the physics governing the "out-of-plane" deformations of a *fluid* lipid bilayer, Canham [3], Helfrich [4] and Evans [5] proposed an important notion of bending elasticity more than twenty years ago. To illustrate the essence of this notion, we consider a single-component fluid lipid bilayer which is in contact with an external reservoir of the lipid molecules. The free energy of this system is given by

$$\mathcal{H}_{el} = \sigma_0 A + \int dA \left[\frac{\kappa}{2}(C_1 + C_2 - C_0)^2 + \kappa_g C_1 C_2 \right], \tag{1}$$

where A is the total area of the bilayer, proportional to the total number of lipid molecules in the bilayer (due to the membrane incompressibility), C_1 and C_2 are the two principal curvatures defined at each point on the bilayer surface. The phenomenological physical parameters are σ_0, κ, κ_g and C_0: σ_0 is proportional to the chemical potential of the external reservoir; κ and κ_g represent quantitative measure of the "flexibility" or "softness" of the bilayer; C_0, often called "bilayer spontaneous curvature", has its origin in *bilayer asymmetry* – the fact that the two monolayers constituting a bilayer are not necessarily identical in terms of their (local or global) chemical or physical states. These physical parameters are the mesoscopic manifestation of all the microscopic interactions and the molecular architecture of the bilayer. One important fact is that κ, the bending rigidity, of phospholipid bilayers typically turns out to be $(5 \times 10^{-13} - 10^{-12})$ erg, or $(5 - 25)k_B T_{\text{room}}$. This energy scale, which is comparable to the thermal energy at physiological temperatures, underlies the observed "softness" of lipid bilayers.

2.2. "In-plane" degrees of freedom and Landau theory

To describe "in-plane" degrees of freedom and their collective ordering processes, we employ a canonical model in condensed matter physics, the Landau-Ginzburg ϕ^4-theory [6]. Let $\phi_1(\vec{x})$ and $\phi_2(\vec{x})$ represent our phenomenological "in-plane" fields of monolayers 1 and 2, respectively. Based on the Landau-Ginzburg theory, the free energy associated with the in-plane fields is given by $F[\phi] = F_1 + F_2$, where

$$F_i = \frac{1}{2} \int dA_i \left\{ \frac{c}{2} (\vec{\nabla} \phi_i)^2 + \frac{t}{2} \phi_i^2 + \frac{g}{4!} \phi_i^4 - \mu \phi_i \right\} . \tag{2}$$

It is a minimal description of the most essential characteristic of an ordering process in a two dimensional system. For example, the generic feature of phase separation in a binary mixture – a region of phase coexistence terminating at a point of critical mixing – is well captured by equation (2), if the following correspondance is established:

$$\phi = \frac{\rho - \rho_c}{\rho_c} , \qquad t \propto \frac{T - T_c}{T_c} , \tag{3}$$

where ρ_c and T_c are the concentration and the temperature, respectively, at the critical mixing point and if g is assumed to take on some positive value for maitaining the thermodynamic stability of the system. $\mu = 0$ corresponds to the situation where the concentration is set at the critical value. When t is positive (above T_c), $\phi = 0$ (the homogeneous state with concentration ρ_c) is the minimum-energy solution; when t becomes negative (below T_c), there exist two minimum-energy solutions of nonzero ϕ, corresponding to the phase coexistence. The first term in equation (2) simply describes the free energy associated with the boundary separating the two coexisting phases. Similarly, off-critical situations (when the concentration takes on values other than the critical one) are described by the model with nonzero values of μ.

The signature of phase separation as described above has been revealed in a study of lipid bilayers of binary mixture DMPC and $DS_{d54}PC$ based on Small-Angle-Neutron-Scattering Technique (SANS) [7]. The presence of coherent domains rich in $DS_{54}PC$ was observed. Furthermore, for the equimolar mixture in equilibrium at high temperatures, the scattering intensity function was shown to have the characteristic Ornstein-Zernike behaviour:

$$I(q) \propto \langle \rho(\vec{q}) \rho(-\vec{q}) \rangle \sim \frac{\xi^2}{1 + (q\xi)^2} \tag{4}$$

with the observed temprature dependence of the coherence-length $\xi \sim (T - T_c)^{-0.5}$. This is consistent with the prediction of the Landau-Ginzburg model that $\xi = \sqrt{c/t}$. Evidence of (critical) demixing or phase separation processes has also been observed in various lipid systems. Addition of Ca^{2+} to two-component lipid bilayers was also found to induce phase separation [8]. Demixing is observed commonly in a range of mixtures of lipids that differ considerably either in their chain lengths or in their head-group structures or charges [9]. Finally, PC-cholesterol mixtures, which are of particular interest to membrane physics, provide another example of systems where critical demixing takes place [2, 10].

Another important type of ordering processes can occur even in single-component lipid bilayers, during which lipid chains collectively undergo transition between a (high-temperature) disordered state and a (low-temperature) ordered state. It is known as the "main transition" or the "chain-melting transition" and involves a considerable change in the molecular densities, ρ_i, in the two monolayers. The main transition can also be approximatively described by equation (2) with

$$\phi_i = (\rho_i - \rho_0)/\rho_0 \,, \quad t < 0 \,, \quad \mu \propto (T - T_m) \,, \tag{5}$$

where T_m is the main-transition temperature [2, 11]. It is important to note that in general we can not expect the free energy density to be symmetric under $\phi \to -\phi$ and $h \to -h$. However, the generics described by equation (2) would not change.

2.3. General model

Finally, based upon the basic descriptions presented in Sections 2.1 and 2.2, we introduce a general model for a bilayer. This model views a bilayer as an entity of two *coupled* monolayers which share a common geometrical conformation ($H = C_1 + C_2$), but can assume different lateral structures (ϕ_1 and ϕ_2). Thus the total free energy of the bilayer is expressed as follows:

$$\begin{aligned} F_{\text{bilayer}} = \int dA \, \Big\{ \sigma_0 + \frac{\kappa}{2}[H^2 - \bar{C}_0(\phi_1 - \phi_2)H] + \frac{t_{12}}{4}\phi_1\phi_2 \\ + \frac{1}{2}\left[\frac{c}{2}(\vec{\nabla}\phi_1)^2 + \frac{t}{2}\phi_1^2 + \frac{g}{4!}\phi_1^4 - \mu\phi_1 + \frac{c}{2}(\vec{\nabla}\phi_2)^2 + \frac{t}{2}\phi_2^2 + \frac{g}{4!}\phi_2^4 - \mu\phi_2\right]\Big\} \,. \end{aligned} \tag{6}$$

This model contains the two lowest-order forms of coupling between the two monolayers, as expressed by the last two terms in the first line of equation (6). The first form is an explicit model expression of the notion of *bilayer spontaneous curvature* and the assumptions involved are based on the following observation of the molecular construction of the bilayer. In general, any two opposing local elements of the two monolayers can be different in their chemical or physical states, leading to a nonzero ($\phi_1 - \phi_2$) or local bilayer asymmetry. The bending flexibility of the bilayer makes it possible for the bilayer to assume different geometrical conformations under a given ($\phi_1 - \phi_2$), corresponding to different bending energy costs. However, the local bending energy will be minimized if the local mean curvature takes on some particular value that expresses this local bilayer asymmetry; and it is assumed that this particular value, the local bilayer spontaneous curvature, is given by $\bar{C}_0(\phi_1 - \phi_2)$, where $\bar{C}_0$ is a phenomenological constant that depends on the material properties of the bilayer. This form of coupling has been proposed in models similar to equation (6) to explore the interplay between different types of "in-plane" degrees of freedom and membrane conformation [12-18].

The second form of coupling between the two monolayers, $t_{12}\phi_1\phi_2$, is introduced to describe the possibility that direct intermonolayer interactions may also depend on the physical or chemical states of the monolayers, although conventional approximations neglect this effect [16,17]. For example, van der Waals interactions between lipid molecules across the bilayer may depend on the conformational states of those molecules. Also in bilayers containing charged lipids, electrostatic interactions may effectively depend on local densities or concentrations of the charged lipids and can therefore be either attractive or repulsive. There are experimental evidences that coupling between two constituent monolayers of this nature is relevant in bilayers formed from lipids with charged head groups [19]. As we will show in the next Section, this coupling term plays a non-trivial role in determining the characteristics of the ordering processes, in particular, lateral structure formations in a bilayer.

An alternative expression of equation (6), which will be more convenient for our study of the model, can be written in terms of $\phi \equiv (\phi_1 - \phi_2)/2$ and $\psi \equiv (\phi_1 + \phi_2)/2$:

$$\begin{aligned} F_{\text{bilayer}} &= \int dA \left\{ \sigma_0 + \frac{\kappa}{2}(H^2 - 2\bar{C}_0\phi H) + \frac{c}{2}(\vec{\nabla}\phi)^2 + \frac{K_\Delta}{2}\phi^2 + \frac{g}{4!}\phi^4 \right. \\ &\left. + \frac{c}{2}(\vec{\nabla}\psi)^2\Delta + \frac{K_\Delta + \Delta_K}{2}\psi^2 + \frac{g}{4!}\psi^4 + \frac{g}{4}\phi^2\psi^2 - \mu\psi \right\}, \qquad (7) \end{aligned}$$

where $K_\Delta \equiv t - t_{12}/2$ and $\Delta_K \equiv t_{12}$.

3. RESULTS AND DISCUSSIONS

Our study of the model, equation (7), is based on so-called "mean-field" considerations which assume that the thermodynamic state of the model is determined by minimizing the free energy, equation (7), with respect to both the "out-of-plane" and the "in-plane" degrees of freedom. In particular, it is assumed that bilayer deformations are small deviations from the flat configuration; and this assumption is justified when σ_0 is "sufficiently large". Our results should also be applicable to systems where the bilayers are confined [17].

Furthermore, we will only consider the situation where the parameter μ is zero, corresponding to "critical demixing" or cases where the main phase transition is nearly critical [20]. The generalization of the results for this case to off-critical situations will be discussed elsewhere [21]. There are still more physical parameters to be dealt with: σ_0, K_Δ, Δ_K, κ, $\bar{C}_0$, c and g. c has the unit of energy, g has the unit of energy density; $\sqrt{c/g}$ thus has the unit of length. Since the precise values of c and g do not influence the qualitative features of the thermodynamic behavior of the model, they are set to 1 for the convenience of computation and presentation. κ and $\bar{C}_0$ are also set to some fixed values, from which we obtain a reduced parameter $r_{\mathrm{sc}} \equiv \kappa \bar{C}_0^2$. It effectively measures the strength of the bilayer spontaneous curvature. We will thus consider situations where σ_0, K_Δ and Δ_K are the relevant control parameters.

The most important effect arising both from the coupling between bilayer deformation and "in-plane" fields and from the presence of surface tension, σ_0, (or any mechanical constraints that stabilize smooth conformations of the bilayer) is a novel type of lateral ordering processes, for which there are no analogs in ordinary fluids. Fluid bilayers can have ordered phases that are often not macroscopically homogeneous, but rather appear spatially modulated, i.e., consisting of domains that alternate between the two coexisting ordered states and have characteristic sizes [13,16,17]. Moreover, the direct interaction between the two monolayers makes such lateral self-organization processes very intricate.

The basic physical forces driving this appearance of ordered domains of well-controlled sizes are in fact apparent to see. There is an important length scale inherent in the system, determined by the two competing energies, the bending rigidity and the surface tension:

$$\xi_{\mathrm{c}}^2 \equiv \kappa/\sigma_0 \, . \tag{8}$$

It defines a length scale for crossover: the bending rigidity κ tends to bend the bilayer when ϕ becomes *ordered* (nonzero) in order to conform with the nonzero bilayer spontaneous curvature and minimize the bending energy, and is more effective on short length scales, while σ_0 effectively controls deformations of long wavelength, tending to keep the bilayer flat on large length scales. ξ_{c} separates these two different regimes of length scales. The smaller σ_0 is, the larger ξ_{c} is. This competition reflected by ξ_{c} is the underlying mechanism for the appearance of the specific length scale characterizing domain size in spatially modulated phases. In fact, lateral ordering of the bilayer into domains of specific size is possible when $0 < \sigma_0 < \sigma_0^*$, where

$$\sigma_0^* = \kappa^2 \bar{C}_0^2 / c \, (= 1) \, , \tag{9}$$

and the sizes of the domains are typically given by

$$q_0^2 = \xi_{\mathrm{c}}^{-2} \left(\sqrt{r_{\mathrm{sc}} \xi_{\mathrm{c}}^2} - 1 \right) . \tag{10}$$

The actual calculations and characterization of the different types of ordered phases and transitions between the phases were carried out by using the following ansatz,

$$\phi = \phi_0 + \phi_1 \cos(\bar{q}_0 x) \, , \quad \psi = \psi_0 + \psi_2 \cos(2\bar{q}_0 x) \, . \tag{11}$$

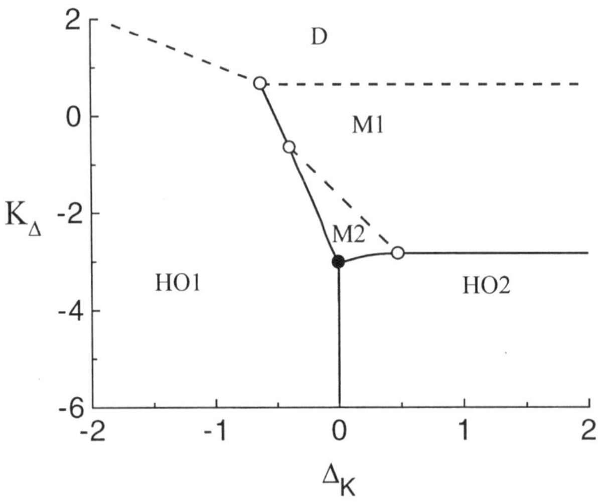

Fig. 1. — A schematic phase diagram showing the principal phases and phase transitions predicted by our model calculation. Solid lines indicate discontinuous phase transitions, and dashed lines represent continuous phase transitions.

The basic reason for proposing this ansatz lies in the observation described in the preceding paragraph. The more specific comments concerning the ansatz will be relegated to a different paper [21]. The results of the calculations are summarized in the following 2-dimensional phase diagrams for which different physical parameters are chosen as the control parameters.

Figure 1 is a phase diagram expressed in terms of K_Δ (effectively, the temperature) and Δ_K (or t_{12}, the strength of the direct interaction between the two monolayers), but for a fixed value of σ_0 that lies between 0 and σ_0^*. The phase diagram is composed of five principal phases:
i) the flat, completely disordered (**D**) phase, characterized by $\phi = 0$, and $\psi = 0$;
ii) the flat, homogeneously ordered (**HO1**) phase, characterized by $\phi = 0$, $\psi = \psi_0 \neq 0$;
iii) the curved, spatially modulated (**M1**) phase I, described by

$$\phi = \phi_1 \cos(\bar{q}_0 x) \ , \ \ \phi_1 \neq 0 \ ; \quad \psi = 0 \ ; \tag{12}$$

iv) the curved, spatially modulated phase (**M2**) phase II, where both ϕ and ψ fields are nonzero and appear modulated, and the period of the modulation of ψ field is half of that of ϕ field, as indicated in the ansatz equation (11);
v) the flat, homogenously ordered (**HO2**) phase, where $\phi_0 \neq 0$, $\phi_1 = 0$, $\psi_2 = 0$.

The physics underlying the appearance of these different phases in different regions of the parameter space is largely intuitive. The **HO1** phase, where the two monolayers are ordered in an identical fashion, appears in the region where

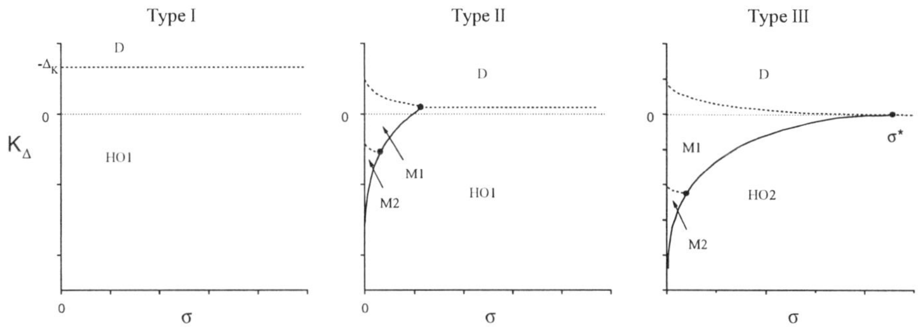

Fig. 2. — A series of schematic phase diagrams expressed in terms of K_Δ and σ_0, for three different regimes of the parameter of the direct interaction between two monolayers, Δ_K. (a) regime of strong attraction: $\Delta_K < -r_{\rm sc}$; (b) regime of intermediate attraction: $-r_{\rm sc} < \Delta_K < 0$; (c) regime of repulsion: $\Delta_K > 0$.

the direct interaction between the two monolayers is attractive and relatively strong, as such interaction favors commensuration of the two monolayers. As this direct interaction becomes less attractive and then repulsive, the energetic requirement for one monolayer to behave commensurately with the other becomes less stringent; consequently, the **M1** and **M2** phases, where the symmetry between the two monolayers is locally broken and the **HO2** phase, where the symmetri is globally broken, appear as the equilibrium phases. Another important outcome of our calculations, illustrated in Figure 1, is that, for bilayers with different degrees of the direct coupling, the lateral ordering process, induced thermally, for example, proceeds via different scenarios, involving different sequences of thermodynamic singularities (*e.g.*, in the specific heat) and different phases.

The representation of the phase diagrams in terms of K_Δ and σ_0 at fixed Δ_K may often be the more convenient one for making connections with experimental results. This representation amounts to tuning temperature or the surface tension (or mechanical constraints) for a given system of lipid bilayers. Figure 2 collects a series of phase diagrams given in this representation, which show three distinct types of topology, corresponding to three regimes of strength of the direct interaction parameter Δ_K. The figures accentuate the complexity of the lateral ordering processes described by the simple model, equation (6), or equation (7).

Although we will not discuss explicitly cases where μ (the chemical potential, or the chain ordering field) is nonzero, let us note, for the sake of completeness, that the distinction between the two modulated phases, **M1** and **M2**, only exists when $\mu = 0$. The presence of a nonzero μ leads to nonzero ψ, thereby determining that a modulated phase should be of the **M2** type [21]. Thus, the **M2** phase is predicted to be the more prevalent form of modulated phases in systems of lipid bilayers, to which our model, equation (6), may be applied.

An important and natural question is, of course, whether the types of basic physical mechanisms that we have just illustrated bear any relevance to real systems of lipid bilayers, i.e., whether those systems will under favorable conditions self-organize into the modulated phases as characterized by our analysis. The following observations indicate that the answer to the question may be a positive one.

First, for lipid bilayers of PhosphatidylCholine (PC) near their main (chain-melting) transition, the relevance of the coupling between bilayer deformations and the monolayer density fields (or, chain conformational states) has been supported by experiments [18]. Thus, in the presence of a nonzero surface tension, or equivalently, a nonzero osmotic pressure difference across a bilayer, or, a confinement potential, one might expect that the main phase transition should be accompanied by an approximately sinusoidal modulation of the bilayer conformation as well as a periodical modulation of the density or thickness profile characteristic of the **M2** phase, since the appropriate Landau-Ginzburg description involves a nonzero ordering field, $\mu \propto (T - T_m)$ (see Eq. (5)). Recent experiments [22] in fact show that upon cooling, PC lipid bilayers transform into a metastable "ripple phase", $P_{\beta'}$(mst), which is stable for hours [23]. This phase is characteristically different from the well-known primary ripple phase, with a larger periodicity (around 270 Å, compared with typical 130 Å of the primary ripple), a symmetric profile of the ripple and an appreciable lateral variation in the bilayer thickness [22]. This would be the characteristics expected of the predicted **M2** phase of the model if the thickness of each monolayer would be thought to depend on its "in-plane" field ϕ_i [2].

Secondly, some systems of bilayers composed of lipid mixtures have been found to self-organize into lateral structures characterized by specific, large (compared with molecular) length scales and by compositional variations [24]. In some cases, structures have been observed to resemble those characteristic of the **M2** phase. For example, in the gel-fluid coexistence region of PC-cholesterol mixtures, domains have been observed, the sizes of which are in the range of $40-60$ nm and show dependence on both temperature and composition [25,26].

We, therefore, conclude that our phenomenological model may be of some experimental relevance in elucidating the potential physical mechanisms driving self-organization processes in lipid bilayers. On the other hand, we are inclined to think that there are probably some physical variables and effects that are yet to be incorporated into the model. Furthermore, we believe that we are still far from providing answers to the ultimate question of whether such physical mechanisms play any active role in membrane-bound biological processes.

ACKNOWLEDGEMENTS

We have profited greatly from discussions with T. Hønger, M. Sabra, T. Gil, and O.G. Mouritsen. This work was supported both by a research fellowship of the Technical University of Denmark and by the Danish Natural Science Research Council.

REFERENCES

[1] For a comprehensive review of physics of biomembranes, refer to *Structure and Dynamics of Membranes: From Cells to Vesicles; Generic and Specific Interactions*, Vols. 1A-B of *Handbook of Biological Physics*, edited by R. Lipowsky and E. Sackmann (North Holland, Amsterdam, 1995). Peliti L. "Amphiphilic Membranes" in *Fluctuating Geometries in Field Theory and Statistical Mechanics* Les Houches Series, edited by F. David and P. Ginsparg (North Holland, Amsterdam) (1996).

[2] Bloom M., Evans E. and Mouritsen O.G., *Quart. Rev. Biophys.* **24** (1991) 293.

[3] Canham P.B., *J. Theor. Biol.* **26** (1970) 61.

[4] Helfrich W., *Z. Naturforsch. Teil C* **28** (1973) 693.

[5] Evans E.A. *Biophys. J.* **14** (1974) 923.

[6] Landau L.D. and Lifshitz E.M., *Statistical physics* Part 1, Vol. 5 in *Course of theoretical physics* 3rd Edn. (Pergamon Press, 1980).

[7] Knoll W., Schmidt G., Ibel K. and Sackmann E., *J. Chem. Phys.* **79** (1983) 3439.

[8] Knoll W., Schmidt G., Rötzer H., Henkel T., Pfeiffer W., Sackmann E., Mittler-Neher S. and Spinke J., *Chem. Phys. Lipids* **57** (1991) 363.

[9] Wu S. and McConnell H.M., *Biochemistry* **14** (1975) 847.

[10] Ipsen H.J., Karlström G., Mouritsen O.G., Wennerström H. and Zuckermann M.J., *Biophys. Biochem. Acta* **905** (1987) 162.

[11] Doniach S., *J. Chem. Phys.* **68** (1978) 4912.

[12] Leibler S., *J. Phys. France* **47** (1986) 507.

[13] Leibler S. and Andelman D., *J. Phys. France* **48** (1987) 2013.

[14] Safran S.A., Pincus P. and Andelman D., *Science* **248** (1990) 354; Safran S.A., Pincus P., Andelman D. and MacKintosh F.C., *Phys. Rev. A* **43** (1991) 1071.

[15] MacKintosh F.C. and Safran S.A., *Phys. Rev. E* **47** (1993) 1180.

[16] Kodama H. and Komura S., *J. Phys. France II* **3** (1993) 1305.

[17] MacKintosh F.C., *Phys. Rev. E* **50** (1994) 2891.

[18] Hønger T., Mortensen K., Ipsen J.H., Lemmich J., Bauer R. and Mouritsen O.G., *Phys. Rev. Lett.* **72** (1994) 3911.

[19] Düzgünes N. Newton C., Fisher K., Fedor J. and Papahadjopoulos D., *Biochem. Biophys. Acta* **944** (1988) 391.

[20] Mouritsen O.G., *Chem. Phys. Lipids* **57** (1991) 179.

[21] Hansen P.L., Miao L. and Ipsen H.J., *Phys. Rev. E* **58** (1998) 2311.
[22] Rappolt M. and Rapp G., *Eur. Biophys. J.* **24** (1996) 381.
[23] Tenchov B., *Chem. Phys. Lipids* **57** (1991) 165.
[24] Sackmann E., "Physical Basis of Self-organization and Function of Membranes: Physics of Vesicles" in reference [1], Vol. 1A.
[25] Mortensen K., Pfeiffer W., Sackmann E. and Knoll W., *Biochim. Biophys. Acta* **945** (1988) 221.
[26] Hicks A., Dinda M. and Singer M.A., *Biochim. Biophys. Acta* **903** (1987) 177.

LECTURE 21

Logical Analysis of Timing-Dependent Signaling Properties in the Immune System

M. Kaufman, F. Andris[1] and O. Leo[1]

Center for Nonlinear Phenomena and Complex Systems, Université Libre de Bruxelles, Campus Plaine CP 231, 1050 Bruxelles, Belgium
[1]Laboratoire de Physiologie Animale, Université Libre de Bruxelles, CP 300, 67 rue des Chevaux, 1640 Rhode St Genèse, Belgium

1. INTRODUCTION

Among the major immunological cell types, helper T lymphocytes have been shown to play a key role in most antibody and cell mediated immune responses. Their activation leads to the production of regulatory factors, called lymphokines, that determine the activation and functioning of other important cells of the immune system. Understanding the mechanisms regulating helper T cell responsiveness is therefore a central issue in immunobiology.

Ligand engagement of the T cell antigen receptor (TCR) triggers a series of biochemical processes within the cell which can lead either to cellular activation, i.e. lymphokine production and cell proliferation, or to the induction of a state of unresponsiveness termed anergy. This anergic state is characterized by a refractoriness to further antigenic stimulation despite normal cell surface TCR expression. In particular, receptor-induced production of the growth lymphokine Interleukin-2 (Il-2), and hence autocrine growth, is inhibited in anergic cells whereas general metabolic functions remain unaffected (Schwartz, 1996).

The conditions determining which of these responses will occur are not completely understood. There is growing evidence that factors influencing the signaling properties are the presence of an additional "costimulatory" signal provided by particular antigen presenting cells (APCs), and the dissociation rate of the ligand. For example, T cell stimulation in vitro in the absence of accessory signals (Jenkins and Schwartz, 1987) or by analogs of immunogenic peptides with faster dissociation rates (Sloan-Lancaster *et al.*, 1993), fails to induce a productive response and frequently results in subsequent unresponsiveness. Recently however, numerous observations performed in vitro and in vivo have indicated that anergy may, in some cases, also develop upon adequate stimulation and following a productive cellular response (Muraille *et al.*, 1995; Andris *et al.*, 1996).

In this paper we present a model for both positive and negative signaling of T lymphocytes with special emphasis on specifying the conditions leading to alternative cellular behavior. Using a simple Boolean formalism developed by Thomas and coworkers (1990), we show how the timing of the signaling events may affect decision-making at branching pathways and determine the properties of receptor signaling and final state of the system.

In Section 2 we briefly recall some properties of the T cell signaling cascade. In Sections 3 and 4 we present the model and shortly describe the Boolean method that has been used to analyze this model. Section 5 is devoted to a discussion of the results of our analysis and their relation to experimental facts and physiological situations. Concluding remarks are given in Section 6.

2. T CELL SIGNALING CASCADE

In normal physiological condition, T cell activation is initiated by the interaction of the TCR with antigen peptides (Ag) bound to self major histocompatibility complex molecules (MHC) on the surface of antigen presenting cells. The TCR is a multimeric complex consisting of the Ag/MHC binding clonotypic $\alpha\beta$ heterodimer, the noncovalently associated invariant CD3 polypeptides and the ζ dimers. The MHC recognizing CD4/CD8 coreceptors may participate in cell stimulation by stabilizing the TCR-Ag/MHC interactions and by contributing to the generation of signals. Among the earliest biochemical events that accompany T cell stimulation via the TCR is an increase in protein tyrosine kinase (PTK) activity which results in tyrosine phosphorylation of several intracellular substrates, including immunoreceptor tyrosine-activation motifs (ITAMs) contained in the cytoplasmic domains of the ζ and CD3 chains of the receptor complex. The TCR itself lacks intrinsic PTK activity but it is physically associated with the nonreceptor tyrosine kinases fyn and lck. Interestingly, both kinases have a site for intermolecular autophosphorylation and phosphorylation at this site markedly enhances their catalytic activity. Fyn and lck are responsible for phosphorylation of the TCR ζ and CD3 ITAM regions. These phosphorylated ITAMs provide docking sites for downstream signaling components and mediate, in particular, the binding of the cytosolic PTK Zap-70. This kinase also has a characteristic autophosphorylation motif which positively regulates its catalytic activity. The series of biochemical processes which is then initiated includes an increase in cytoplasmic free calcium concentration and the activation of protein kinase C, activation of Ras, and activation of several serine/threonine kinases and phosphatases. These events ultimately lead to gene transcription and to functional responses such as lymphokine production, new surface receptor expression and proliferation (Berridge, 1997).

In addition to TCR recognition of antigen, a second signal delivered independently of the Ag-specific signal and involving another receptor on the T cell and another signaling molecule on the APC is required for effective Il-2 production and subsequent cell division. Which mechanisms are at work in costimulation is not precisely known, but additional or increased transcriptional activity and stabilization of the Il-2 messenger RNA are implicated (Umlauf *et al.*, 1995). One effect, however, of the absence of costimulation is that the cell is driven into the anergic state.

3. THE MODEL

Our model incorporates the following major features and assumptions:

(i) Autophosphorylation is an important feature of the early events of TCR signaling. As extensively discussed in Kaufman *et al.* (1996), this autocatalytic step may lead to a bistable behavior characterized by a low and high phosphorylative activity of the receptor-associated TKs. Moreover, this bistability allows for sustained TK activity after removal of the stimulus. In agreement, it has recently been observed that stimulation of T helper clones leads to an elevated activity of the fyn kinase after ligand removal (Gajewski *et al.*, 1994).

(ii) Tyrosine phosphorylation is considered to both stimulate and inhibit the signaling process (Kaufman *et al.*, 1996). In particular, residual TK activity is assumed to be responsible for a deficient signal transduction capacity of the TCR system. Support for this view is provided by the observation that increased levels of several tyrosine-phosphorylated substrates are associated with defective signaling in *lpr* and *gld* mutant mouse strains (Samelson *et al.*, 1986), and in anergic T helper murine clones (Gajewski *et al.*, 1994).

(iii) In agreement with the observation that signaling through the Il-2 receptors and/or cell proliferation reverse anergy (Beverly *et al.*, 1992), we consider that positive signaling inhibits, through a yet undefined mechanism, the persistent activity of the TKs.

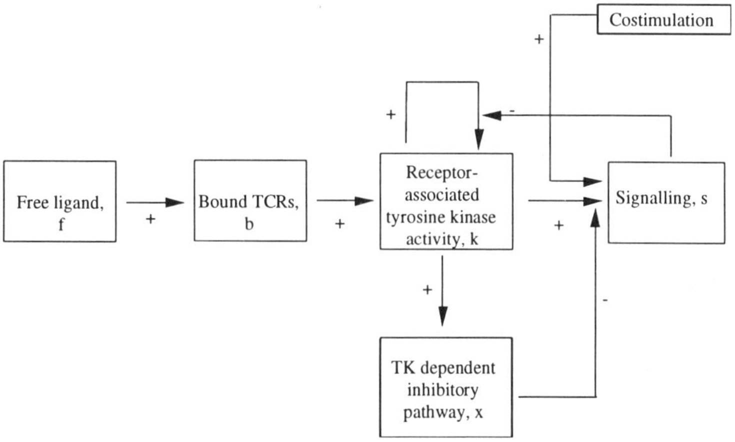

Fig. 1. Schematic interaction diagram. f = free ligand; b = TCRs bound to ligand; k = receptor-associated tyrosine kinase activity; x = tyrosine kinase dependent inhibitory pathway; s = metabolic and mitogenic response. Positive and negative interactions are indicated by a plus and minus sign, respectively.

(iv) Costimulation does not act to suppress the negative signal but rather enhances and accelerates the positive signaling process (Beverly *et al.*, 1992; Schwartz, 1996). This assumption is supported by the finding that the kinetics of Il-2 production is influenced by the level of costimulatory molecules expressed by APCs (Muraille *et al.*, 1996).

The model shown in Figure 1 consists of a series of interconnected events each of which requires a characteristic time to be realized. Binding of free ligand to the TCRs activates the receptor-associated tyrosine kinases. TK activation induces signaling but also activates an inhibitory pathway which negatively affects the signal. Costimulation participates in signal generation. The positive action of k on itself indicates that once receptor-associated TK activity is established, it remains sustained even in the absence of ligand. This autocatalytic maintenance mechanism is suppressed by Il-2 linked signaling and proliferation. Except for the initial increase in TK activity, the signal includes here both the early (Ca rise, PKC and Ras activation...) and late transduction events (Il-2 production, upregulation of Il-2 receptors, cell proliferation). A separation between early and late events leads to quantitative rather than qualitative differences in the description and will not be considered here.

4. LOGICAL DESCRIPTION

4.1. Logical variables and logical equations

Except for costimulation, we associate to each element of the system a logical variable which takes the logical values 0 or 1. Justification for treating the interactions as "on" or "off" comes from the observation that the effect of biological regulators typically changes from inefficient to efficient over a rather small range of concentration and short period of time. Thus, f = 1 means a significant level of free ligand available, otherwise f = 0; b = 1 means a significant fraction of receptors bound to ligand, otherwise b = 0; k = 1 means that the TK are significantly activated, otherwise k = 0; and so on. Costimulation is considered to accelerate the signaling process and is, at this stage of our analysis, accounted for in an implicit way as will appear below.

To describe the dynamics of the system one has to consider transition rules which reflect the evolution of the state variables under the influence of the signaling interactions. These logical functions are built in order to take the value 1 for the conditions in which the homonymous variable tends to 1 or remains at 1, *i.e.* for the values of the state variables which impose an efficient activation of the element considered. For our model we define the following set of logical equations:

$$B = f \tag{1}$$

$$K = b + k.\bar{s} \tag{2}$$

$$X = k \tag{3}$$

$$S = b.k.\bar{x} \tag{4}$$

where we use the logical notations: $\bar{z}$ = NOT z (logical complement); z . y = z AND y (logical product); z + y = z OR y (inclusive OR).

Equation (1) states that the TCRs will become bound (B = 1) if a significant level of free ligand is available (f = 1); otherwise B = 0. Equation (2) expresses that receptor-associated tyrosine kinase activity is established if the receptors are bound to ligand (b = 1). It also involves a maintenance mechanism, through autophosphorylation, which persists in the absence of ligand-bound TCRs. This maintenance mechanism is suppressed by the occurrence of Il-2 linked signaling events and proliferation (s = 1). Thus, K = 1 if either or both b = 1, $k.\bar{s}$ = 1; otherwise K = 0. Equation (3) states that the receptor-associated tyrosine kinases activate, directly or indirectly, an inhibitory pathway. X = 1 if k = 1; otherwise X = 0. Finally, equation (4) means that signaling will occur (S = 1) if the receptors are bound (b = 1) and the receptor-associated kinases are activated (k = 1), and if their positive action on signaling is not inhibited (x = 0); otherwise S = 0. For sake of simplicity we do not include an explicit description of the evolution of the concentration of free ligand and treat free ligand here as an input variable.

There is a temporal relation between logical functions and variables: if the value of a logical function and its corresponding homonymous variable disagree, there is a commitment for the variable to align its value with the value of the function. This command will effectively be executed after a time delay, unless there is a counter order before the delay has elapsed. Moreover we adopt a fully asynchronous description in which all the time delays are different, except for accidental coincidence (Thomas and D'Ari, 1990; Thomas, 1991). The variables of our system are thus updated according to the transition rules as asynchronous automata.

4.2. State tables

The state of the system is described by the combined logical values 0 or 1 of the state variables b, k, x and s, presented in this defined order. From the logical equations one derives state tables which provide, at any time, the value of the function vector for each value of the state vector (Tab. I).

Free ligand f, has the value 0 initially (ligand absent, left table); it takes the value 1 when the ligand is added (ligand present, right table); and then after association either remains at the value 1 in the continuous presence of free ligand, or returns to the value 0 if free ligand is removed.

A plus or minus sign over the logical value of a state variable indicates that that variable is being driven upward (0 $\rightarrow$ 1) or downward (1 $\rightarrow$ 0) by the interactions. Parentheses or circles around a state indicate that the state is stable under the given input conditions. In such regime states all the functions and corresponding variables have the same logical value.

Note that when a state vector has more than one switch command, the function vector or "image" is usually not the next state and may even never be reached because of the occurrence of counter orders. It is only if an exact simultaneity of the different transitions is considered that the image vector coincides with the next state, but this rules out any possibility of branching and choice for the system.

Table I. State tables in the absence (f = 0) and presence (f = 1) of free ligand.

f = 0

b	k	x	s	B	K	X	S
(0	0	0	0)	0	0	0	0
0	0	0	$\bar{1}$	0	0	0	0
0	0	$\bar{1}$	$\bar{1}$	0	0	0	0
0	0	$\bar{1}$	0	0	0	0	0
(0	1	1	0)	0	1	1	0
0	$\bar{1}$	1	$\bar{1}$	0	0	1	0
0	$\bar{1}$	$\overset{+}{0}$	$\bar{1}$	0	0	1	0
0	1	$\overset{+}{0}$	0	0	1	1	0
$\bar{1}$	1	$\overset{+}{0}$	$\overset{+}{0}$	0	1	1	1
$\bar{1}$	1	$\overset{+}{0}$	1	0	1	1	1
$\bar{1}$	1	1	$\bar{1}$	0	1	1	0
$\bar{1}$	1	1	0	0	1	1	0
$\bar{1}$	$\overset{+}{0}$	$\bar{1}$	0	0	1	0	0
$\bar{1}$	$\overset{+}{0}$	$\bar{1}$	$\bar{1}$	0	1	0	0
$\bar{1}$	$\overset{+}{0}$	0	$\bar{1}$	0	1	0	0
$\bar{1}$	$\overset{+}{0}$	0	0	0	1	0	0

f = 1

b	k	x	s	B	K	X	S
$\overset{+}{0}$	0	0	0	1	0	0	0
$\overset{+}{0}$	0	0	$\bar{1}$	1	0	0	0
$\overset{+}{0}$	0	$\bar{1}$	$\bar{1}$	1	0	0	0
$\overset{+}{0}$	0	$\bar{1}$	0	1	0	0	0
$\overset{+}{0}$	1	1	0	1	1	1	0
$\overset{+}{0}$	$\bar{1}$	1	$\bar{1}$	1	0	1	0
$\overset{+}{0}$	$\bar{1}$	$\overset{+}{0}$	$\bar{1}$	1	0	1	0
$\overset{+}{0}$	1	$\overset{+}{0}$	0	1	1	1	0
1	1	$\overset{+}{0}$	$\overset{+}{0}$	1	1	1	1
1	1	$\overset{+}{0}$	1	1	1	1	1
1	1	1	$\bar{1}$	1	1	1	0
(1	1	1	0)	1	1	1	0
1	$\overset{+}{0}$	$\bar{1}$	0	1	1	0	0
1	$\overset{+}{0}$	$\bar{1}$	$\bar{1}$	1	1	0	0
1	$\overset{+}{0}$	0	$\bar{1}$	1	1	0	0
1	$\overset{+}{0}$	0	0	1	1	0	0

4.3. Transition diagram

From the state tables, one can now derive all the possible temporal sequences of logical states.

Let us start from the virgin state (0000) which is stable in the absence of ligand. We add free ligand to the system and the receptors start to bind (B = 1, b = 0). When a significant fraction of the receptors are bound (B = 1, b = 1), we remove free ligand from the system. The receptor-ligand complexes will now tend to dissociate (B = 0, b = 1). Depending on whether ligand dissociation or tyrosine kinase activation occurs first two different pathways will be followed. If tyrosine kinase activity appears before unbinding of the receptors, again different branches will be followed depending on which next event is realized first, and so on. This defines a whole network of branching pathways (Fig. 2). Note that the formulation which is presented here amounts to model the signaling behavior resulting from a single ligand-binding event per receptor. This situation is likely to occur at low ligand concentration and vast excess of receptors. The effect of repeated rebinding of ligand to the *same* receptors during the course of the signaling process (*e.g.* at high ligand concentration) will be discussed elsewhere.

Two groups of pathways may be distinguished. Pathways 1 to 3 reach a final stable state without the appearance of positive signaling. Pathways 4 to 12 all involve a stage where positive signaling is present. Furthermore, the model predicts that two different stable final states can be reached after ligand dissociation:

(i) state (0000) where all the state variables are back to their initial zero value. This state corresponds to full immunocompetence.

(ii) state (0110) in which tyrosine kinase activity is on and inhibition present. As shown in Figure 3a, this state is characterized by an unresponsiveness to further stimulation and corresponds to the anergic state. Unresponsiveness is however reversed by external addition of the signaling component s (Fig. 3b), in agreement with the experimental observation that anergy is reversed after addition of exogeneous Il-2.

Each event in our logical network requires a characteristic time delay to be realized. In agreement with the observation that costimulatory functions accelerate the signaling process, we consider here that the effect of costimulation is to decrease the time delay t_s for switching the positive signal on. Thus, $t_s < t_x$ describes the presence of costimulation, whereas $t_s > t_x$ corresponds to the absence of costimulation.

Taking into account that it is the branch requiring the least time which will be followed, the conditions for selecting alternative routes at each branching point in the network are given by inequalities among the different time delays. Using the tools of Boolean algebra and combinatorial logic, overall conditions can then be deduced for traversing branches leading to a particular cellular response or final situation.

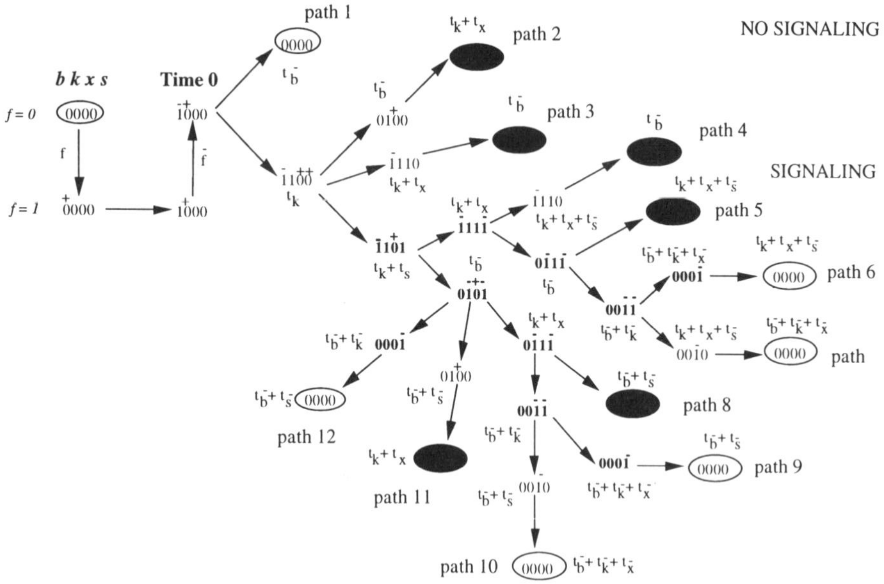

Fig. 2. Transition diagram showing the sequences of states and possible pathways. The total time to reach a given state, relative to time 0, is written next to each state in terms of the characteristic transition times (delays) of the state variables. The on and off delays corresponding to variable b are designated by t_b and $t_{\bar{b}}$, respectively, and similarly for the other state variables.

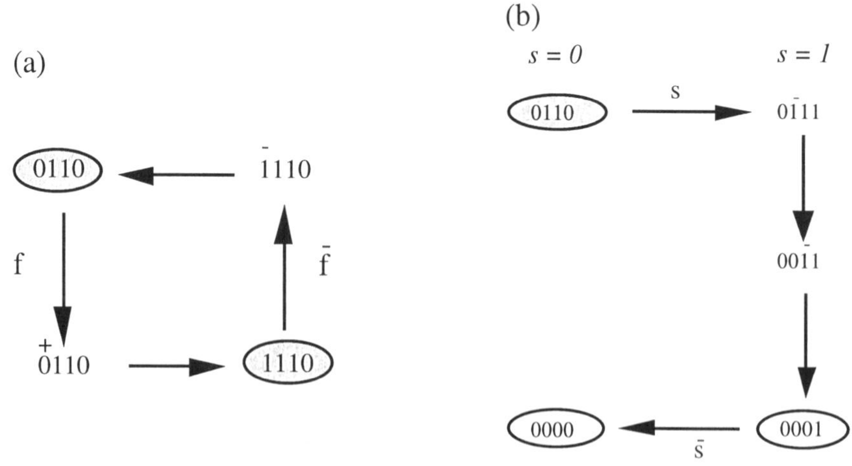

Fig. 3. (a) Addition of free ligand to the anergic state fails to evoke a productive response. After ligand dissociation the systems returns to the unresponsive state. (b) Reversal of anergy. Addition of the signaling component s restores the fully responsive state.

5. RESULTS

5.1. Null ligands

Pathway 1 is followed if ligand dissociation precedes kinase activation ($t_{\bar{b}} < t_k$) and corresponds to what is observed for "null" or inactive ligands. Dissociation is extremely rapid and no detectable changes are observed within the cells (Lyons *et al.*, 1996).

5.2. Altered Peptide Ligands

Pathway 2 is followed if ligand dissociation is slower than kinase activation but faster than activation of the signaling or inhibitory pathways ($t_k < t_{\bar{b}} < \text{Min}\ \{t_k + t_s, t_k + t_x\}$). Upon restimulation with an activatory ligand in the presence of costimulation, signaling is prevented by the already activated inhibitory pathway. This situation accounts for stimulation with altered peptide ligands (APL). These variants of antigenic TCR ligands, characterized by strongly increased off-rates (Lyons *et al.*, 1996), generally only lead to unproductive initial phosphorylation reactions and drive the cells into the anergic state, even in the presence of the necessary costimulatory molecules (Sloan-Lancaster *et al.*, 1993; Jameson and Bevan, 1995).

APL's are extensively studied for therapeutical reasons. Through anergy induction such ligands may provide a way to specifically control or modify unwanted immune responses such as autoimmune responses or immunopathological inflammatory immune responses. The principle of weak TCR-ligand interactions might also be involved in positive selection of the T cells in the thymus. On the contrary, mutations leading to such inhibitory characteristics may be exploited by viruses for escaping from immune response, resulting in persistent infections.

5.3. Absence of costimulation

Pathway 3 is followed if $t_{\bar{b}} > t_k + t_x$ and $t_s > t_x$. Here, the receptor-associated kinases are activated but inhibition precedes signal transmission so that there is no significant positive signaling. Again, the system ends up in the unresponsive state and restimulation in optimal conditions does not lead to Il-2 production and proliferation. This pathway accounts for stimulation with an activatory ligand in the absence of costimulatory signals. Such stimulation conditions have been shown to elicit only unproductive transduction events and to lead to subsequent anergy (Jenkins and Schwartz, 1987).

Physiologically, T cell anergy induced by the absence of costimulation appears to be a mechanism by which potentially autoreactive T cells are inactivated in the periphery, thus maintaining peripheral tolerance to self antigens. Therapeutically, the selective induction of anergy by preventing costimulation could be very useful for avoiding graft rejection. In contrast, anergy due to the absence of costimulation is a potential mechanism by which tumours may escape immune surveillance and pathogens may evade destruction.

5.4. Positive signaling

The general conditions for positive signaling are: $t_s < t_x$ and $t_{\bar{b}} > t_k + t_s$. Signaling should precede inhibition and ligand dissociation has now to be slower than the time needed to activate the kinases and to transmit the signal. These conditions correspond to paths 4 to 12 and account for signaling after stimulation with an activatory ligand in the presence of costimulation.

After signaling the system may end up in anyone of two stable steady states, corresponding either to recovery of responsiveness (0000) or to anergy (0110). Using again the tools of combinatorial logic the necessary and sufficient conditions to reach each of these two states can be determined, independently of the precise pathway that is followed. These conditions are:

(i) For signaling with recovery of responsiveness:

$$(t_{\bar{s}} > t_{\bar{k}}) \text{ AND } (t_{\bar{b}} < t_k + t_x \text{ OR } t_{\bar{b}} < t_k + t_x + t_{\bar{s}} - t_{\bar{k}}).$$

The activatory phase will thus be followed by recovery of immunocompetence if *both* the ligand does not bind too strongly and positive signaling does not decay before inactivation of the kinases.

(ii) For signaling followed by anergy:

$$(t_{\bar{s}} < t_{\bar{k}}) \text{ OR } (t_{\bar{b}} > \mathrm{Max}\{t_k + t_x, t_k + t_x + t_{\bar{s}} - t_{\bar{k}}\}).$$

The activatory phase will be followed by anergy if *either or both* the signal decays rapidly or ligand dissociation is very slow.

Note that these two conditions are the logical complement of each other, indicating that all the logical possibilities have been considered. The duration of the positive signaling phase can also easily be calculated.

The first situation is related to vaccination : the number of cells that are able to react to antigen will increase. The second situation corresponds to "activation-induced anergy". Transient T cell activation preceding the induction of unresponsiveness has been observed in vitro and in vivo in response to anti-CD3 antibodies (Andris *et al.*, 1996). Similarly, exposure of mature T cells in vivo to viral and bacterial superantigens leads to massive T cell activation followed by a long phase of anergy (MacDonald *et al.*, 1993). Such activation-induced anergy may represent a mechanism for limiting excessive inflammatory responses which represent a potential threat for the organism.

6. CONCLUDING REMARKS

The present model emphasizes the dual role of receptor-associated tyrosine kinases in the regulation of cellular competence. The autophosphorylative activity of these kinases leads to a bistable behavior thereby endowing the system with a memory of its previous antigenic experience. Residual tyrosine kinase activity is responsible for actively maintaining the inhibitory pathway at work, thus leading to anergy.

Our logical description is based on the approach developed by Thomas and coworkers, which emphasizes the importance of feedback loops and asynchronous switching in regulatory networks. This formalism has already been applied to other biological examples such as the decision between lysis and lysogenization in bacteriophage λ (Thomas and Van Ham, 1974; Thomas *et al.*, 1976), or the choice between memory and paralysis in the humoral immune response (Kaufman *et al.*, 1985). See also Thomas and D'Ari (1990). More recently, it has been used to illustrate the influence of hormone residence time on the insulin receptor in determining metabolic vs mitogenic signaling (Shymko *et al.*, 1996; 1997).

For our simple model, the logical analysis provides conditions for the selection of specific branches in the signaling pathway expressed in terms of inequalities among the time delays required for the activation or deactivation of the components of the pathway.

The conditions for anergy induction are given in terms of the ligand average residence time on the receptors, which is directly related to the kinetic dissociation rate. Anergy is induced if ligand dissociation is either rapid or very slow, suggesting that in addition to inadequate stimulation unresponsiveness may also develop as a consequence of an excessive stimulation. Positive signaling requires an acceleration of the signaling process by costimulation, and will be followed by recovery of T cell responsiveness if the ligand average residence time is in the required range. In addition, our approach may account for the fact that some ligands are inhibitory at low concentrations or for some cell types and become activatory at higher concentrations or for other cell types.

Several experimental (Lyons *et al.*, 1996; Valitutti and Lanzavecchia, 1997) and theoretical papers (McKeithan, 1995; Rabinowitz *et al.*, 1996) have stressed the influence of ligand dissociation rate for T cell activation. In particular, the kinetic discrimination model proposed by Rabinowitz *et al.* (1996) relates T cell activation to a ratio of complete (positive) and incomplete (negative) signals, thus allowing for a discrimination between agonist and antagonist ligands. In the framework of our model, based on autophosphorylative protein tyrosine kinase activity, we extend these ideas to cover a broader context and wider variety of situations.

Our analysis shows that both the rate of dissociation of ligand from T cell receptors and the rate of some key intracellular biochemical processes may determine the activatory or inhibitory properties of the ligand. Although our description does not fully characterize all the processes involved in T cell signaling, it accounts for a large body of data and allows to integrate different experimental observations into a common framework.

ACKNOWLEDGEMENTS

We thank R. Thomas and J. Urbain for helpful discussions. This work has been supported by the Belgian program on interuniversity attraction poles, initiated by the Belgian State, Prime Minister's Office Policy Programming. The scientific responsibility is assumed by its authors.

REFERENCES

Andris F., Van Mechelen M., De Mattia F., Baus E., Urbain J. and Leo O., *Eur. J. Immunol.* **26** (1996) 1187-1195.

Berridge M.J., *Critical Rev. Immunol.* **17** (1997) 155-178.

Beverly B., Kang S.-M., Lenardo M.J. and Schwartz R.H., *Int. Immunol.* **4** (1992) 661-671.

Gajewski T.F., Quian D., Fields P. and Fitch F.W., *Proc. Natl. Acad. Sci. USA* **91** (1994) 38-42.

Jameson S.C. and Bevan M.J., *Immunity* **2** (1995) 1-11.

Jenkins M.K. and Schwartz R.H., *J. Exp. Med.* **165** (1987) 302-319.

Kaufman M., Urbain J. and Thomas R., *J. Theor. Biol.* **114** (1985) 507-534.

Kaufman M., Andris F. and Leo O., *Int. Immunol.* **8** (1996) 613-624.

Lyons D.S., Lieberman S.A., Hampl J., Boniface J.J., Chien Y-H., Berg L.J. and Davis M.M., *Immunity* **5** (1996) 53-61.

MacDonald H.R., Lees R.K., Baschieri S., Herrmann T. and Lussow A.R., *Immunol. Rev.* **133** (1993) 105-117.

McKeithan T.W., *Proc. Natl. Acad. Sci. USA* **92** (1995) 5042-5046.

Muraille E., De Smedt T., Urbain J., Moser M. and Leo O., *Eur. J. Immunol.* **25** (1995) 2111-2114.

Muraille E., Devos S., Thielemans K., Urbain J. Moser M. and Leo O., *Immunology* **89** (1996) 245-249.

Rabinowitz J.D., Beeson C., Lyons D.S., Davis M.M. and McConnell H.M., *Proc. Natl. Acad. Sci. USA* **93** (1996) 1401-1405.

Samelson L.E., Davidson W.F., Morse H.C. and Klausner R.D., *Nature* **324** (1986) 674-676.

Schwartz R.H., *J. Exp. Med.* **184** (1996) 1-8.

Shymko R.M., De Meyts P. and Thomas R., *Exp. Clin. Endocrinol. Diabetes* **104** (1996) 72-73.

Shymko R.M., De Meyts P. and Thomas R., *Biochem. J.* **326** (1997) 463-469.

Sloan-Lancaster J., Evavold B.D. and Allen P.M., *Nature* **363** (1993) 156-159.

Thomas R. and Van Ham P., *Biochimie* **56** (1974) 1529-1547.
Thomas R., Gathoye A.M. and Lambert L., *Eur. J. Biochem.* **71** (1976) 211-227.
Thomas R. and D'Ari R., Biological Feedback (CRC Press Inc., Boca Raton, Florida, 1990).
Thomas R., *J. Theor. Biol.* **153** (1991) 1-23.
Umlauf S.W., Beverly B., Lantz O. and Schwartz R.H., *Mol. Cell. Biol.* **15** (1995) 3197-3205.
Valitutti S. and Lanzavecchia A., *Immunol. Today* **18** (1997) 299-304.

LECTURE 22

A Water Channel Network in Cell Membranes of the Filter Chamber of Homopteran Insects

P. Bron, V. Lagrée, A. Froger, I. Pellerin, S. Deschamps, J.-F. Hubert, C. Delamarche, A. Cavalier, J.-P. Rolland, J. Gouranton and D. Thomas

Canaux et Récepteurs Membranaires, UPRES-A 6026 du CNRS, Biologie Cellulaire et Reproduction, Université de Rennes1, Campus de Beaulieu, 35042 Rennes Cedex, France

1. INTRODUCTION

Water is the most ubiquitous molecule in the living cell and movement of water across the cell membrane accompanies fundamental cell functions. All biological membranes exhibit some water permeability as a result of diffusion through the lipid bilayer and osmotic gradients constitute the driving force for water flow. Osmotic water permeability is therefore of the highest relevance. However, some cells have the ability to transport water across their cell membrane at greatly accelerated rates, for example mammalian red blood cells, epithelial cells of the renal proximal tubules. Water permeability in such cells is simply too high to be accounted for by lipid-mediated diffusion, thus leading biophysicists to predict that water-selective channels must exist. The search for water channel began not surprisingly in tissues that had been already identified from physiological studies as having high water permeabilities. But the molecular basis of water channels remained elusive for a long time, since many attempts to determine its structure by biochemical approaches and expression cloning were unsuccessful. The reasons of this failure were linked to the inability of the water channel to be labeled by its substrate, the lack of highly specific inhibitors and the basal diffusional permeability of cell membranes.

The prediction was borne out by a serendipitous event with the discovery of CHIP28 (channel-forming integral membrane protein of 28 kDa) in the membranes of red blood cells. The isolation of CHIP28 complementary DNA (cDNA) and the analysis of its product function using *Xenopus* oocytes as expression systems allowed for the first time the demonstration of the existence of a protein water channel (Preston *et al.*, 1992). CHIP28, renamed AQP1, became the archetype of water channels (Agre *et al.*, 1993). Other aquaporins have further been functionally identified. They were isolated from bacteria and from various mammalian, amphibian or plant tissues. All aquaporins are 25 – 30 kDa membrane proteins with similar primary sequences that class them in the MIP family (Gorin *et al.*, 1984), named from its archetype MIP26, the major intrinsic protein of bovine lens fibers. MIP-related proteins all have six stretches of hydrophobic sequence and share an internal repeat which both contain a highly conserved motif Asn-Pro-Ala (NPA), suggesting the duplication of an ancestral gene (Reizer *et al.*, 1993).

We contributed to this quest for the "wet" channel in studying a model which could be favorable for isolating water channel proteins. We investigated an epithelial complex found in the digestive tract of some homopteran insects feeding on plant sap. In this complex, called the "filter chamber", a significant water transfer is believed to occur down a transepithelial osmotic gradient (Gouranton, 1968). In a structural study on the filter chamber of *Cicadella viridis*, we observed that epithelial cell membranes exhibit a network of intramembrane particles, which cover the whole surface of the membranes (Fig. 1). By polyacrylamide gel electrophoresis, we have shown that the major constituent of the filter chamber purified membranes is a 25 kDa hydrophobic polypeptide (P25) (Hubert *et al.*, 1989). In this review we will report the strategy we followed to demonstrate that P25 is a water channel.

Fig. 1. Electron micrograph of a freeze-fracture preparation carried out on the filter chamber of *Cicadella viridis*. The cell membrane exhibits a network of intramembrane particles. Each particle has a size of about 9 nm (× 100 000).

2. P25 BELONGS TO THE MIP FAMILY PROTEIN (Beuron *et al.*, 1995)

As a result of its extremely high representation in the plasma membranes, it appeared very likely that P25 takes an important part in the constitution of the regular array within the native membranes and is involved in the water transport function of the filter chamber epithelia. We hypothesized that it could be a water channel and thus belongs to the MIP family. Besides functional studies of P25, we focused our work on cloning the cDNA encoding P25, associated with its structural determination.

Using polymerase chain reaction, we isolated a 360-base pair cDNA, named *cic*, from RNA of the filter chamber. *cic* encodes a 119-amino acid polypeptide (CIC), whose homologies with MIP26 is 38%. Then, using a specific antibody raised against a 15 amino-acid peptide derived from the CIC sequence, we were able to conclude that CIC and P25 are identical entities and that therefore P25 belongs to the MIP family.

Having demonstrated that P25 is a member of MIP family and considering the rarity of MIP structural organization, it appeared interesting to investigate its fine structure, The two-dimensional membrane crystal constituted in the filter chamber by the extraordinary abundant P25 was analyzed using electron microscopy and image processing. The basic motif of the lattice was found to be composed of 4 elongated bilobed domains arranged around a central pit (Fig. 2).

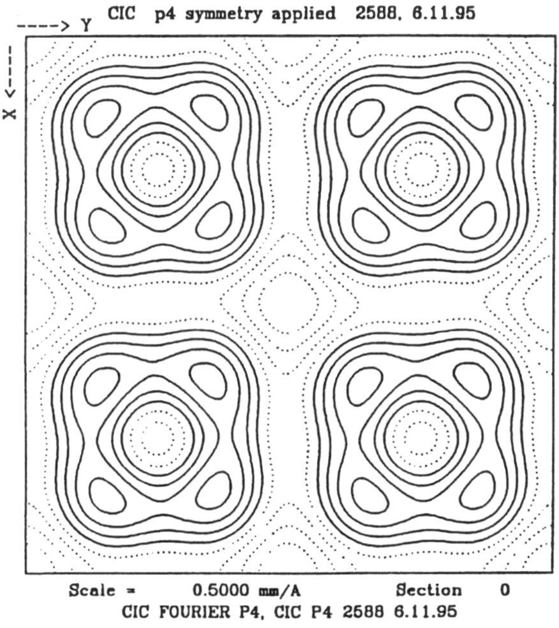

Fig. 2. Two-dimensional projection of 4 membrane protein tetramers calculated from the fourier transform of electron images of isolated membrane. The unit cell dimensions are 9.6 × 9.6 nm; 90°.

Combined with mass informations provided by Scanning Transmission Electron Microscopy in dark-field mode (Thomas *et al.,* 1994) these images were interpreted as tetramers of P25. Thus, we proposed an informative two-dimensional projection map of the P25 tetramer, consistent with the two-dimensional structure reported for AQP1 (Walz *et al.*, 1994).

3. INCORPORATION OF PROTEINS INTO *XENOPUS OOCYTES* BY PROTEOLIPOSOME MICROINJECTION (Le Cahérec *et al.*, 1996a)

A direct demonstration that a protein being involved in rapid osmotic flux across membranes can be performed by the injection of messenger RNA or *in vitro* transcribed RNAs into defolliculated *Xenopus oocytes*. After 2-3 days incubation, oocytes are exposed to hypotonic buffer. Oocytes expressing water channels show a significant increase (> 15-fold) in osmotic water permeability. This increase is inhibited by addition of mercurials, but is restored with reducing agents, indicating the presence of a critical sulfhydryl group in the pore.

Oocyte water permeability is measured from the time course of oocyte swelling in response to an hypotonic extracellular buffer. The swelling is monitored using a microscope equipped for video recording. The oocyte equatorial plane is focused and volume variations are recorded at regular intervals for 5 min. Results are expressed as the relative oocyte volume (V/V_0) computed from the relative oocyte area (A/A_0) in the focal plane.

$$V/V_0 = (A/A_0)^{3/2}.$$

The osmotic water permeability coefficient (Pf, in cm/s) is calculated from the oocyte-surface area (S = 0.045 cm^2), the initial volume ($V_0 = 9 \times 10 - 4$ cm^3), the molecular volume of water (V_w = 18 cm^3/mol), and the initial rate of swelling $d(V/V_0)/dt$, by means of the equation:

$$Pf = V_0 \times d(V/V_0)/dt/[S \times V_w \times (Osm_{out} - Osm_{in})]$$

where Osm_{out} = 176 mosmol and Osm_{in} = 34 mosmol.

However, in the case of a lack of cloned cDNA, the procedure of injecting cDNA transcripts cannot be used. To overcome this problem, we proposed an alternative system allowing functional insertion of exogenous proteins into the plasma membrane of *Xenopus* oocytes. We microinjected proteoliposome suspensions into the cytoplasm and then analyzed membrane protein function. The proteins used were: human AQP1, MIP26 from bovine eye lens and P25 from *Cicadella*. The subsequent insertion of these proteins into the plasma membrane of oocytes was demonstrated by immunocytochemistry (Fig. 3). Oocytes microinjected with either AQP1 or P25-proteoliposomes exhibited significantly increased osmotic membrane water permeabilities compared to those measured for oocytes injected with liposomes alone or with MIP26-proteoliposomes. These effects were inhibited

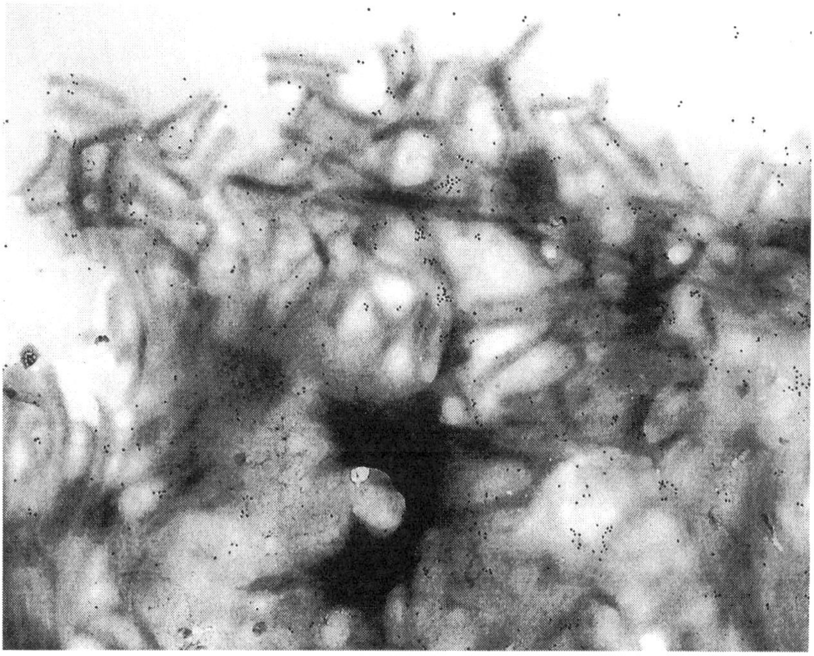

Fig. 3. Subcellular localization of protein injected into oocytes by immuno-electron microscopy 10 nm gold particles indicate the insertion of P25 into the microvilli of the oocyte plasma membrane (× 30 000).

by $HgCl_2$ in a reversible manner. From this study, we demonstrated that P25, from *Cicadella viridis* filter chamber, exhibits in this system water channel properties and therefore is an aquaporin.

4. MOLECULAR CLONING AND FUNCTIONAL CHARACTERIZATION (Le Cahérec *et al.*, 1996b)

Among the unanswered questions about aquaporins, the nature of the water channel is of importance. Comparison of sequence-related functional differences between several MIP is thus a field of great interest to identify the molecular motifs involved in the water-specificity of aquaporins. The number of sequences available for this purpose is regularly increasing, however the nature and the precise localization of the water pore remain undetermined.

We have isolated and cloned a full length cDNA clone from a filter-chamber library of *Cicadella viridis*. This cDNA encodes for a 26 kDa protein whose sequence is 43% identical to AQP1. Translation of the corresponding RNA into *Xenopus* oocytes generated a polypeptide that was specifically recognized by polyclonal antibodies raised against the native protein p25 and insertion of the protein into the membrane oocyte was checked by immunocytochemistry. Expression of the protein in oocytes led to a 15-fold increase of osmotic membrane permeability (Fig. 4). We called this protein AQPcic for *Cicadella* aquaporin.

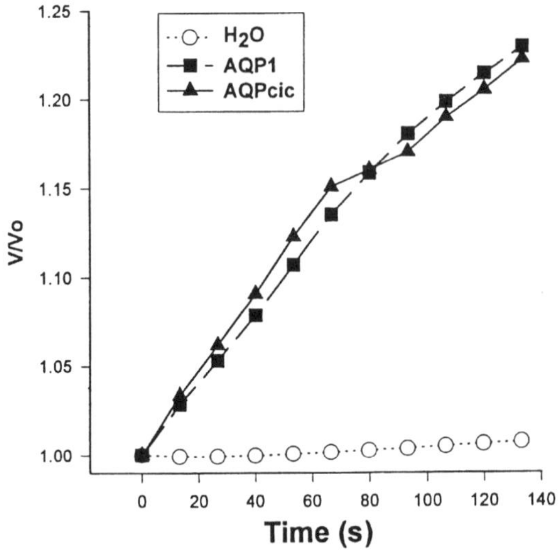

Fig. 4. Time-course of osmotic swelling of oocytes expressing AQP1 or AQPcic in response to a 5-fold dilution of the extracellular buffer.

Like for AQP1, the increase of membrane permeability was inhibited by $HgCl_2$. However, oocytes expressing AQPcic were less sensitive to mercury than oocytes expressing AQP1 (Fig. 5). In order to identify the cysteine(s) involved in permeability inhibition by $HgCl_2$, we have constructed mutants of AQPcic by site-directed mutagenesis. The complementary RNAs were injected into oocytes and water permeability was measured in the presence or absence of $HgCl_2$. One mutant for a cysteine appeared to be insensitive to $HgCl_2$, thus allowing us to identify a cysteine residue that is situated close to the pore.

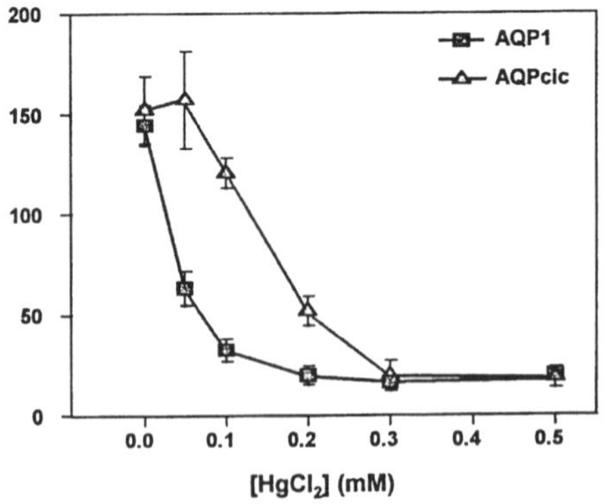

Fig. 5. Mercury effect on osmotic water permeability of oocytes expressing AQP1 or AQPcic.

5. FUTURE PROSPECTS

Electron crystallography. Sequence analysis of the MIP family predicted that aquaporins as well as other MIP proteins would be α-helical transmembrane proteins with six membrane-spanning helices. Studies on the membrane topology of AQP1 by epitope insertion and by site-directed mutagenesis led to the "hourglass model" in which the two hydrophilic loops containing the NPA sequences are folded into the lipid bilayer, forming the aqueous pore (Jung *et al.*, 1994). Electron crystallographic studies on AQP1 two-dimensional crystals generated a projection map of the water at about 6 Å resolution where 6 to 8 density peaks were resolved, corresponding probably to alpha-helices (Jap *et al.*, 1995; Mitra *et al.*, 1995; Walz *et al.*, 1995). Two years later, the same groups published a three-dimensional structure of AQP1 at 6 – 7 Å resolution. Each AQP1 monomer has six transmembrane tilted α-helices which form a barrel surrounding a central density region. This central mass could be formed by the extended loops B and E which carry the functionally important NPA boxes (Li *et al.*, 1997; Walz *et al.*, 1997; Cheng *et al.*, 1997). These works provide a basic structural design for AQP1 water channel and are clearly a step toward an atomic resolution structure which is required to apprehend the molecular mechanism of the water channel.

Since these works were essentially conducted on AQP1, which was the only aquaporin, so far, purified and crystallized, we thought that the production, purification, and crystallization of a different aquaporin could deserve the general knowledge of the aquaporin family. We have successfully produced in yeast cells a functionally recombinant AQPcic in sufficient amount to obtain the protein in a high purified state. Crystallization experiments are presently under progress to produce large crystals suitable for electron crystallography.

Sequence analysis and site-directed mutagenesis: Sequence analysis of the MIP family proteins lead to major conclusions: (1) MIP proteins probably arose by intragenic duplication (2). All MIP family members contain highly conserved amino acids and can be identified from a signature sequence. Presently the MIP family contains two functional types which have been well characterized: the specific water transport (aquaporins) and the small solutes transport like glycerol (glycerol facilitators). To discover the discriminant amino acids which should be responsible for the specificity of the function, we have used a statistic approach to compare 42 MIP sequences representative of all groups of organisms. Our main results have shown that one can differentiate the aquaporins from the glycerol facilitators by looking at a very small number of amino acids. However, the physiological significance of these observations needs to be experimentally confirmed using mutagenesis experiments. To be able to transform an aquaporin molecule into a glycerol facilitator will be one challenge for the understanding of the molecular basis of water and glycerol transport.

REFERENCES

Agre P., Preston G.M., Smith B.L., Jung J.S., Raina S., Moon C., Guggino W.B. and Nielsen S., *Am. J. Physiol.* **265** (1993) F463-F476.

Beuron F., Le Cahérec F., Guillam M.T., Cavalier A., Garret A., Tassan J.P., Delamarche C., Schultz P., Mallouh V., Rolland J.P., Hubert J.F., Gouranton J. and Thomas D., *J. Biol. Chem.* **270** (1995) 17414-17422.

Cheng A., van Hoek A.N., Yeager M., Verkman A.S. and Mitra A.K., *Nature* **387** (1997) 627-630.

Gorin M.B., Yancey S.B., Cline J., Revel J.P. and Horwitz J., *Cell.* **39** (1984) 49-59.

Gouranton J., *J. Microsc. Paris* **7** (1968) 559-574.

Hubert J.F., Thomas D., Cavalier A. and Gouranton J., *Biol. Cell.* **66** (1989) 155-163.

Jap B.K.and Li H., *J. Mol. Biol.* **251** (1995) 413-420.

Jung J.S., Preston G.M., Smith B.L., Guggino W.B. and Agre P., *J. Biol. Chem.* **269** (1994) 14648-14654.

Li H., Lee S. and Jap B.K., *Nature Struct. Biol.* **4** (1997) 263-265.

Le Cahérec F., Bron P., Verbavatz J.M., Garret A., Morel G., Cavalier A., Bonnec G., Thomas D., Gouranton J. and Hubert J.F., *J. Cell. Sci.* **109** (1995a) 1285-1295.

Le Cahérec F., Deschamps S., Delamarche C., Pellerin I., Bonnec G., Guillam M.T., Thomas D., Gouranton J. and Hubert J.F., *Eur. J. Biochem.* **241** (1995b) 707-715.

Mitra A.K., van Hoek A.N., Wiener M.C., Verkman A.S. and Yeager M., *Nature Struct. Biol.* **2** (1995) 726-729.

Preston G.M., Caroll T.P., Guggino W.B. and Agre P., *Science* **256** (1992) 385-387.

Reizer J., Reizer A. and Saier M.H., *Crit. Rev. Biochem. Mol. Biol.* **28** (1993) 235-257.

Thomas D., Schultz P., Steven A.C. and Wall J., *Biol. Cell.* **80** (1994) 181-192.

Walz T., Smith B.L., Zeidel, Engel A. and Agre P., *J. Biol. Chem.* **269** (1994) 1583-1586.

Walz T., Typke D., Smith B.L., Agre P. and Engel A., *Nature Struct. Biol.* **2** (1995) 730-732.

Walz T., Hirai T., Murata K., Heymann J.B., Mitsuoka K., Fujiyoshi Y., Smith B.L., Agre P. and Engel A., *Nature* **387** (1997) 624-627.

LECTURE 23

Creative Genomic Webs

E. Ben-Jacob

School of Physics and Astronomy, Raymond & Beverly Sackler Faculty of Exact Sciences, Tel-Aviv University, Tel-Aviv 69978, Israel

1. INTRODUCTION

In a previous manuscript [1] I have presented a new view of bacterial colonies as smart entities. Here I start with a new picture of the genome as an adaptive cybernetic unit with self-awareness which is presented in Sections 2 through 4. The genome, as I see it, is not merely a storage device, but a sophisticated cybernetic entity well beyond a universal Turing machine [2]. Metaphorically speaking, it includes a user, a computational unit, and a hardware engineer and technicians. The computational unit itself supersedes the universal Turing machine, since during computations the structure is dynamic and changes adaptively according to the needs dictated by the computations. The crucial component is the "user" which can recognize difficulties imposed by the environment and formulate problems requiring solution. This "user" possesses information about the past and present abilities of the system, which it can apply when searching for a solution to a current problem. It also has the potential for interpreting and assigning meaning to the computations. I further assume that the genome has self-awareness. In Section 3, I propose that the following requirements must be fulfilled by a system to possess self-awareness: 1) It has to be a cell composed of agents (*vs.* sets composed of elements). 2) It has to evolve in time. 3) It has to be an open system – constantly exchanging energy and information with the environment. 4) It has to have an advanced language (with self-reference to sentences and to its grammar).

At first it seems that the assumption that the genome is an adaptive cybernetic unit with self-awareness will suffice to explain evolution (Sect. 4). However, this is not so. A lemma extended from Gödel's theorem sets limitations on self-improvement. Naively phrased, it says that a system cannot design another system which is more complex than itself.

In Section 6, I use the distinction between Kuhn's normal science (problem solving within the scientific paradigm) and scientific revolutions (creation of a new scientific paradigm) as a metaphor to define horizontal genomic changes *vs.* vertical genomic leaps. The extension of Gödel's theorem would imply that the genome is not capable of performing genomic leaps. Yet these may be the most relevant changes in evolution.

Are we back to random mutations? Could it be that the horizontal genomic changes are self designed changes in response to the environment, and the more dramatic vertical genomic leaps are due to Darwinian evolution? I don't think so, and suggest the new picture of cooperative evolution as an alternative.

First, I propose that the vertical genomic leaps are in response to an existing paradox. In Section 6, I present the creativity paradox, and my picture in which the paradoxes are the gears of creativity, serving as the new principles on which the new paradigm is established. Next (Sect. 7), based on the contemporary knowledge of genetic communication in a stressed colony, I propose that the colony forms a genomic web (I use the term web instead of network to emphasize that the building blocks are self-aware agents and not elements). The genomic web is a "super-mind" relative to the individual genome. Thus, a paradox for the genome is a problem for the web, which in turn exercises its creativity on the genome level. Or, in other words, the web is far more complex than the individual genome, so it can design a new and more advanced genome which would represent a vertical leap beyond the previous genomic version. This is why I use the terms cooperative self-improvement or cooperative evolution.

I conclude in section 8 with some implications of the new picture, and speculate about its validity to eukaryotes.

2. THREE LEVELS OF INFORMATION TRANSFER AND THE CONCEPT OF CYBERNATORS (CYBERNETIC AGENTS)

Morphotype transitions during growth of bacterial colonies have led Ben-Jacob *et al.* to propose a new cybernetic framework [3-7]. The latter is also motivated by the experimental findings about adaptive mutagenesis and is based on the contemporary knowledge about the genetic agents (*e.g.* plasmids, transposons, phages, as well as other dynamic agents) discussed in the previous manuscript [1]. I have mentioned that these autonomous genetic agents, which can perform genetic changes in the host cell, can have their own "self-interests" and their own direct communication channels to the conditions outside the cell.

In the new picture we designate autonomous genetic agents whose function is regulated by holoparameters (*i.e.* colony parameters such as growth kinetics, cellular density, density of metabolic byproducts, level of starvation, *etc.*) as cybernators. I emphasize that an agent here is not necessarily a specific single macromolecule. It could be a combination of units or even a collective excitation of the genome performing the specific function. In other words, generally it should be viewed as a conceptual unit, although specifically it might be one macromolecule or a collection of molecules. The crucial point is that, since the cybernators' activity is regulated by holoparameters, it can produce changes in the genome's activity and structure that modify the individual cells in a manner beneficial to the colony as a whole. Thus the bacteria possess a cybernetic capacity which serves to regulate three levels of interactions: the cybernator, the cell and the colony. The "interest" of the cybernator serves the "purpose" of the colony by readjusting the genome of the single cell. The cybernator provides a singular feedback mechanism as the colony uses it to induce changes in the single cell, thus leading to consistent adaptive self-organization of the colony.

3. GENOMIC ADAPTATION AND GENOMIC LEARNING

I proceed here with a detour to present two additional ideas that emerged from the observations of the morphotype transitions. Clearly, the potential to perform the transitions from one morphotype to another in response to environmental conditions is available within the bacteria, as well as the capacity for "deciding" to go through the transition (although the "decision" can be a collective action of many bacteria in the colony).

Whether the mechanism is based on activation of cybernators (as is proposed in Refs. [3-6]), or is "ordinary" epigenetics [8] or phase variations [9], or another mechanism yet to be revealed, the important point is that it provides the colony with the potential to select the preferred morphotype according to the environmental conditions.

To emphasize the special nature of such morphotype transition, to distinguish it from ordinary reversible phenotypic adaptation and to draw upon the possible relations with adaptive mutagenesis, I refer to it as cooperative genomic adaptation. The advantage of the latter (over phenotypic adaptation) is probably with respect to more severe but less frequent changes in the environmental conditions, *e.g.* soft soil *vs.* hard soil in different seasons of the year. Indeed, the concept of time has a major role in the process of genomic adaptation, as will be discussed in reference [10].

The possession at present of the potential for genomic adaptation means that it had to be acquired by the genome sometime in the past during its course of evolutionary history. I refer to the above process of acquisition as genomic learning, to emphasize my assumption that it is not a result of Darwinian evolution (although in principal one can construct a complicated explanation based on the contemporary picture of evolution). For the genome to perform learning in the sense of "learning from experience", the following requirements must be fulfilled:

1. Exposure of the bacteria to several cycles of alternating environmental conditions (*e.g.* wet and dry soil).
2. Stored information about past environmental conditions.
3. Self-information: Information about past and present abilities of the genome.
4. Means for the genome to recognize difficulties and formulate problems and for problem solving according to the collected and processed information, both about internal state and external conditions (including state of other bacteria).

5. Cybernetic capacity: means for the genome to change itself according to solutions to problems. Ranging from reorganization and restructuring [11-13] (*e.g.* activation and deactivation of genes, replication of genes, moving genes, *etc.*) to actual interlacing of new sequences which we propose [6].

The acquisition of the potential for the morphotype transitions is only one out of many examples which can be assumed as genomic learning (see Ref. [10] for more examples). In all cases, if genome learning is assumed the above requirements must be fulfilled.

4. THE GENOME AS AN ADAPTIVE CYBERNETIC UNIT WITH SELF-AWARENESS

Back in 1992 [3] we referred to our observations of complex colonial patterning as an example of adaptive self-organization, and proposed that "the genome can be viewed as an adaptive cybernetic unit". We have concluded that "along the above assumptions, the colony organization (being the environment) can directly affect the genetic metamorphosis of the individuals. Hence, we expect to observe synchronized, autocatalytic and cooperative genetic variations of the colony, either spontaneous or in response to imposed growth conditions".

In a follow up publication submitted a year later, [4] we suggested a possible mechanism based on the concept of cybernators to provide the singular feedback between the colony and the individual bacteria. The two publications (Refs. [3, 4]) were primarily devoted to report our experimental observations, and the new picture remained somewhat fuzzy. Now it is time to clarify things. I devote this section to elaborate on the new proposed picture of the genome, and I try to clarify what are the known facts and what are our new assumptions and conclusions.

As is described in the previous manuscript [1], there is a vast amount of knowledge about the structure and functions of the extra-chromosomal elements in the genome. The introduction of the new terminology of cybernators is to emphasize the new interpretation of their role as part of the cybernetic capacity of the genome.

It is well known that the genome can change itself. We have proposed that the changes are neither random nor automatic but rather self-designed by the environment. Trivially, the ability to design changes requires computational capabilities. Indeed, Shapiro [13] proposed "... thinking of genomes as complex interactive information systems, in many ways comparable to those involving computers".

We have referred to the genome as an adaptive cybernetic unit [3, 4] in order to emphasize that, in our view, it is beyond a universal Turing machine [14]. As I mention in the introduction, metaphorically speaking, the genome includes a user with a computational unit and a hardware engineer with a team of technicians for continuous design and implementation of changes in the hardware. Such a complex is beyond a universal Turing machine. In the latter, the structure is static and is decoupled from the input/output and the computation process. The genome is a dynamic entity. If its structure changes adaptively according to the performed computations, it implies that the genome is capable of self reference, has self-information and, most crucially, has self-awareness. The user represents the ability of the genome to recognize that it faces a difficulty (imposed by the environmental conditions), formulate the problem associated with the difficulty and initiate a search for its solution. As discussed in Section 3, the genome employs its past experience in the process. The user also represents the ability of the genome to interpret and assign meaning to the outcome of its computations and compare it with its interpretation of the environmental conditions.

It might seem that I have been carried away from facts to fantasy. So before going on I would like to emphasize that it is not necessarily so. We know that the genome can change itself, and in some cases we know that it is done purposely, in order to adapt to the environment. If we combine this information with the assumption that the acquisition of the potential for transformations is via genomic learning, it directly implies that the genome is an adaptive cybernetic unit with self awareness (*i.e.*, it has all the features described above).

To refer to the genome as being self-aware is a very strong statement with farreaching implications. The issue will be presented in a forthcoming publication [10]. I briefly describe here the main points needed for this presentation. Our logic and mathematics are based on the notion of a set composed of elements. Implicitly, the set is closed and static, the elements have a fixed identity (it does not change due to the fact that they are part of the set) and they either do not have internal structure or, if they do, it is not relevant to the definition of the set. The set is defined by an external observer, *i.e.*, it is not a result of self-assembly of the elements under a common goal. The elements, being passive and of no structure, do not have any information about the set. The definition of sets leads to logical paradoxes (Russel-type, like the famous barber paradox) when we try to include a notion of self-reference. Russel and others have devoted much effort to construct formal axiomatic systems free of inherent logical paradoxes. Gödel's theorem [15, 16] proved that they all have to be "incomplete", including the Principa Mathematica of Russell and Whitehead. It is important to emphasize that Gödel's theorem applies to closed systems which are also fixed in time. I propose that one has to take an entirely different approach and not start with the notion of sets of elements. I believe that here is exactly where the reductionist approach fails. We cannot reach self-awareness starting from passive elements, no matter how intricate their assembly is. I propose to replace elements by agents, that possess internal structure, purpose and some level of self-interest, and whose identity is not fixed. The notion of a set is replaced by a cell, which refers to a collection of agents with a common goal and mutual dependence. It also implies that the system of agents is open, *i.e.*, it exchanges energy and information with the environment. I argue that, in order for a cell of agents to be self-aware, it must have an advanced language, *i.e.*, a language which permits self-reference to sentences and to its grammar. The language also enables the individual agents to have information about the entire system. In addition the cell has strong coupling with the environment. The "self" is emerged through this coupling. There is no meaning of "self" in a closed system.

5. GÖDEL'S THEOREM AND THE LIMITATIONS OF SELF-IMPROVEMENT

Gödel's theorem (paraphrased by Hofstadter into a more "digestible" form) states [16]:

All consistent axiomatic formulations of number theory include undecidable propositions.

The great achievement of Gödel was the connection of the idea of self-referential statements in language with number theory. Clearly, mathematical statements in number theory are about the properties of whole numbers, which by themselves are not statements, nor are their properties. However, Gödel had the insight that a statement of number theory could be about a statement of number theory (even about itself, *i.e.* self-reference).

For this, numbers should be mapped (one to one mapping) to statements, by a certain code, and Gödel has indeed constructed one. This coding trick enables statements of number theory to be understood on two different levels: 1. as statements of number theory; and 2. as statements about statements of number theory [16].

Using his code, Gödel teleported the Epimenides paradox ("This statement is false": true-false-true-...) into number theory in a version "This statement of number theory does not have any proof in the system of Principia Mathematica (or any fixed axiomatic system). One implication of Gödel's theorem is that no fixed axiomatic system, no matter how complex, could even represent the complexity of the whole numbers. Gödel's theorem cannot be directly applied to the genome. One can do the same trick and map the DNA sequence either to the whole numbers or to statements in language. However, Gödel's theorem deals with infinite systems while the genome is finite [17, 18].

To apply Gödel's theorem, another mapping should be considered – that performed in Nature. This is the mapping from DNA sequence to proteins. The proteins define a finite set of "words" in an infinite "language" [19]. Functional combinations of proteins are then sentences, and the interactions between them are the grammar. This picture might be supported by studies of correlations in DNA sequences and applications of Zipf's tests [20, 21]. Once we have an infinite language, Gödel's theorem can be applicable. To escape the limitations posed by the theorem, the sequence must change in time.

Let me elaborate on this point. The set of all possible environmental conditions poses an infinite number of problems which cannot be solved by any given language. Luckily, at a given instance of time the organism faces only a finite number of relevant problems. So there should be a version of the language which provides solutions to the current problems.

We have proposed that the genome is capable of performing self-designed genomic changes. Thus, at first it seems that assuming the genome to be an adaptive cybernetic unit with self-awareness is sufficient to explain evolution. This is not the case. A lemma extended from Gödel's theorem sets limitations on self-improvements (Ref. [10]). Simply put, it would state that "a system cannot self-design another system which is more advanced than itself" (this is in contradiction to the claim in Ref. [22]).

In the next section I define two types of genomic changes – horizontal changes *vs.* vertical leaps. The individual genome is capable of performing the first kind, but not the second. Only a genomic n etwork is capable of vertical leap, which are creative events.

6. PROBLEMS *vs.* PARADOXES AND HORIZONTAL GENOMIC CHANGES *vs.* VERTICAL GENOMIC LEAPS

It is customary to borrow ideas from the picture of evolution of organisms to describe the evolution of scientific theories. Here I engage in the reverse intellectual exercise. For reasons to be clarified below, I draw on the metaphor of the advancement of scientific ideas and propose to distinguish between two types of genetic changes. The identification is done according to the level of difficulties faced by the bacteria, the nature of the means required to cope with the difficulties and the type of genetic changes performed to cope with the difficulties.

Kuhn identifies two types of scientific progress – "normal science" and "scientific revolutions". Most scientific activities belong to the category of normal science. This proceeds by solving problems within a well defined conceptual plane or within a given theoretical framework with specified "rules of the game". The problems are also formulated within the conceptual plane of the present paradigm. To the other category belong the rare events of scientific revolutions that transcend science from a given theoretical framework to a new one. Scientific revolutions are initiated when scientists encounter a paradox, that is a problem which cannot be solved within the conceptual boundaries of the current paradigm. To solve a paradox, a new paradigm must be created, with an enlarged conceptual space and new "rules". The paradox is both the motive to the event of the creation of a new paradigm and the conceptual gear connecting the old paradigm to the new one. The paradox itself becomes the core principle upon which the new theoretical framework is constructed.

What relevance does the above bear to genetic changes of real living organisms? Organisms face at times difficulties best characterized as problems, and at times ones that could only be regarded as paradoxes. By problem I mean here a difficulty or existential hazard the solution to which can be obtained by using the tools at the disposal of the organism. A trivial example would be exposure to antibiotic for which the bacterium has a silent (inactiveable) gene that the bacterium must activate. Adaptive mutagenesis is another example of problem solving. The nature of the genetic changes performed to cope with these difficulties is such that an organism undergoing them may still be considered the same organism, though an improved one. So I propose to refer to changes resulted from "problem solving" as "horizontal genomic changes". I have in mind a picture of these changes as a trajectory on a plane defined by the organism, in analogy with "normal science" which is a trajectory on a plane defined by the paradigm.

At present I do not have a good definition for the plane of the organism, and must rely on intuition. In the future we intend to use Gödel's approach (Sect. 5) to reach a definition.

Genetic changes which move the organism a step higher on the evolutionary axis represent "vertical genomic leaps" which are transitions from one plane to another. In analogy to scientific revolutions, I expect the "vertical genomic leap" to be a solution to a paradox, not a problem. A paradox here would be a difficulty to which the genome cannot find a solution using its own tools, since the solution is a new genome which is more advanced in comparison to the original one. For example, I believe that the emergence of sporulating bacteria is a "vertical genomic leap". What paradox could have led to such a solution? Sporulation enables bacteria to survive otherwise lethal conditions. The "decision" to sporulate (which is reached collectively) is based on the prediction that conditions will become lethal. The need to learn from lethal conditions could have been the paradox that forced the bacteria to come up with a vertical leap in order to survive.

7. THE COLONIAL WISDOM: GENOMIC WEBS AND EMERGENCE OF CREATIVITY

According to the extension of Gödel's theorem, the genome can design and perform horizontal genomic changes but not vertical leaps involving paradox solving. Are we back to random mutations? Could it be that the simpler horizontal changes resulted from designed changes and the more relevant (for evolution) vertical leaps are the outcome of random mistakes? The dilemma is solved when we assume cooperative behavior.

Say you would like to design a new, more advanced computer for a certain task. The best strategy would be to construct a network of computers to do the designing. Even though each individual computer is less advanced than the new computer, their network can, in principle, be superior to it. This is, for example, how Intel486 was designed: A network of Intel386 was employed for the task.

Back to the bacteria. It is known that in a stressed colony, some of the bacteria become competent by rendering their membrane more permeable to genetic material, while other bacteria go through lysis: break open and deposit their genetic material in the media [9, 23]. In addition, direct genetic connections between the bacteria are formed by means of conjugation or transduction [9, 23]. We propose that these features indicate that the stressed colony turns into a genetic network, which is the highest level of colony cooperation. To emphasize that the network is composed of agents (each genome is by itself a cybernetic agent) I refer to it as a "genomic web". I further assume, that in order to establish the genomic web, the bacteria produce (or activate) special cybernators enhancing the efficient and sophisticated genomic communication. Once formed, the genomic web is a "super-mind" relative to the individual genome. Thus, a paradox for the genome is a solvable problem for the web. The web, being more complex than the individual genome, can design and construct a new and more advanced genome relative to the original ones, *i.e.*, perform a vertical genomic leap. Such a leap is best described as a cooperative self-improvement or cooperative evolution.

The formation of a creative web is far from being trivial and requires very special environmental conditions. Not every assembly of agents leads to a more sophisticated entity. As we well know from daily experience, a committee composed of very intelligent individuals can be a fairly dumb entity. It depends on the balance between the agents' self-interest and their level of awareness of the new entity. In other words, the environmental conditions should be such that the individual bacteria will give up most of their awareness as individual entities [10].

In principle, the genome is capable of solving problems on its own, but it is more efficient to solve problems cooperatively. Hence I expect that genomic webs are also employed for the task of problem solving. The harder the problem, the more advanced the genomic web formed. Indeed, as I mention in the introduction, we now have evidence that adaptive mutagenesis requires cooperation of the bacteria.

The picture of a creative genomic web is very appealing. Yet a nagging conceptual difficulty is left. We would expect the colonies of new bacteria which are the outcome of a vertical genomic leap to be more advanced than the colonies of the original bacteria. But, if we truly regard the colony as a multicellular organism, it will be in contradiction with the extension of Gödel's theorem; if we regard the colony as our system, it would imply that a system is capable of designing a system more advanced than itself. I believe that the colony of the new bacteria can only be improved relative to the original colonies, and not more complex. In order to keep the picture consistent, we have to assume that genetic communication between many colonies of the same bacteria, or a number of colonies of different bacteria, are required for the design of a vertical leap on the colonial level (we discuss this point in Ref. [10]).

8. POSSIBLE IMPLICATION OF THE NEW PICTURE AND DARWINIAN EVOLUTION *vs.* COOPERATIVE EVOLUTION

The new picture I have presented here has many potential implications, both practical and philosophical. For exemple, at present, the bacteria seem to be winning the war we fight against them with antibiotics, developing drug resistance as fast as we develop new drugs, or faster. In order to outsmart them, we must first realize how smart they are, and accordingly develop new strategies for treatment. If, as I claim, the strength of the bacteria lies in colonial communication and cooperation, then a way to go would be to blackout and jam their communication rather than (or along with) disable the individual bacteria [24].

All along, I was referring to bacterial colonies and drawing conclusions from observations of bacteria. However, I believe that the idea of the emergence of creative web under stress is universal. I believe that eukaryotes have not lost the option of genetic communication in the course of evolution from prokaryotes, and that under stress, colonies of single cells eukaryotes establish genetic networks in very much the same way bacteria do. Some initial hints that this might be the case are provided by observations of adaptive mutagenesis in yeast.

In multicellular eukaryotes I expect continuous exchange of genetic information between cells. There are fragments of knowledge which, put together, could support a picture of genetic communication in multicellular organisms. However, in the absence of a proper theory, some were discarded as meaningless and others were studied separately. They were never put together and considered as parts of one picture.

There are reports from the seventies about circulating nucleic acids in higher organisms [25] and from the sixties and the seventies about released DNA segments from cells of eukaryotes [26-29]. These observations met with strong skepticism and, as they were not considered to be of any importance, have not been tested again.

It is well known that cancer cells can emit genetic material which induces other cells to become cancerous. This is clearly a case of transfer of genetic information between cells. Another recently studied phenomenon that involves such transfers is the death of cells. It is known that the dying processes of cells are very complicated and involve restructuring of the DNA into packed units which are deposited into the blood stream when the cell dies.

I assume genetic communication in multicellular organisms with the hope that in the future the fragments of knowledge can be collected to provide a solid proof.

If there is indeed genetic communication in eukaryotes, then the state of the eukaryote can directly affect genetic changes in its individual cells, in the same manner that the state of the colony affects genetic changes in the individual bacterium. I would like to emphasize that indeed macro to micro singular feedback should exist for efficient control. In this regard, I believe that there are cells specialized in producing cybernators. The latter affect germ cells, thus providing a plausible mechanism for designed changes in eukaryotes, changes brought about by the creative acts of genomic webs established within the organism.

A collection of eukaryotes can establish a web whose basic element is an individual eukaryote. Any means of communication between the organisms, if it is capable of affecting the state of the organism, indirectly affects the genetic level of each one. Hence a genetic web of eukaryotes can be formed.

I expect strong coupling between the genetic webs of different species which are functionally coupled. This coupling will cause induction of genetic changes from one web to another which can provide a plausible mechanism for the observed avalanche effects in evolution [30].

To conclude, I hope I was successful in convincing the reader that Vitalism is not the only alternative to Darwinism. I propose a new option, that of cooperative evolution based on the formation of creative webs. The emergence of the new picture involves a shift from the pure reductionistic point of view to a rational holistic one, in which creativity is well within the realm of Natural sciences.

ACKNOWLEDGMENTS

The ideas presented here are part of an intellectual endeavor I began in 1988. My students I. Cohen and A. Tenenbaum are important participants in this endeavor.

I thank I. Cohen for his invaluable help in the preparation of this manuscript. I would also like to thank Y. Aharonov, M. Azbel, H. Bloom and M. Sternberg for many discussions and mainly for their encouragements. I thank Inna Brains for technical assistance. I was lucky to receive support from the program for alternative thinking at Tel-Aviv University at the embryonic stages of the project. Currently it is supported in part by a Grant from the BSF No. 95-00410 and a Grant from the Israel Science Foundation, and the Sigel Prize for research.

REFERENCES

[1] Ben-Jacob E., Bacterial wisdom, *Physica A* (in press).
[2] Penrose R., *Shadows of The Mind* (Oxford University Press, Oxford, 1994).
[3] Ben-Jacob E., Shmueli H., Shochet O. and Tenenbaum A., Adaptive self-organization during growth of bacterial colonies, *Physica A* **187** (1992) 378-424.
[4] Ben-Jacob E., Tenenbaum A., Shochet O. and Avidan O., Holotransformations of bacterial colonies and genome cybernetics, *Physica A* **202** (1994) 1-47.
[5] Ben-Jacob E., Shochet O., Cohen I., Tenenbaum A., Czirók A. and Vicsek T., Cooperative strategies in formation of complex bacterial patterns, *Fractals* **3** (1995) 849-868.
[6] Ben-Jacob E. and Cohen I., Cooperative formation of bacterial patterns, edited by J.A. Shapiro and M. Dworkin, *Bacteria as Multicellular Organisms* (Oxford University Press, New-York, 1997) pp. 394-416.
[7] Ben-Jacob E. and Cohen I., Adaptive self-organization of bacterial colonies, edited by F. Schweitzer, *Self-Organization of Complex Structures: From Individual to Collective Dynamics* (Gordon and Breach Science Publ., London, 1997) pp. 243-256.
[8] Jablonka E. and Lamb M.J., *Epigenetic Inheritance and Evolution* (Oxford University Press, 1995).
[9] King R.C. and Stanfield W.D., *A Dictionary of Genetics* (Oxford University Press, 4th edition, 1990).
[10] Ben-Jacob E., Cohen I. and Tenenbaum A., Paradoxes as the origin of cooperative self-improvement in creative evolution, unpublished.
[11] Shapiro J.A., Observations on the formation of clones containing *arab-lacz* cistron fusions, *Mol. Gen. Genet.* **194** (1984) 79-90.
[12] Shapiro J.A. and Trubatch D., Sequential events in bacterial colony morphogenesis, *Physica D* **49** (1991) 214-223.
[13] Shapiro J.A., Natural genetic engineering in evolution, *Genetica* **86** (1992) 99-111.
[14] Siegelmann H.T., Computation beyond the turing machine, *Science* **268** (1995) 545-548.
[15] Gödel K., *On Formally Undecidable Propositions* (Basic Books New-York, 1962).
[16] Hofstadter D.R., *Gödel, Escher, Bach: an eternal golden braid* (Basic Books, New-York, 1979).

[17] Stent G., Explicit and implicit semantic content of the genetic information. In *Proceeding of the 4th International Conference on The Unity of Sciences* (New-York, 1975).

[18] Steiner G., *After Babel: Aspects of Language and Translation* (Oxford University Press, New-York, 1975).

[19] Jones S., *The Language of The Genes* (Flaming, Glasgow, 1993).

[20] Peng C.K., Buldyrev S., Goldberger A., Havlin S., Sciortino F., Simons M. and Stanley H.E., Long-range correlations in nucleotide sequences, *Nature* **356** (1992) 168-171.

[21] Czirók, Mantegna R.N., Havlin S. and Stanley H.E., Correlations in binary sequences and generalized zipf analysis, *Phys. Rev. E* **52** (1995) 446-452.

[22] Laing R., Artifial organisms: history, problems, directions, edited by C. Langton, *Artificial Life* (Addison-Wesley Publishing Company, New-York, 1989) pp. 49-61.

[23] Joset F. and Guespin-Michel J., *Prokaryotic Genetics* (Blackwell Scientific Publishing, London, 1993).

[24] Ben-Jacob E., Cohen I. and Gutnick D.L., Cooperative organization of bacterial colonies: From genotype to morphotype, *Ann. Rev. Microbiol.* (in press).

[25] Stroun M., Anker P., Maurice P. and Gahan P.B., Circulating nucleic acids in higher organisms, *Int. Rev. Cyto.* **51** (1977) 1-48.

[26] Ledoux L., *Informative Molecules in Biological Systems* (North-Holland Publ., Amsterdam, 1971).

[27] Bendich, Wilczok T. and Borenffreund E., *Science* **148** (1965) 374.

[28] Roosa R.A. and Bailey E., *J. Cell. Physiol.* **75** (1970) 137.

[29] Reid L. and Blackwell P.M., *Aust. J. Med. Technol.* **2** (1970) 44.

[30] Szathmary E. and Maynard Smith J., The major evolutionary transitions, *Nature* **374** (1995) 227-232.

Neuronal Networks

LECTURE 24

Hebbian Learning of Temporal Correlations: Sound Localization in the Barn Owl Auditory System

J.L. van Hemmen and R. Kempter

Physik-Department der TU München,
85747 Garching bei München, Germany

1. INTRODUCTION

Until recently an unresolved paradox existed in auditory and electrosensory neural systems [1, 2]: they encode behaviorally relevant signals in the range of a few microseconds with neurons whose time constants are at least one order of magnitude bigger. In this paper we will explain how the paradox can be resolved. We take the barn owl's auditory system as an example and present results of a modeling study [3] of a neuron in the laminar nucleus that combines inputs from both ears:

(i) A single neuron can be quite a good coincidence detector despite its receiving stochastic input [20]. The reason is that it is driven by *many* presynaptic signals that arrive more or less coherently and that these signals are, to a decent approximation, stochastically independent of each other.

(ii) Coherence presupposes the "right" underlying hardware which operates with a μs precision. A simple proposal were to generate the hardware by genetic coding but that is hard to believe since thousands of axons require a immense amount of genes that were not able to adapt themselves to fluctuating circumstances such as food. In our opinion, the way out is provided by an unsupervised Hebbian learning rule that selects synapses and, hence, axons with the "right" delays and suppresses the rest. This is – in a sense – an "evolutionary" process with a "survival of the fittest".

(iii) Combining two groups of input, say from the left and the right ear, one gets the very same tuning. Evaluating the output of many laminar neurons through a population code one arrives at the final μs precision.

In the following we will implement and illustrate these ideas. We do not want to bother the reader by technical details, for which (s)he should consult the literature, in particular [3, 20]. Instead we would like to focus on the essentials and maybe stick to half of the truth, if it leads to an intuitive understanding of the whole truth.

2. TEMPORAL CODING

The importance of temporal coding in neural information processing has been debated vigorously [3]. One of the central issues in this debate is the question of whether neuronal firing can be more precise than the time constants of the neuronal processes involved [6]. We address this problem and study sound localization as it is performed by the barn owl's auditory system.

Barn owls are night hunters, despite the fact that their vision is excellent. Not so good, though, that they can see at night. The barn owl therefore needs a substitute, viz., sound localization. Its typical prey is the field mouse, which is fairly nourishing so that, under normal circumstances, five mice per night suffice. In spring, however, the masculine animal has to feed his offspring, numbering six to seven. That is to say, it has to catch a mouse every ten minutes – and it does so.

When a mouse moves through the grass it produces a broadband spectrum and the barn owl samples the sound between 2 and 8 kHz so as to localize the mouse. Why kiloherz (10^3 Hz)? The speed of sound v is about 330 meters per second (m s^{-1}). Denoting the wave length by λ and the frequency by f, we have $v = f\lambda$ so that for 8 kHz λ amounts to about 4 cm, the size of the owl's head. Sitting in a tree, the barn owl has to perform both horizontal and vertical localization so as to fix the direction of its prey. Vertical localization is done by measuring the *intensity difference* between left and right ear. To this end the ears are positioned somewhat asymmetrically and one is "looking" downwards, the other upwards. For horizontal or azimuthal sound localization, the object of our study, the owl determines the *interaural time difference* or, in short, the ITD.

A barn owl can locate sounds and, hence, prey in a horizontal plane with a precision of $1 - 2$ degrees, a capability which requires a time resolution of less than 5 μs. Figure 1 depicts an interaural time difference and Figure 2 is an early explanation due to Jeffress [7] of how a neuron could represent a specific direction by being a meeting point of two inputs, one coming from the right, the other one from the left. This idea exploits in an essential way that *single* spikes carry information.

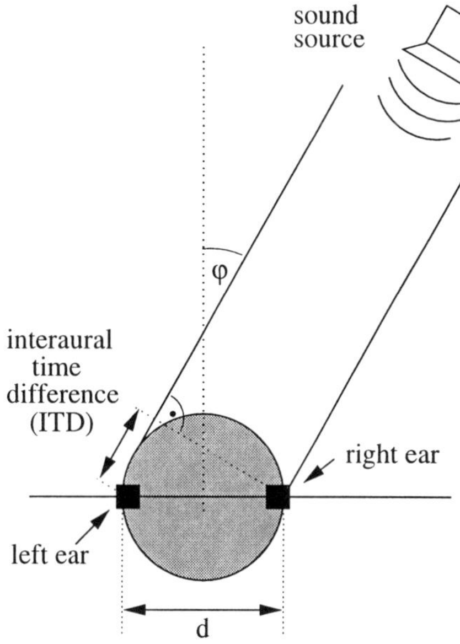

Fig. 1. — Interaural time difference. Sound coming from a source to the right of the head arrives at the left and right ear at different times. The interaural time difference (ITD) is determined by the azimuthal angle φ, the head's diameter $d \approx 4\,\mathrm{cm}$ and the speed $c \approx 330\,\mathrm{m/s}$ of sound. For small angles φ we obtain ITD $= \varphi\, d/c$. The barn owl's angular accuracy $\Delta\varphi \approx 1 - 2°$ of sound localization corresponds to a temporal precision $\Delta\mathrm{ITD} = \Delta\varphi\, d/c < 5$ microseconds ($1\ \mu\mathrm{s} = 10^{-6}$ s).

After passing through the cochlea, a kind of "inverse" piano([1]), sound information splits into two pathways, intensity and ITD. The latter continues via the magnocellular nucleus, which produces a phase-locked output (plus a stochastic jitter), and then proceeds to the laminar nucleus, where signals from both ears meet for the first time. The key process in the auditory system is "phase locking": spikes occur preferentially at a certain phase, termed "mean phase", of a stimulating tone of frequencies up to 8 kHz. In order to understand the high degree of temporal precision in phase locking, we have to consider both the typical duration τ_s of synaptic input current and the membrane time constant τ_m. For auditory neurons in chicken [10], time constants of synaptic input are already in the range of 200 μs. Even though the passive membrane time

([1]) A piano is an instrument where by pushing a specific key one produces a specific tone. Conversely, by "inverse piano" we mean a fictitious instrument that "pushes" the different keys according to the various tones that come in. In this way one obtains a spectral decomposition of the incoming sound.

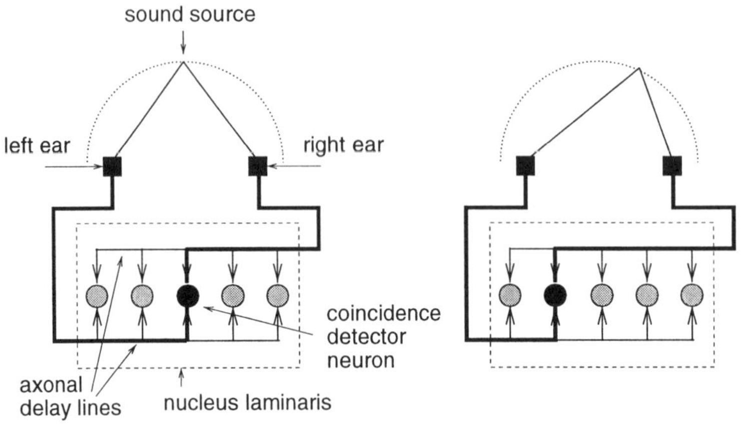

Fig. 2. — Jeffress model [7] suggesting a neural circuit in the barn owl's nucleus laminaris (NL) for processing interaural time differences (ITDs). By means of such a circuit an ITD is transformed into a characteristic spatial activity pattern of the neurons (horizontal row of grey/black disks). Starting from the sound source the sound reaches both ears with a azimuth-dependent delay. In the ears the sound is transformed into action potentials (or spikes). These travel along axons that serve as delay lines to the NL. Axons from the left and the right ear converge in the NL at neurons which act as coincidence detectors. If spikes arrive at a coincidence detector neuron "simultaneously", the neuron shows up high activity (dark disk). Temporally displaced inputs cause a weaker response (brighter disks). A comparison of both figures shows that coincident input requires a compensation of a shorter delay from the sound source to one ear by a shorter axonal delay from the other ear. Depending on the azimuthal position of the sound source, there is a characteristic spatial pattern of neuronal activity, which is shown above and can be processed by "higher" brain areas. Though the Jeffress scheme is a nice idea, it has – as most cybernetic schemes – a few drawbacks: (i) there is not a *single* delay line but there are hundreds (altogether many thousands) of delay lines, and (ii) one still has to show that the hardware does what the scheme wants it to do. We will face both problems.

constant is about 2 ms, the *effective* membrane time constant of magnocellular and laminar neurons is less than 200 μs because there is an outward rectifying current which is activated above resting potential. This kind of outward rectifying current is commonly found in phase-locking neurons of the auditory system. As a result of the short time constants, the width of single excitatory postsynaptic potentials (EPSPs) in chicken is about 500 – 800 μs at half of the maximum amplitude [10]. We think it is reasonable to assume that laminar neurons in an auditory specialist such as the barn owl are twice as fast as those in chicken and model EPSPs by taking a width of 250 μs.

Let $\mathcal{V}$ be the membrane voltage at the soma of a laminar neuron. When $\mathcal{V}$ reaches the threshold ϑ, a neuron fires. As for the neuronal dynamics, one has to integrate the differential equation $d\mathcal{V}/dt = -\mathcal{V}/\tau_m + I(t)$ with an *effective* membrane time constant $\tau_m = 100\ \mu$s. An input spike arriving at time t_j^f at a synapse j evokes an input current $I_j^f(t) = (1/\tau_s)\exp[-(t - t_j^f)/\tau_s]$ for $t > t_j^f$ with $\tau_s = 100\ \mu$s. The total input is $I(t) = \sum_{j,f} J_j I_j^f(t)$ where J_j is the efficacy of synapse j and the sum runs over all synapses and firing times preceding t. For each input channel j, firing times are generated with a periodic probability density $P(t) = (\nu T/\sigma\sqrt{2\pi})\sum_{m=-\infty}^{\infty}\exp[-(t - mT - \Delta_j)^2/2\sigma^2]$ where $\nu = 1$ kHz is a rate, $T = 200\ \mu$s is the period of one of the tones of the stimulus([2]), Δ_j is the transmission delay from the ear to a neuron in the laminar nucleus, and $\sigma = 40\ \mu$s is a measure of the temporal jitter. We impose absolute refractoriness through a minimal interspike interval of 0.5 ms.

If the transmission delays Δ_j are drawn from the broad, *mono*modal distribution of Figure 3a, the one the young barn owl starts with after hatching, then spike arrival is incoherent. If on the other hand the delay distribution is that of Figure 3c, *i.e.*, a *multi*modal one with the maxima at distances T, then spike arrival is coherent. To fix the effective time constant τ_m of the integrate-and-fire model, we started from a more detailed model [3] with an explicit description of the outward rectifying current [10].

We concentrate on a single frequency channel. The stimulus is a pure tone of, *e.g.*, 5 kHz (period $T = 200\ \mu$s). The input from magnocellular neurons to our laminar integrate-and-fire neuron is a more or less periodic sequence of spikes. Presynaptic spikes occur preferentially around a specific mean phase of the tone [9]. The jitter of $\sigma = 40\ \mu$s takes care of internal sources of variability and noise along the auditory pathway as well as of the bandwidth of frequency tuning of auditory neurons. The mean firing rate is relatively low (667 Hz) so that each presynaptic spike train skips most of the cycles of the 5 kHz tone.

EPSPs of spikes from more than 100 presynaptic neurons that arrive coherently with a common mean phase lead to an oscillating membrane potential $\mathcal{V}$, that builds up, reaches the threshold in the *rising* stage of the cycle and in this way produces a slightly better phase locking than that of the input. The upshot turns out to be (see Sect. 3) that a single neuron can reach a precision of about 25 μs, which is not bad in view of the time constants involved.

If presynaptic spikes arrive with random phases, however, $\mathcal{V}$ shows aperiodic fluctuations and the output spikes have a *uniform* phase distribution (Fig. 3a). We expect this to be the initial, embryonic, condition before the connections between the magnocellular and the laminar nucleus are tuned during a sensitive period [11]. In adult owls, transmission delays from the ear to the laminar nucleus differ greatly. Their mean value is between 2 and 3 ms, and the standard deviation is about $200 - 240\ \mu$s. Without a tuning of the delays, any phase information would be lost. A Hebbian learning rule [3] proves to be an efficient tuning mechanism.

([2]) The cochlea operates like an "inverse piano" and we now consider one of its "keys".

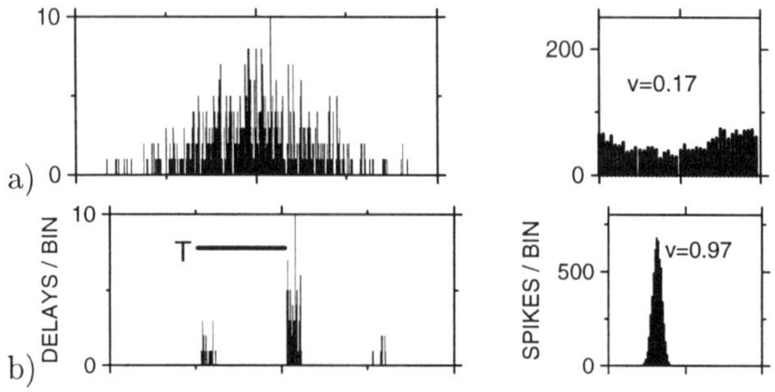

Fig. 3. — Hebbian learning. In a–c, the left-hand graphs show synapses binned as a function of the signal transmission delay Δ. On the right-hand side the distribution of output phases is shown in period histograms (bin width 5 μs). (a) Before learning, there are 600 synapses with signal transmission delays (left) drawn from a broad Gaussian distribution (mean 2.5 ms, width 0.6 ms). The output of the laminar model neuron shows *no* phase locking to a 2 kHz (not shown) or a 5 kHz signal (right). (b) After a period of learning while being stimulated by a 2 kHz signal. The 105 synapses which survive learning have delays which differ by *multiples* of the period $T = 500$ μs (scaling bar). The output spikes exhibit phase locking with vector strength [12] $v = 0.97$ corresponding to a temporal precision of 20 μs (right).

3. HEBBIAN LEARNING

The setup of the previous section implicitly assumed that phase-locked spikes with some jitter impinge upon a laminar neuron. Three weeks after hatching, however, there seems to be a broad continuous distribution of delays and, thus, no phase locking. Genetic coding as a way out is not plausible since there are many thousands of axons that have to be tuned and genetics cannot adapt to a changing environment. How, then, does a barn owl get the right "hardware"? We suggest by Hebbian learning or, sloppily formulated, "practice makes perfect".

In Hebbian learning, synaptic strength is changed by a small amount ϵ, if presynaptic spike arrival and postsynaptic firing "coincide". This simultaneity constraint is implemented by a learning window $W(s)$ where $s = t_j^f - t_n$ is the difference between the arrival time of a presynaptic spike t_j^f at synapse j and the postsynaptic firing t_n. In our model, $W(s)$ has two regimes; see Figure 3d and [13]. For $s < 0$, $W(s)$ is positive. Thus the efficacy of synapses which are repeatedly active shortly *before* a postsynaptic spike occurs is increased [14]. The weights of synapses that are active shortly *after* the postsynaptic spike originated are *de*creased [15]. (Our theory was earlier, though, than its experimental verification:-) Since depolarization is known to induce potentiation of active synapses [14] and the neuron is depolarized most of the time

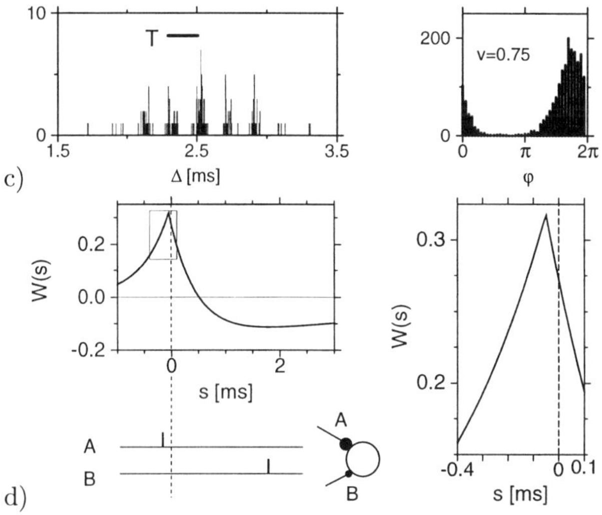

Fig. 3. — (c) Same plot as in b, after learning a 5 kHz input signal. 154 saturated synapses survive. The output spikes exhibit phase locking ($v = 0.75$) corresponding to a temporal precision of 25 μs. Neither this nor the 20 μs under b live up to the required 5 μs; see Section 4. (d) Learning window W, the key notion of the present paper, as a function of the delay $s = t_j^f - t_n$ between a postsynaptic firing time t_n and presynaptic spike arrival t_j^f at synapse j. The graph on the right-hand side shows the boxed region around the maximum on an expanded scale. If $W(s)$ is positive (negative) for some s, the synaptic efficacy is increased (decreased). The postsynaptic firing occurs, *e.g.*, at $s = 0$ (vertical dashed line). Learning is most efficient, if presynaptic spikes arrive shortly *before* the postsynaptic neuron starts firing as in the "fat" synapse A. The "meager" synapse B, which fires *after* the postsynaptic spike originated, is weakened. Before learning, all synapses have identical weights $J_j = 1$. During learning weights are changed according to $\Delta J_j = \epsilon \sum_f [\gamma + \sum_n W(t_j^f - t^n)]$ with factors $\epsilon = 0.002, \gamma = 0.1$. The sum runs over all spike arrival times $t_j^f < t$ and all postsynaptic firings $t^n < t$. We take $W(s) = 0.3 \exp[(s + 0.05)/0.5]$ for $s < -0.05$ and $W(s) = 0.5 \exp[-(s + 0.05)/0.5] - 0.2 \exp[-(s + 0.05)/5]$ for $s \geq -0.05$ where s is a time in ms. Learning occurs during 3000 seconds in time steps of 5 μs. The synaptic strength saturates at a maximum of 3. A synapse whose weight vanishes is removed. The efficiency of phase locking is quantified by the vector strength. Spike input and electrical time constants are as described in the main text.

between two spikes, we also strengthen each active synapse by a small amount γ, even if no postsynaptic spike occurs. The procedure of continuously strengthening and weakening the synapses automatically (sic!) leads to a normalization of the total input strength feeding into a postsynaptic neuron in a competitive self-organized process.

As a result of learning, a clearly structured distribution of synapses evolves

(Figs. 3b,c). Delays of the remaining synapses differ approximately by multiples of the period T of the stimulating tone. We emphasize that the period was not previously known to the cell. As is shown by Figures 3b and c, the final synaptic pattern of a neuron stimulated by a 2 kHz signal during learning is different from that of a 5 kHz neuron.

The evolution of synaptic weights during learning does *not* depend on the specific shape of the learning window W but only on some generic properties. In particular, the learning window can be nonzero in a domain which extends over several milliseconds (*cf.* [15]) and therefore is large as compared to the period T of the sound (Fig. 3d). Efficient learning depends on the temporal relationship between the process of strengthening the synaptic weights and that of weakening them. The maximum of the window function $W(s)$ should be at $\tilde{s} \approx -\tau_s/2$ where τ_s is the rise time of the postsynaptic potential (Fig. 3d). This can be understood through a self-consistency argument. Let us assume that learning has led to a sharply peaked distribution with all synapses having a common delay Δ (mod T). Coherent input arriving at the synapses can trigger a postsynaptic spike with a mean delay of approximately $\tau_s/2$. Because of learning, all synapses which are active slightly *before* the postsynaptic firing will be strengthened. If $\tilde{s} = -\tau_s/2$, the maximal increase of synaptic weights occurs at those synapses which are already strongest.

So far we have considered monaural input, but laminar neurons receive input from both ears and exhibit a sensitivity to the *interaural* time difference (ITD). We divide the synapses into two groups, viz., with input from the left and the right ear. During learning, both ears are stimulated by the same signal and with a fixed ITD. The learning rule selects synaptic connections so that spikes arrive coherently for exactly this ITD. If the same ITD is used *after* learning, the neuron is driven optimally and emits phase-locked spikes. If the ITD does not match, phase locking of the output spikes breaks down and the mean firing rate decreases; see Figures 4a and b.

4. POPULATION CODING

Temporal information conveyed by a *single* laminar neuron is limited. The temporal precision of phase locking is about 20 – 25 μs (Fig. 3) and the ITD tuning curve is only weakly modulated (Fig. 4a). Nevertheless, barn owls achieve a behavioral performance corresponding to a temporal precision of 5 μs. How can we understand this remarkable accuracy? The barn owl needs a reaction time of about 100 ms before it starts a head movement [8]. During 100 ms, the firing pattern of a *population* of laminar neurons contains enough information to resolve ITDs with a precision of 5 μs, as we will now show.

Let us consider a *group* of laminar neurons with ITD tuning curves as in Figure 4a, but shifted along the ITD axis so that the optimal responses occur at different ITDs. Neurons are stimulated by a tone with fixed, but unknown, ITD. We estimate the ITD from the neuronal firing pattern by a "population

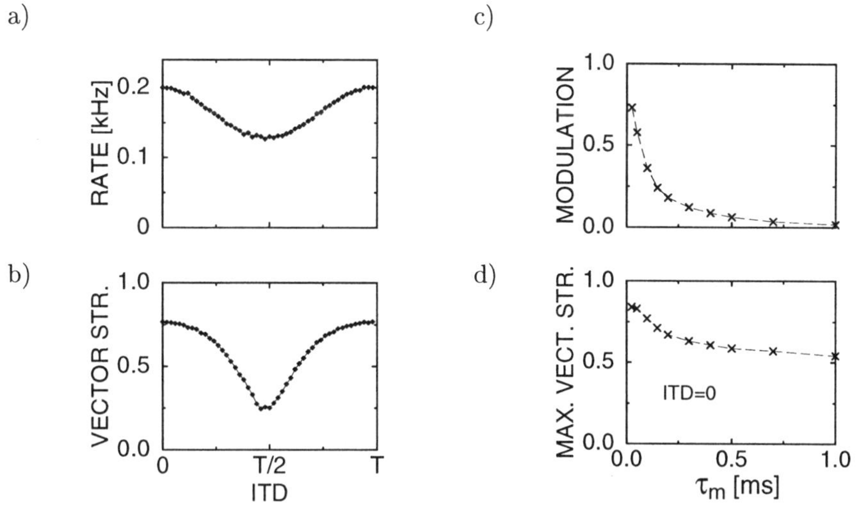

Fig. 4. — a) and b). Conversion of spike code (input) into rate code (output) [20] through tuning to interaural time difference (ITD). The output rate a) and the vector strength b) of a laminar neuron are shown as a function of the ITD. The neuron has been tuned to a 5 kHz signal ($T = 200\ \mu$s) as described in the text and the legend of Figure 3. Half of the 154 synapses which survive learning receive signals from the left ear, the others from the right ear. The neuron exhibits best phase locking and maximal output rate $f_{\max}$ for the ITD $= 0$ used during learning. The rate has a minimum $f_{\min}$ for ITD $= T/2$. c) and d): effect of τ_m. We define the modulation to be $(f_{\max} - f_{\min})/f_{\max}$ and show it as a function of the effective membrane time constant τ_m in c). The maximal vector strength at ITD $= 0$ is shown as a function of τ_m in d).

vector" decoding scheme [16]: each neuron "votes" for the ITD corresponding to its optimal response and the votes are weighted according to the number of spikes that occur in a time window of 100 ms. We find that the activity of about 100 independent neurons provides enough information to estimate the ITD with a precision of 5 μs (Fig. 5) – apart from ambiguities that are due to the periodicity of the signal. Weak correlations from common inputs rescale the absolute values but do not alter the conclusions.

In Figure 5 we took N neurons labeled by $1 \leq k \leq N$ with ITD tuning curves similar to Figure 4a and approximated by $f_k(x) = \{164 + 36\cos[(x - x_k)2\pi/T]\}$ Hz where x is the ITD and T the period of the stimulating tone, x_k is the optimal ITD of neuron k, and f_k its mean firing rate in Hz. The values of x_k have a uniform distribution between 0 and T. Spikes of neuron k are generated by a Poisson process with mean rate f_k. In an additional simulation (data not shown) we have confirmed that Poisson statistics give a reasonable assumption

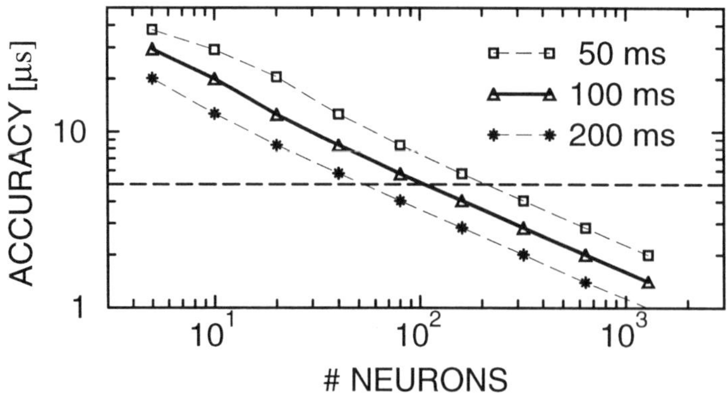

Fig. 5. — What population coding is good for. The temporal precision that can be achieved in a population of N independent neurons is shown as a function of N under the assumption of population vector coding [16, 19]. The ITD is estimated based on spikes occurring in a time window of $\tau = 50$ ms (squares), 100 ms (triangles), or 200 ms (stars). The accuracy is defined as the average deviation of the estimate from the actual ITD of the stimulus. The horizontal line at 5 μs is the boundary given by the behavioral performance of 1-2 degree of azimuthal angle [8]. For large N, the accuracy scales as $1/\sqrt{\tau N}$ as expected from the central limit theorem.

and the results of Figure 5 do not change, if we generate spikes by an integrate-and-fire neuron as described above. The ITD of the stimulus is estimated [16] to be $x^{\text{est}} = (T/2\pi) \arg[\sum_k n_k \, e^{i2\pi x_k/T}]$ where n_k is the number of spikes of neuron k measured in a time window of length τ. We plot the accuracy $\langle |x^{\text{est}} - x|^2 \rangle^{1/2}$ averaged over 20 000 trials. We see that 100 μs averaging time needs about 100 neurons as input to get a precision of 5 μs. In reality one finds about 150 neurons, leaving room for additional correlations among the neurons.

The firing precision of $20 - 25$ μs of *single* neurons has been achieved despite the fact that input spikes exhibit a jitter and evoke EPSPs which are ten times broader than the precision just mentioned. In an additional set of simulations, we have systematically varied the effective membrane time constant τ_m and, hence, the width of the EPSP. We find that the temporal precision of neuronal spiking depends only weakly on τ_m (Fig. 4d). This is possible because firing always occurs during the *rise* time of a postsynaptic potential. On the other hand, for a 5 kHz signal, ITD tuning breaks down rapidly, if τ_m exceeds 0.1 ms (Fig. 4c). Thus modulation of ITD tuning does indeed require short EPSPs. In fact, very short time constants have been measured by Reyes *et al.* [10]. With the benefit of hindsight we now also see that the stimulating frequency cannot be arbitrarily high. In practice, the barn owl does not go beyond 10 kHz – completely consistent with the above estimates.

5. DISCUSSION

Our results show that the temporal precision of output spikes does not have the membrane time constant itself as an upper bound. Spike timing can achieve a resolution shorter than the rise time of an EPSP, given *coherent* spike arrival – and spike arrival times can be tuned accurately by time-resolved Hebbian learning. The key notion of our theory is the *learning window* describing how a synapse is to change if the presynaptic spike arrives before/after the postsynaptic one and specifying what time scales are involved. It is quite natural that W is positive/negative depending on the presynaptic spike coming too early/too late. If it is slightly earlier, it favors the firing of the postsynaptic neuron so that the synapse is to be strengthened. Coming too late, that's no good and the synapse is to be weakened. This way the mechanism leads to a stable equilibrium. Instead of weeding out synapses by means of W one could think of mechanisms that allow the synapses to speed up/slow down or the axons to shrink/grow depending on the presynaptic spikes being too late/early. New phenomena do not occur.

The barn owl's neurons reaching a μs precision one may wonder: why is Nature that good? In the case of a laminar neuron, because of three reasons, all simple biology. First, the neuron receives input through many synapses so that noise is averaged out. Second, it uses a time-resolved Hebbian learning rule singling out the "right" synapses. In a sense, this is nothing but evolution on a microscale, the "right" synapses representing the "survival of the fittest". Third, combining many neurons performing the same calculation one attains a population code, which is better than that of the individual neurons.

It is tempting to apply the same ideas to information processing in the cerebellum [17], the hippocampus [4], or the cortex [5]. If we multiply time constants by a factor of about 100 and reinterpret our results, then we are led to the conclusion that in areas where effective membrane time constants are in the range order of $10-20$ ms [18], a temporal code with an accuracy of $1-3$ ms should be, and seems to be, possible.

ACKNOWLEDGMENTS

R.K. has been supported by the DFG (He 1729/8-1).

REFERENCES

[1] Carr C.E. *Annu. Rev. Neurosci.* **16** (1993) 223-243.

[2] Heiligenberg W. *Neural Nets in Electric Fish* (MIT Press, Cambridge, MA, 1991).

[3] Gerstner W., Kempter R., van Hemmen J.L. and Wagner H. *Nature* **383** (1996) 76-78 and references quoted therein.

[4] Hopfield J.J., *Nature* **376** (1995) 33-36.
[5] Abeles M., in *Models of Neural Networks II*, edited by E. Domany, J.L. van Hemmen and K. Schulten (Springer, New York, 1994) pp. 121-140.
[6] Softky W. and Koch C., *J. Neurosci.* **13** (1993) 334-350.
[7] Jeffress L.A., *J. Comp. Physiol. Psychol.* **41** (1948) 35-39.
[8] Knudsen E.I., Blasdel G.G. and Konishi M., *J. Comp. Physiol.* **133** (1979) 1-11.
[9] Sullivan W.E. and Konishi M., *J. Neurosci.* **4** (1984) 1787-1799.
[10] Reyes A.D., Rubel E.W. and Spain W.J., *J. Neurosci.* **16** (1996) 993-1007; **14** (1994) 5352-5364.
[11] Carr C.E., in *Advances in Hearing Research*, edited by G.A. Manley, G.M. Klump, C. Köppl, H. Fastl and H. Oeckinghaus (World Scientific, Singapore, 1995) pp. 24-30.
[12] Goldberg J.M. and Brown P.B., *J. Neurophysiol.* **32** (1969) 613-636.
[13] Van Hemmen J.L., Gerstner W., Herz A.V.M., Kühn R. Vaas M., in *Konnektionismus in Artificial Intelligence und Kognitionsforschung*, edited by G. Dorffner (Springer, Berlin, 1990) pp. 153-162.
[14] Bliss T.V.P. and Collingridge G.L., *Nature* **361** (1993) 31-39.
[15] Markram H., Lübke J., Frotscher M. and Sakmann B., *Science* **275** (1997) 213-215.
[16] Salinas E. and Abbott L.F., *J. Comput. Neurosci.* **1** (1994) 89-107.
[17] Braitenberg V., *Network* **4** (1993) 11-17.
[18] Bernander Ö., Douglas R.J., Martin K.A.C. and Koch C., *Proc. Natl. Acad. Sci. USA* **88** (1991) 11569-11571.
[19] Georgopoulos A.P., Schwartz A.B. and Kettner R.E., *Science* **233** (1986) 1416-1419.
[20] Kempter R., Gerstner W., van Hemmen J.L. and Wagner H., *Neural Comput.* **10**/8 (1998) in press.

Application of Physical Models and Phenomena to Biological Systems

LECTURE 25

Structures of Supercoiled DNA and their Biological Implications

T.R. Strick, J.-F. Allemand, A. Bensimon[1], D. Bensimon and V. Croquette

Laboratoire de Physique Statistique de l'ENS, CNRS, associé aux Universités Paris VI et VII, 24 rue Lhomond, 75231 Paris Cedex 05, France
[1]Laboratoire de Biophysique de l'ADN, Département des Biotechnologies, Institut Pasteur, 25 rue du Dr. Roux, 75724 Paris Cedex 15, France

1. INTRODUCTION

The double-helical structure of DNA rapidly adapts to changes in the molecule's degree of supercoiling. The overwound or underwound double-helical axis can assume exotic forms such as plectonemes (the braided forms that appear on twisted phone cords) [1], and supercoiling-induced denaturation of certain DNA sequences can allow the formation of stem-loop structures or cruciforms [2]. In the thirty-odd years since DNA supercoiling was discovered [3], it has been shown that supercoiling is involved in or affected by biological processes such as DNA transcription [4, 5], DNA recombination [6], DNA replication [7] and the packaging of eukaryotic genomes [8]. Until today, it was only possible to control the supercoiling of circular plasmid DNAs using intercalators (such as ethidium bromide) or commercially available topoisomerases. These techniques have several disadvantages; they do not allow for real-time modification and analysis of DNA supercoiling, nor do they allow for precise, controllable and reversible DNA supercoiling. We have established a new technique based on the tools of DNA micromanipulation which gives us the possibility of executing precise, quantitative and reversible supercoiling of an individual *linear* DNA molecule in real time [9]. Here we will describe our experimental setup and the properties of supercoiled DNA, and then present preliminary experiments concerning the supercoiling-assisted hybridization of homologous DNA sequences and its possible implications for genetic recombination.

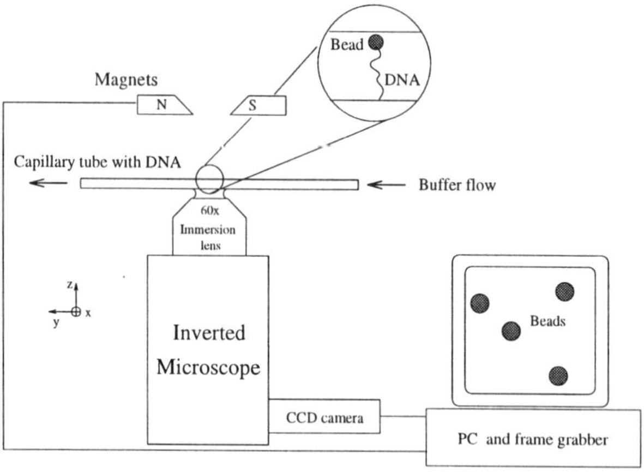

Fig. 1. — An overview of the experimental setup. The $\sim$ 16 μm-long DNA molecule is bound to a glass surface (the bottom of a 1 mm $\times$ 1 mm $\times$ 50 mm capillary tube) at one end by digoxigenin/anti-digoxigenin links and at the other end to a superparamagnetic bead (3 μm diameter) by biotin/streptavidin links. The magnetic field used to pull on the bead and control its rotation is generated by Co-Sm magnets and focused by asymetric polar pieces with a 1.7 mm gap. This magnetic device can rotate about the optical axis, inducing synchronous rotation of the magnetic bead, and it can be lowered (raised) to increase (decrease) the stretching force. The samples are observed on a Nikon Diaphot-200 inverted microscope with a $\times$60 immersion oil objective. Video data relating the Brownian motion of the magnetic bead is generated by a square pixel XC77CE Sony camera connected to a Cyclope frame grabber (timed on the pixel clock of the camera) installed in a 486 – 33 MHz PC.

1.1. Overview of setup

We begin by binding single λ-DNA (50.5 kbp, or 16.8 μm long) molecules by one end to a treated glass surface, while the other end is bound to a small magnetic bead (Fig. 1). The binding is achieved by digoxigenin–anti-digoxigenin and biotin-streptavidin bonds. By allowing for multiple anchoring sites between the DNA and the bead and the DNA and the glass surface, we are able to exert torsional control on the DNA molecule. A rotating magnetic field is used to induce rotation of the magnetic bead, and reversibly over- and underwind the DNA molecule. By controlling the number of turns n added to or removed from the DNA, we can change its degree of supercoiling σ. The magnetic field gradient (along the z-axis of the experiment) is used to increase or decrease the stretching force exerted by the magnetic bead on the DNA molecule. The degree of supercoiling of the molecule can therefore be quantitatively controlled and monitored, while video-microscopy based tethered-particle motion analysis [9, 10] allows us to measure the extension of the molecule as well as the stretching force acting on the DNA.

1.2. Basics of DNA supercoiling

The topology of torsionally-constrained DNA molecules can be described by a few simple quantities [11, 12]. The first is the twist (Tw) of the molecule, the number of times the two strands that make up the double helix twist around each other. The second is the writhe (Wr) of the molecule. The writhe represents the wrapping of the molecule's axis with itself (the braided structures known as plectonemes are highly writhed). If one constrains the ends of the DNA molecule, then the total number of times that the two strands of the helix cross each other (either by twist or writhe) becomes a topological invariant of the system known as the linking number: Lk. A mathematical theorem due to White [11] states that:

$$Lk = Tw + Wr. \tag{1}$$

In the case of an anchored DNA, rotating one end of the DNA allows one to access and modify Lk without damaging the molecule. The linking number of a torsionally relaxed (linear or circular) DNA is written as $Lk_0 = Tw_0 + Wr_0$. Tw_0 is the number of helical repeats in B-DNA: 10.5 base-pairs per turn. Assuming that DNA has no significant spontaneous curvature with which it could form coils or loops then $Wr_0 = 0$. The relative difference in linking number between the supercoiled and relaxed forms of DNA is called the excess linking number, σ:

$$\sigma = \frac{Lk - Lk_0}{Lk_0}. \tag{2}$$

The molecule is overwound when σ is positive, underwound when it is negative.

2. BASIC EXPERIMENTS

2.1. Results

Our system thus allows us to control two parameters: the DNA's degree of supercoiling σ (to within about 0.02%) and the stretching force F that pulls on the molecule (to within $\sim$ 10%). Moreover, we can measure the DNA's extension l with a precision of 100 nm. We can thus obtain l *versus* σ curves (at constant force) or F *versus* l curves (at constant supercoiling). Let us first consider the case where the DNA is under no torsional stress ($\sigma = 0$); the force-extension diagram in Figure 2 shows that a stretching force of about 5 picoNewtons (pN) is sufficient to bring the molecule to $\sim$ 95% of its crystallographic length. In order to get a feel for the scale of the forces involved, motor proteins such as T4 RNA polymerase generate forces on the order of 10 pN [13], and the traction of chromosomes during meiosis and mitosis involves forces greater than $\sim$ 500 pN

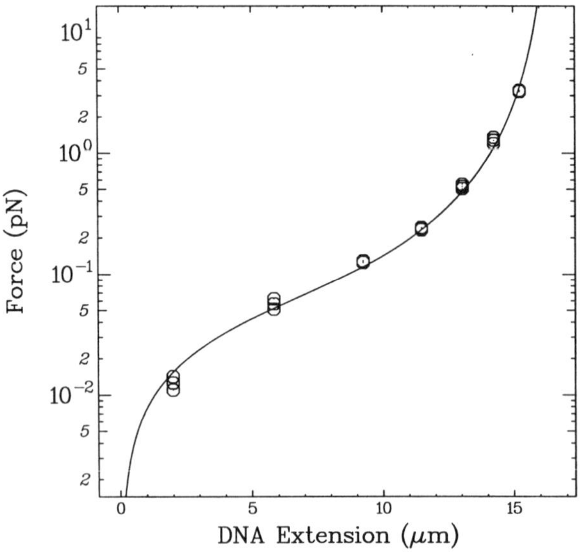

Fig. 2. — Force *versus* extension curve for a single DNA molecule at $\sigma = 0$ in a 10 mM phosphate buffer. The curve is well-described (see fit) by a worm-like chain (WLC) model with a persistence length of $\sim$ 53 nm.

[14, 15]. The torsionally-relaxed force-extension curve is fit (solid curve) by the worm-like chain (WLC) model, which describes DNA as a continuous, semi-flexible rod characterized by a persistence length $\xi \sim 50$ nm [16] (in solutions containing at least 10 mM of a monovalent cation).

Let us now set the stretching force applied to the DNA and measure the molecule's extension l as a function of its supercoiling σ (Fig. 3). At a low force ($F = 0.1$ pN) the elastic behavior of DNA is symmetric under positive or negative supercoiling. Like a phone cord the molecule continuously "contracts", as each added turn allows it to form plectonemes and "shorten". The symmetry axis of this low-force l *versus* σ curve defines the torsionless ($\sigma = 0$) state of the molecule. At an intermediate force ($F = 1.3$ pN), the chiral nature of the molecule is evident. The extension of the molecule does not change as it is underwound, whereas it continues to contract when overwound. Finally at high forces (here $F = 8$ pN) the DNA's extension depends only slightly on the degree of supercoiling and is similar to that expected from a torsion-free worm-like chain.

These three regimes are also evident in the F *versus* l plots at fixed σ (Figs. 4(A) ($\sigma < 0$) and (B) ($\sigma > 0$)):

- At low forces ($F < 0.5$ pN) under- and overwound DNA at the same

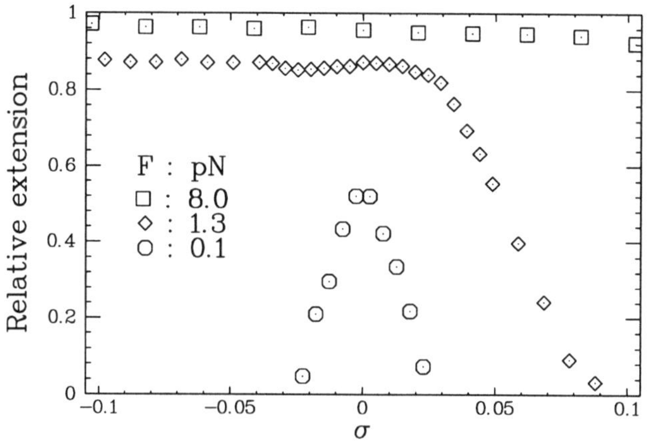

Fig. 3. — Extension *versus* supercoiling curves for a single DNA molecule in a 10 mM phosphate buffer. The force is held constant at 0.1, 1.3 and 8 pN. The three forces represented here were chosen to emphasize the three regimes observed in this type of experiment. In the low-force regime $F < 0.4$ pN, the DNA molecule responds in a symmetric manner to positive or negative supercoiling by forming plectonemes. These plectonemes grow with $|\sigma|$, reducing the molecule's extension. At intermediate forces 0.5 pN $< F <$ 3 pN, the extension of negatively supercoiled DNA is relatively insensitive to changes in the molecule's linking number. Positively supercoiled DNA, on the other hand, contracts as the excess linking number grows. In the high force $F >$ 3 pN regime, the extensions of both positively and negatively supercoiled DNA are relatively independent of changes in the linking number.

value of $|\sigma|$ have the same extension for a given force. Their rigidity, the force required to stretch the molecule to a given length, increases with $|\sigma|$.

- At intermediate forces ($0.5 < F < 3$ pN), DNA behaves very differently if it is positively or negatively coiled. Indeed, at a force $F = F_c^- \sim 0.5$ pN, underwound DNA undergoes an abrupt transition to an extension similar to that of a torsionless molecule. On the other hand overwound DNA extends continuously as the stretching force is increased.
- At higher forces (3 pN $< F <$ 10 pN) DNA, whether under- or overwound, has a force *versus* extension behaviour similar to that of a torsionless DNA. Indeed, at $F = F_c^+ \sim 3$ pN positively coiled DNA with $\sigma > 0.1$ undergoes an abrupt transition to an extended state, in a manner similar to that observed for underwound molecules at $F_c^- \sim 0.5$ pN.

The solid lines in Figures 4 are the force-extension diagrams for supercoiled DNA as calculated by Marko and Siggia [1].

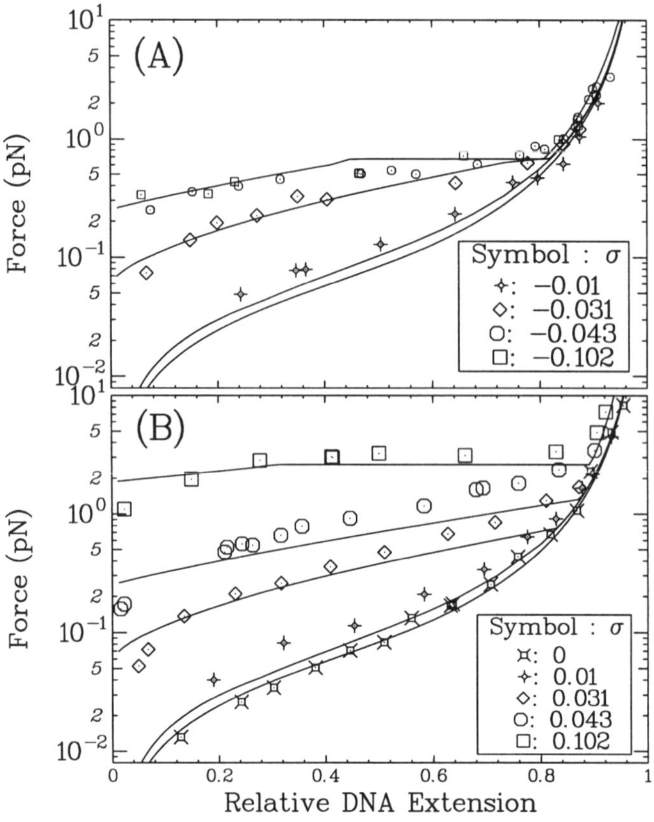

Fig. 4. — Force *versus* extension curves for negatively (A) and positively (B) supercoiled DNA in 10 mM phosphate buffer. The $|\sigma| = 0$ curve was fitted by a WLC with a persistence length of 53 nm. The other solid curves are fits generated by Marko and Siggia [1]. At low forces ($F < 0.3$ pN) the curves are similar for positive and negative supecoiling, whereas at ${F_c}^- \sim 0.5$ pN the negatively supercoiled molecule undergoes an abrupt transition to an extended state which behaves like a molecule with $\sigma = 0$. The positively supercoiled DNA undergoes a similar abrupt transition to an extended state when $\sigma > 0.1$ and $F = {F_c}^+ \sim 3$ pN.

2.2. Interpretation

Biological experiments indicate that negatively supercoiled DNA molecules can alleviate excess torsional stress by denaturing locally. It has been shown, for example, that negative supercoiling reduces the melting temperature of DNA by favoring the local separation of the two constituent strands [17]. These biochemical studies indicate that the alternative DNA structure which appears in stretched, unwound DNA is a denaturation bubble. We propose that the abrupt lengthening of the underwound molecule at the critical force F_c^- is due to the simultaneous destruction of plectonemic structures and local denaturation of the DNA. The separated strands (with $\sigma_{bubble} = -1$) thus absorb the molecule's linking number deficit ($\sigma < 0$) and permit the remainder of the DNA to adopt a nearly normal B-form conformation ($\sigma \sim 0$). A similar argument can be made for the transition observed at F_c^+, in which the molecule's overwinding ($\sigma > 0$) would be absorbed by a local structural transition to a form with a helical pitch shorter than that of B-DNA (10.5 bp/turn).

3. FURTHER EXPERIMENTS AND CONCLUSION

In order to conclusively show that negatively supercoiled DNA responds to stress by locally denaturing, we have attempted to reproduce an experiment inspired by Beattie *et al.* [18]. They showed that negatively supercoiled plasmids, when heated, are capable of removing from a solution fragments of single-stranded homologous DNA, simply by hybridizing with them. By analyzing the sedimentation rates of these plasmids, it was found that the negatively supercoiled plasmids go from a compact form (before hybridization with the homologous DNA) to a relaxed form (after hybridization). The compact, pre-hybridization form is due to the presence of plectonemes, whereas the relaxed form is due to the stabilization of a denaturation bubble that occurs when single-stranded DNA hybridizes onto this bubble.

We therefore used the polymerase chain reaction (PCR) to generate 1 kb single-stranded fragments of λ-DNA. This was then added to an experiment where a linear λ-DNA was mechanically unwound by $n = -500$ turns from its $\sigma = 0$ state and then stretched with a force $F > 0.5$ pN. After waiting for an hour, we lowered the stretching force to $F < 0.5$ pN and re-measured the molecule's extension *versus* supercoiling curve (see Fig. 5). We found that the curve obtained was "broadened" by about 100 turns relative to the initial pre-hybridization curve. The probe was then dehybridized by overwinding the duplex and stretching it: the molecule then recovered its initial behavior. No such change in the molecule's behavior was observed when non-homologous single-stranded DNA was used. These preliminary experiments indicate that negatively supercoiled, stretched DNA forms a denaturation bubble capable of hybridizing to homologous single-stranded DNA. Moreover, the hybridization could be reversed by mechanically closing this denaturation bubble. Additional experiments have been performed to verify that as the molecule is progressively unwound denaturation begins in AT-rich regions before advancing into GC-rich regions [19].

In conclusion, we see that stretching a supercoiled DNA molecule can induce a reversible local melting of the molecule. The interplay between coiling and stretching can thus regulate strand exchange between a duplex DNA and a homologous single-stranded DNA. As enzymes (topoisomerases) are known to

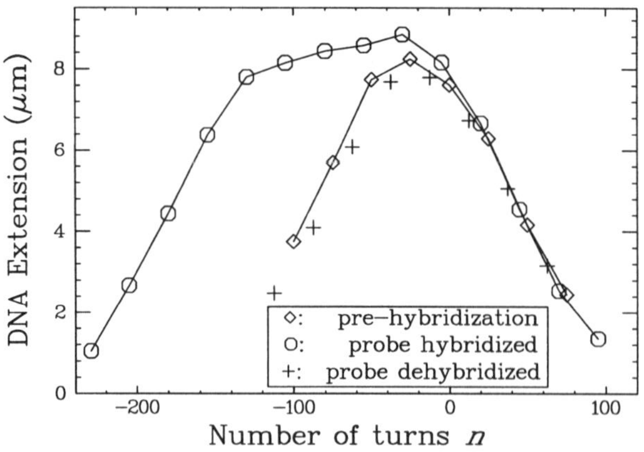

Fig. 5. — Hybridization-associated changes in the DNA's extension *versus* supercoiling behavior. The experiments were carried out at 42 degrees celsius in a 10 mM phosphate buffer supplemented with 150 mM NaCl. The extension *versus* supercoiling curves were taken at the same force ($F \sim 0.1$ pN) before hybridization, after hybridization and after dehybridization of 1000 bases of single-stranded homologous DNA. After hybridization, the curve is "broadened" by about 100 turns relative to the initial pre-hybridization curve. After dehybridization is performed (by rewinding and stretching the DNA) the molecule's initial behavior is recovered.

control the degree of supercoiling of DNA and molecular motors (and crowding) are capable of stretching DNA with the rather weak forces reported here, this mechanism might be used *in vivo* by the cell to regulate recA-mediated strand exchange [6] during recombination as well as other DNA/protein interactions.

REFERENCES

[1] Marko J.F. and Siggia E.D., Statistical Mechanics of Supercoiled DNA, *Phys. Rev. E* **52** (1995) 2912–2938.

[2] Palecek E., Local Supercoil-Stabilized Structures, *Crit. Rev. Biochem. Mol. Biol.* **26** (1991) 151–226.

[3] Vinograd J., Lebowitz J., Radloff R., Watson R. and Laipis P., The twisted circular form of polyoma virus DNA, *Proc. Natl. Acad. Sci. USA* **53** (1965) 1104–1111.

[4] Wu H.-Y., Shyy S., Wang J.C. and Liu L.F., Transcription Generates Positively and Negatively Supercoiled Domains in the Template *Cell.* **53** (1988) 433–440.

[5] Dunaway M. and Ostrander E.A., Local domains of supercoiling activate a eukaryotic promoter *in vivo*, *Nature* **361** (1993) 746–748.

[6] Stasiak A. and Di Capua E., The helicity of DNA in complexes with Rec-A protein, *Nature* **299** (1982) 185–187.

[7] Murray A.W. and Szostak J.W., Chromosome segregation in mitosis and meiosis, *Annu. Rev. Cell. Biol.*, **1** (1985) 289–315.

[8] Patterton H.-G. and von Holt C., Negative supercoiling and nucleosome cores. I The Effect of Negative Supercoiling on the Efficiency of Nucleosome Core Formation *in vitro*, *J. Mol. Biol* **229** (1993) 623–636.

[9] Strick T.R., Allemand J.F., Bensimon D., Bensimon A. and Croquette V., The elasticity of a single supercoiled DNA molecule, *Science* **271** (1996) 1835–1837.

[10] Gelles J., Schnapp B.J. and Sheetz M., Tracking kinesin-driven movements with nanometre-scale precision, *Nature* **331** (1988) 450–453.

[11] White J.H., Self-linking and the Gauss integral in higher dimensions, *Am. J. Math.* **91** (1969) 693–728.

[12] Calladine C.R. and Drew H.R., Understanding DNA (Academic Press, 1992).

[13] Yin H., Wang M.D., Svoboda K., Landick R., Block S.M. and Gelles J., Transcription against an applied force, *Science* **270** (1995) 1653–1656.

[14] Bensimon D., Simon A.J., Croquette V. and Bensimon A., Stretching DNA with a receding meniscus: Experiments and Models, *Phys. Rev. Lett.* **74** (1995) 4754–4757.

[15] Duplantier B., Jannink G. and Sikorav J.-L., Anaphase chromatid motion: involvement of type II DNA topoisomerases, *Biophys. J.* **69** (1995) 1596–1605.

[16] Bustamante C., Marko J.F., Siggia E.D. and Smith S., Entropic Elasticity of λ-Phage DNA, *Science* **265** (1994) 1599–1600.

[17] Kowalski D., Natale D. and Eddy M., Stable DNA unwinding, not breathing, accounts for single-strand-specific nuclease hypersensitivity of specific A+T-rich sequences, *Proc. Natl. Acad. Sci.* **85** (1988) 9464–9468.

[18] Beattie K.L., Wiegand R.C. and Radding C.M., Uptake of Homologous Single-stranded Fragments by Superhelical DNA, *J. Mol. Biol.* **116** (1977) 783–803.

[19] Strick T.R., Bensimon D. and Croquette V., Homologous Pairing in Stretched Supercoiled DNA, *Proc. Natl. Acad. Sci.* (1998) to be published.

AUTHORS

J.-F. Allemand, Laboratoire de Physique Statistique de l'ENS, CNRS, associé aux Universités Paris VI et VII, 24 rue Lhomond, 75231 Paris Cedex 05, France

K.I. Anderson, Institute of Molecular Biology, Austrian Academy of Sciences, Billrothstr. 11, A-5020 Salzburg, Austria

F. Andris, Laboratoire de Physiologie Animale, Université Libre de Bruxelles, CP 300, 67 rue des Chevaux, 1640 Rhode St Genèse, Belgium

M.-C. Bellissent-Funel, Laboratoire Léon Brillouin, CEA-CNRS, CEA Saclay, 91191 Gif-sur-Yvette, France

A.-M. Benoliel, Laboratoire d'Immunologie, INSERM U387, BP. 29, 13274 Marseille Cedex 09, France

E. Ben-Jacob, School of Physics and Astronomy, Raymond & Beverly Sackler Faculty of Exact Sciences, Tel-Aviv University, Tel-Aviv 69978, Israel

A. Bensimon, Laboratoire de Biophysique de l'ADN, Département des Biotechnologies, Institut Pasteur, 25 rue du Dr. Roux, 75724 Paris Cedex 15, France

D. Bensimon, Laboratoire de Physique Statistique de l'ENS, CNRS, associé aux Universités Paris VI et VII, 24 rue Lhomond, 75231 Paris Cedex 05, France

A. Ben-Ze'ev, Department of Molecular Cell Biology, The Weizmann Institute of Science, Rehovot 76100, Israel

D.A. Beysens, Département de Recherche Fondamentale sur la Matière Condensée, CEA Grenoble, 17 rue des Martyrs, 38054 Grenoble, France

P. Bongrand, Laboratoire d'Immunologie, INSERM U387, BP. 29, 13274 Marseille Cedex 09, France

P. Bron, Canaux et Récepteurs Membranaires, UPRES-A 6026 du CNRS, Biologie Cellulaire et Reproduction, Université de Rennes 1, Campus de Beaulieu, 35042 Rennes Cedex, France

A. Cavalier, Canaux et Récepteurs Membranaires, UPRES-A 6026 du CNRS, Biologie Cellulaire et Reproduction, Université de Rennes 1, Campus de Beaulieu, 35042 Rennes Cedex, France

D.L. Charest, Department of Medicine, Faculty of Medicine, University of British Columbia, 1779 W., 75th avenue Vancouver, B.C., V6P 6P2, Canada

V. Croquette, Laboratoire de Physique Statistique de l'ENS, CNRS, associé aux Universités Paris VI et VII, 24 rue Lhomond, 75231 Paris Cedex 05, France

M. Daoud, Laboratoire Léon Brillouin, CE Saclay, 91191 Gif-sur-Yvette, France

C. Delamarche, Canaux et Récepteurs Membranaires, UPRES-A 6026 du CNRS, Biologie Cellulaire et Reproduction, Université de Rennes 1, Campus de Beaulieu, 35042 Rennes Cedex, France

S. Deschamps, Canaux et Récepteurs Membranaires, UPRES-A 6026 du CNRS, Biologie Cellulaire et Reproduction, Université de Rennes 1, Campus de Beaulieu, 35042 Rennes Cedex, France

D. Drasdo, Max-Planck-Inst. f. Kolloid-und Grenzflächenforschung, Kantstr. 55, 14513 Teltow, Berlin, Germany

J. Engel, Biozentrum, Dept. Biophysical Chemistry, Klingelbergstrasse 70, 4056 Basel, Switzerland

G. Forgacs, Department of Physics and Biology, Clarkson University, Potsdam, NY 13699-5820, USA

E. Frey, Institut für Theoretische Physik und Institut für Biophysik, Physik-Department der Technischen Universität München, James-Franck-Strasse, 85747 Garching, Germany

A. Froger, Canaux et Récepteurs Membranaires, UPRES-A 6026 du CNRS, Biologie Cellulaire et Reproduction, Université de Rennes 1, Campus de Beaulieu, 35042 Rennes Cedex, France

J.A. Glazier, Department of Physics, University of Notre Dame, Notre Dame, IN 46556-5670, USA

P.J. Goldschmidt-Clermont, The Heart and Lung Institute at the Ohio State University 420 W, 12th Avenue Columbus, Ohio 43210, USA

J. Gouranton, Canaux et Récepteurs Membranaires, UPRES-A 6026 du CNRS, Biologie Cellulaire et Reproduction, Université de Rennes 1, Campus de Beaulieu, 35042 Rennes Cedex, France

P.L. Hansen, Department of Chemistry, The Technical University of Denmark, Building 206, 2800 Lyngby, Denmark

J.-F. Hubert, Canaux et Récepteurs Membranaires, UPRES-A 6026 du CNRS, Biologie Cellulaire et Reproduction, Université de Rennes 1, Campus de Beaulieu, 35042 Rennes Cedex, France

D.E. Ingber, Departments of Pathology and Surgery, Harvard Medical School and Children's Hospital, Boston, MA 02115, USA

J.H. Ipsen, Department of Chemistry, The Technical University of Denmark, Building 206, 2800 Lyngby, Denmark

D. Isabey, INSERM U492, Physiopathologie et Thérapeutique Respiratoires, Hôpital Henry Mondor, 94010 Créteil Cedex, France

K. Itoh, Department of Microbiology and Molecular Genetics, Harvard Medical School and Molecular Medicine Unit, Beth Israel Deaconess Medical Center, 330 Brookline Ave., Boston, MA, USA

P.A. Janmey, Experimental Medicine Division, Brigham and Women's Hospital, Harvard Medical School, 221 Longwood Ave., Boston, MA 02115, USA

M. Kaufman, Center for Nonlinear Phenomena and Complex Systems, Université Libre de Bruxelles, Campus Plaine CP 231, 1050 Bruxelles, Belgium

R. Kempter, Physik-Department der TU München, 85747 Garching bei München, Germany

K. Kroy, Institut für Theoretische Physik und Institut für Biophysik, Physik-Department der Technischen Universität München, James-Franck-Strasse, 85747 Garching, Germany

A. Kuspa, Department of Biochemistry, Baylor College of Medicine, Houston, TX 77030, USA

V. Lagrée, Canaux et Récepteurs Membranaires, UPRES-A 6026 du CNRS, Biologie Cellulaire et Reproduction, Université de Rennes 1, Campus de Beaulieu, 35042 Rennes Cedex, France

O. Leo, Laboratoire de Physiologie Animale, Université Libre de Bruxelles, CP 300, 67 rue des Chevaux, 1640 Rhode St Genèse, Belgium

W.F. Loomis, Center for Molecular Biology, UCSD, La Jolla, CA 92093, USA

L. Moldovan, The Heart and Lung Institute at the Ohio State University 420 W, 12th Avenue Columbus, Ohio 43210, USA

L. Miao, Department of Chemistry, The Technical University of Denmark, Building 206, 2800 Lyngby, Denmark

S.A. Newman, Department of Cell Biology and Anatomy, New York Medical College, Valhalla, NY 10595, USA

C. Oddou, Laboratoire de Mécanique Physique, Université Paris 12-Val-de-Marne, France

S.L. Pelech, Department of Medicine, Faculty of Medicine, University of British Columbia, 1779 W., 75th avenue Vancouver, B.C., V6P 6P2, Canada

I. Pellerin, Canaux et Récepteurs Membranaires, UPRES-A 6026 du CNRS, Biologie Cellulaire et Reproduction, Université de Rennes 1, Campus de Beaulieu, 35042 Rennes Cedex, France

E. Planus, INSERM U492, Physiopathologie et Thérapeutique Respiratoires, Hôpital Henry Mondor, 94010 Créteil Cedex, France

F. Richelme, Laboratoire d'Immunologie, INSERM U387, BP. 29, 13274 Marseille Cedex 09, France

J.-P. Rolland, Canaux et Récepteurs Membranaires, UPRES-A 6026 du CNRS, Biologie Cellulaire et Reproduction, Université de Rennes 1, Campus de Beaulieu, 35042 Rennes Cedex, France

E. Sackmann, Institut für Theoretische Physik und Institut für Biophysik, Physik-Department der Technischen Universität München, James-Franck-Strasse, 85747 Garching, Germany

J.V. Shah, Experimental Medicine Division, Brigham and Women's Hospital, Harvard Medical School, 221 Longwood Ave., Boston, MA 02115, USA

G. Shaulsky, Center for Molecular Biology, UCSD, La Jolla, CA 92093, USA

S.Y. Sokol, Department of Microbiology and Molecular Genetics, Harvard Medical School and Molecular Medicine Unit, Beth Israel Deaconess Medical Center, 330 Brookline Ave., Boston, MA, USA

T.R. Strick, Laboratoire de Physique Statistique de l'ENS, CNRS, associé aux Universités Paris VI et VII, 24 rue Lhomond, 75231 Paris Cedex 05, France

J.X. Tang, Experimental Medicine Division, Brigham and Women's Hospital, Harvard Medical School, 221 Longwood Ave., Boston, MA 02115, USA

D. Thomas, Canaux et Récepteurs Membranaires, UPRES-A 6026 du CNRS, Biologie Cellulaire et Reproduction, Université de Rennes 1, Campus de Beaulieu, 35042 Rennes Cedex, France

A. Upadhyaya, Department of Physics, University of Notre Dame, Notre Dame, IN 46556-5670, USA

J.L. van Hemmen, Physik-Department der TU München, 85747 Garching bei München, Germany

N. Wang, Center for Molecular Biology, UCSD, La Jolla, CA 92093, USA

S. Wendling, Laboratoire de Mécanique Physique, Université Paris 12-Val-de-Marne, France

J. Wilhelm, Institut für Theoretische Physik und Institut für Biophysik, Physik-Department der Technischen Universität München, James-Franck-Strasse, 85747 Garching, Germany

Imprimé en France. Jouve, 733 rue Saint-Léonard, BP. 3, 53101 Mayenne Cedex, France
Dépôt légal : octobre 1998